Ius Gentium: Comparative Perspectives on Law and Justice

Volume 53

More information about this series at http://www.springer.com/series/7888

Deborah Cao • Steven White
Editors

Animal Law and Welfare - International Perspectives

Springer

Editors
Deborah Cao
Law Futures Centre
Griffith University
Brisbane, QLD, Australia

Steven White
Griffith Law School
Griffith University
Brisbane, QLD, Australia

ISSN 1534-6781 ISSN 2214-9902 (electronic)
Ius Gentium: Comparative Perspectives on Law and Justice
ISBN 978-3-319-26816-3 ISBN 978-3-319-26818-7 (eBook)
DOI 10.1007/978-3-319-26818-7

Library of Congress Control Number: 2016930403

Springer Cham Heidelberg New York Dordrecht London

Printed on acid-free paper

Springer International Publishing AG Switzerland is part of Springer Science+Business Media (www.springer.com)

Contents

Contributors

David Bilchitz Department of Public Law, University of Johannesburg, Johannesburg, South Africa

Donald M. Broom Centre for Animal Welfare and Anthrozoology, Department of Veterinary Medicine, University of Cambridge, Cambridge, UK

Deborah Cao Law Futures Centre, Griffith University, Brisbane, QLD, Australia

David Favre Animal Legal & Historical Center, Michigan State University College of Law, East Lansing, MI, USA

Jed Goodfellow Department of Law, Macquarie University, Sydney, NSW, Australia

Nicki McGrath Centre for Animal Welfare and Ethics, School of Veterinary Science, University of Queensland, Gattton, QLD, Australia

Clive J.C. Phillips Centre for Animal Welfare and Ethics, School of Veterinary Science, University of Queensland, Gatton, QLD, Australia

Joan E. Schaffner George Washington University Law School, Washington, DC, USA

Tagore Trajano de Almeida Silva Tiradentes University, Aracaju, Sergipe, Brazil

Joy M. Verrinder Centre for Animal Welfare and Ethics, School of Veterinary Science, University of Queensland, Gatton, QLD, Australia

Paul Waldau Canisius College, Buffalo, NY, USA

Steven White Grifith Law School, Griffith University, Brisbane, QLD, Australia

Yossi Wolfson Attorney-at-Law, Jerusalem, Israel

About the Authors

David Bilchitz is a professor at the University of Johannesburg and director of the South African Institute for Advanced Constitutional, Public, Human Rights and International Law (SAIFAC). He is also secretary-general of the International Association of Constitutional Law (IACL). He has a BA (Hons) LLB from Wits University and graduated with an MPhil and PhD from the University of Cambridge. He publishes widely in the field of constitutional and fundamental rights law and has authored the monograph *Poverty and Fundamental Rights: The Justification and Enforcement of Socio-Economic Rights* (Oxford University Press, 2007). He has published several pioneering articles on animals and the law in South Africa and is also involved actively in promoting efforts to improve the protections for animals in the laws of South Africa through law reform, litigation and campaigning. He has helped in the campaigns to prevent elephant culling and to strengthen the legal framework governing performing animals.

Donald M. Broom is emeritus professor of animal welfare, Cambridge University, Department of Veterinary Medicine, and has developed concepts and methods of scientific assessment of animal welfare and studied cognitive abilities of animals, the welfare of animals in relation to housing and transport, behaviour problems, attitudes to animals and ethics of animal usage. He has published over 300 refereed papers, lectured on animal welfare in 43 countries and served on UK (FAWC, APC, Seals) and Council of Europe committees. He has been chairman or vice-chairman of EU Scientific Committees on Animal Welfare (1990–2009) and a member of the European Food Safety Authority (EFSA) Scientific Panel on Animal Health and Welfare until June 2012. He chaired the O.I.E. group on Welfare of Animals during Land Transport. His books include *Stress and Animal Welfare*, *The Evolution of Morality and Religion*, *Domestic Animal Behaviour and Welfare* and *Sentience and Animal Welfare*.

Deborah Cao is a professor at Griffith University, Australia. She is a linguist and legal scholar. She has published in areas including legal theory, legal semiotics, legal translation, the philosophical and linguistic analysis of Chinese law and legal culture and animal law. She is also editor of the *International Journal for the Semiotics of Law*. Her books include *Chinese Law: A Language Perspective* (Ashgate, 2004), *Translating Law* (Multilingual Matters, 2007), *Animals Are Not Things* (China Law Press, 2007), *Animal Law in Australia and New Zealand* (Thomson Reuters, 2010), *While the Dog Gently Weeps* (China Jilin University Press, 2010), *Animal Law in Australia* (2nd ed. Thomson Reuters, 2015), *Animals in China: Law and Society* (Palgrave Macmillan, 2015) and *Law and Language in China* (Rowman & Littlefield, forthcoming). She was named one of the top 200 most influential blog writers in China in 2012.

David Favre is a professor of law at Michigan State University College of Law, USA. Over the past 30 years, Professor Favre has written several articles and books dealing with animal issues including such topics as animal cruelty, wildlife law, the use of animals for scientific research, respectful use and international control of animal trade. His books include the case book *Animal Law: Welfare, Interest, and Rights (2nd ed.)*, *Animal Law and Dog Behavior* and *International Trade in Endangered Species*. He introduced the concept of "Living Property" which was developed in a number of law review articles over the past decade. He created and is editor in chief of the largest animal legal web resource, www.animallaw.info. Now residing on a farm in lower Michigan, Professor Favre shares his space with sheep, chickens and the usual assortment of dogs and cats.

Jed Goodfellow is a PhD candidate within the Legal Governance Concentration of Research Excellence at Macquarie Law School, Australia. His research concerns the animal welfare regulatory framework within the Australian agricultural sector. Jed also teaches animal law at Macquarie University and works as a policy officer for RSPCA Australia with a focus on legislative and regulatory issues affecting animal welfare. Previously, Jed practised as a prosecutor for RSPCA South Australia and a solicitor for a leading commercial law firm. Jed also served as an animal cruelty inspector for RSPCA Queensland throughout his undergraduate studies.

Nicki McGrath is currently studying for PhD in chicken vocalisations and their interpretation with the Centre for Animal Welfare and Ethics at the University of Queensland's School of Veterinary Science, Australia. She has a master's in applied animal behaviour and welfare from Edinburgh University, for which she conducted a study of public understanding of animals' experiences of grief in Brisbane. She also spent 2 years setting up and running a wildlife research expedition in the Amazon, where she was responsible for bird call research. She has 11 years' experience in the corporate workplace.

Clive J. C. Phillips studied agriculture at Reading University and obtained a PhD in dairy cow nutrition and behaviour from the University of Glasgow. He lectured in farm animal production and medicine at the Universities of Cambridge and Wales and conducted research into cattle and sheep welfare. As the inaugural holder of the University of Queensland Chair in Animal Welfare, Australia, he is now involved in research in animal welfare and ethics and the development and implementation of Australian State and Federal government animal welfare policies. He has written widely on animal welfare and management in scientific journals, blogs and books, and he edits a new journal in the field, *Animals*, and a series of books on animal welfare for Springer.

Joan E. Schaffner, is an associate professor of law at the George Washington University Law School, USA. She received her BS in mechanical engineering (magna cum laude) and JD (Order of the Coif) from the University of Southern California and her MS in mechanical engineering from the Massachusetts Institute of Technology. In addition to teaching civil procedure, sexuality and the law and remedies, she directs the George Washington University Animal Law Program and has presented on animal law panels at conferences worldwide. She is the author of *Introduction to Animals and the Law* (Palgrave Macmillan, 2011), a co-author and editor of *A Lawyer's Guide to Dangerous Dog Issues* (ABA, 2009) and *Litigating Animal Law Disputes: A Complete Guide for Lawyers* (ABA, 2009) and author of several book chapters including "Canine Profiling" in *Global Guide to Animal Protection* (Univ. Illinois Press, 2013), "Animal Cruelty and the Law: Permitted Conduct" in *Animal Cruelty: A Multidisciplinary Approac*h (Carolina Academic Press, 2013) and "Laws and Policy to Address the Link of Family Violence" in *The Link Between Animal Abuse and Humane Violence* (Sussex Academic Press, 2009). She is active in various organisations including the past chair of ABA TIPS Animal Law Committee, founding chair of the American Association of Law Schools Section on Animal Law and fellow of the Oxford Centre for Animal Ethics. In 2013 she received the Excellence in the Advancement of Animal Law Award from the American Bar Association, Tort Trial & Insurance Practice Section.

Tagore Trajano de Almeida Silva is a practising lawyer in Brazil and president of the Institute for Animal Abolitionist (www.abolicionismoanimal.org.br). He received his postdoctorate in law from Pace Law School and his PhD and master's in public law from the Federal University of Bahia. He has been a visiting researcher at the University of Science and Technology of China, serves as the editor of the *Brazilian Animal Rights Review* (www.animallaw.info#international) and has coordinated several world conferences on bioethics and animal rights. He is an expert on issues relating to Brazilian constitutional and environmental law as well as bioethics and animal law and has published articles in Brazil and the USA on comparative law issues. In addition to teaching postgraduate courses in environmental law at the Federal University of Bahia, Professor Trajano is a founding member of the Asociación Latinoamericana de Derecho Ambiental. His research papers can be viewed on SSRN: http://ssrn.com/author=1855392.

Joy M. Verrinder is currently studying for a PhD in animal ethics education with the Centre for Animal Welfare and Ethics at the University of Queensland's School of Veterinary Science, Australia. She has a master's in professional ethics and governance and a master's in business administration focusing on public policy and strategic planning. For the past 10 years as strategic development officer of the Animal Welfare League Queensland, she has been progressing legislation and policy changes at local, state and national levels to prevent the killing of abandoned cats and dogs in pounds and shelters in Australia. Joy has served on government animal ethics committees and boards of various animal organisations. She has also been an educator and deputy principal in secondary schools.

Paul Waldau is an educator, scholar and activist working at the intersection of animal studies, law, ethics, religion and cultural studies. A professor at Canisius College in Buffalo, New York, Paul has been the lead faculty member for the master of science graduate programme in anthrozoology since its founding in 2011 and director of the programme since September of 2015. Paul has also taught animal law at Harvard Law School since 2002. He also teaches Harvard's summer term course "Animals: Religion and Ethics". The former director of the Centre for Animals and Public Policy, Paul taught veterinary ethics and public policy at Tufts University School of Veterinary Medicine for more than a decade. Paul has completed five books, the most recent of which are *Animal Studies: An Introduction* (Oxford University Press, 2013) and *Animal Rights* (Oxford University Press, 2011). He is also coeditor of *A Communion of Subjects: Animals in Religion, Science, and Ethics* (Columbia University Press, 2006) and *An Elephant in the Room: The Science and Well-Being of Elephants in Captivity* (Center for Animals and Public Policy, 2008). His first book was *The Specter of Speciesism: Buddhist and Christian Views of Animals* (Oxford University Press, 2001).

Steven White is a lecturer at Griffith University, Australia. He has published widely on animal law, including coediting *Animal Law in Australasia* (Federation Press, 2nd ed., 2013). He continues to teach one of the first undergraduate courses offered on animal law in Australia and for a number of years has been a consultant to Brisbane law firm Couper Geysen – Family and Animal Law. Steven will be completing a PhD in 2016 on the regulation of the protection of companion and farm animals in Queensland, Australia.

Yossi Wolfson is an Israeli attorney-at-law and an animal liberation activist. His legal work has included representation of human and nonhuman animal rights organisations in courts in constitutional, administrative and civil cases. He has been involved in the animal liberation movement since the 1980s and took part in some of the principal campaigns and legal cases for animals in Israel, as well as in legislation processes.

Chapter 1
Introduction: Animal Protection in an Interconnected World

Steven White and Deborah Cao

1.1 Introduction

In our increasingly interconnected and wired world, some of the biggest global stars have been nonhuman animals. On blogs, on Facebook and all around the Internet, claws and clicks go hand in hand or paw in paw, with the furry claiming cyberspace. In 2014, one of the most emailed stories on *The New York Times*'s website was about the biology of cats.[1] According to media reports, there has been a sharp rise in the proportion of American dog and cat owners with provisions in their wills for their pets, with nearly one in every ten now making such arrangements.[2] One of the most fervently embraced documentaries of 2013 was *Blackfish*, shown over and over on CNN. And these are not just "feel good" stories about cute and cuddly animals. They are about animal suffering, animal science, animal intelligence and cognition, animal behaviour and social life, animal welfare and law, and above all, animal dignity, rights and justice. For instance, in a *New York Times's* op-ed piece, the commentator writes about "according animals dignity".[3] These topics are not academic jargon but increasingly entering the popular cyber parlance. In the meantime, apart from stories and images of animals going viral in traditional and social

[1] http://www.nytimes.com/2014/01/07/science/cat-sense-explains-what-theyre-really-thinking.html

[2] http://www.nytimes.com/2014/01/14/opinion/bruni-according-animals-dignity.html?_r=0

[3] http://www.nytimes.com/2014/01/14/opinion/bruni-according-animals-dignity.html?_r=0

S. White (✉)
Griffith Law School, Griffith University,
Brisbane, QLD, Australia
e-mail: steven.white@griffith.edu.au

D. Cao
Law Futures Centre, Griffith University, Brisbane, QLD, Australia
e-mail: d.cao@griffith.edu.au

D. Cao, S. White (eds.), *Animal Law and Welfare - International Perspectives*,
Ius Gentium: Comparative Perspectives on Law and Justice 53,
DOI 10.1007/978-3-319-26818-7_1

media around the world, significant legal battles are being fought on behalf of animals, for instance, in the International Court of Justice in the Hague and in the courtrooms of New York. At the end of 2013, a team of lawyers was filing writs of habeas corpus in New York on behalf of four captive chimpanzees as part of the Nonhuman Rights Project.[4] At the other end of the world, in Australia, a group of animal lawyers, scientists and scholars was gathering to discuss animal law and animal welfare, the result of which is this edited book.

The symposium, held on the Gold Coast in Australia in early-December 2013, was designed to provide a small number of animal law and animal welfare science scholars the time to meet for 2 days of intensive discussion and debate. The symposium, by contrast with the usual conference format of a short paper and sometimes desultory question time, allowed for extended, collegial discussion across a range of inter-related animal law and welfare science issues. The impetus for the symposium, and for the chapters in this book, derive from a sobering reality: despite the developments in animal protection law over the last 200 years, and, in particular, the developments in the legal front over the last three decades, animal cruelty is not decreasing but increasing world-wide. We are witnessing the globalization of animal cruelty. This is so despite laws for protecting animals being accompanied and increasingly informed by the deepening of knowledge in animal welfare science, and despite the recognition of the legal and moral wrong of acts of cruelty against animals in both law and in the general community in most countries. Animal cruelty is increasing in terms of scale and in more varied forms, for both domestic animals and, increasingly, for wildlife. Factory farming, which originated in the West, has now been introduced to developing countries and is expanding rapidly; wildlife is being used and abused for various kinds of human consumption on an unprecedented scale, especially in Asia and we are facing the real possibility that African elephants and rhinos may become extinct in the next decade; indiscriminate killing of different species of animals occurs every so often in different countries on a massive scale due to health scares and panic fuelled by a fear of the spread of disease. Given this expanding cruelty, what are the possible solutions in terms of animal law and welfare? In an age of globalization, a global solution through international cooperation and communication of animal matters is essential to deal with animal cruelty. Such a response must be based on a multi-jurisdictional understanding of the phenomenon of animal protection, addressing both the particular forms it takes in different jurisdictions and, at a broader level, the potential for sharing knowledge and resources to improve the living conditions for animals across all jurisdictions. While there are several texts that address animal law from international perspectives (for instance, *Globalization and Animal Law* by Kelch (2011) and *A Worldview of Animal Law* by Wagman and Liebman (2011)), this book is distinctive in bringing together the disciplines of animal law and animal welfare science, as well as others such as ethics and criminology, in a way which allows for an integrated exploration of animal protection across a range of jurisdictions. The range of jurisdictions canvassed in the

[4] For original filings and subsequent developments of the chimpanzee case, see http://www.nonhumanrightsproject.org/category/courtfilings/.

book reflects not only the work of those who participated in the symposium, but also of others who subsequently and generously agreed to contribute a chapter.

The chapters included in this book address some of the most topical and important issues in animal law and animal welfare in different jurisdictions in the world. International law related to animals is examined, including E.U. animal welfare law, the latest World Trade Organization (W.T.O.) ruling on seal products and the E.U. ban, the *Blackfish* story and U.S. law for cetaceans, wildlife trafficking and related crimes concerning Africa and China, an examination of both historical and current animal protection laws in Australia, and the laws concerning animal protection in Israel, South Africa and Brazil. The book deliberately seeks to draw out perspectives across a range of jurisdictions which mostly have not been the subject of extensive consideration in the animal protection literature. This is particularly so for those jurisdictions which have received comparatively little consideration in the English language literature. Questions which have been explored extensively elsewhere in the literature, including in the areas of moral philosophy and ethics concerning animals, feminist approaches to animal protection, animal protection activism and the doctrinal law of animal cruelty, are not specifically addressed here, although these and other aspects of animal protection are drawn on in various places through the book.

1.2 Perspectives on Animal Law and Welfare Science: General Principles and Specific Issues

The book is divided into two parts. In Part I, the chapters address general issues, in animal law and welfare science, which cut across all jurisdictions. Building on this base, Part II explores developments in animal legal protection in different jurisdictions from five different continents: Australia, Africa, Asia, Europe, North America and South America.

In Chap. 2, **Paul Waldau** addresses the need for "a second wave of animal law". Animal Law is now taught in many law schools around the world. However, Waldau argues that it is time to examine the future direction of animal law, including its place in the broader realm of Animal Studies. He poses four key questions for guiding this process: (1) how should animal law courses in law school proceed given the obvious interdisciplinary reach of the larger field of Animal Studies?; (2) what is the place in second wave animal law for answers to the root question "who, what are 'animals'?"; (3) why is the field of "Animal Studies" important to legal education and, more broadly, education as a whole?; and, (4) what roles have personal connection with and meeting of nonhuman animals played in first wave animal law, and what roles might they play in second wave animal law, the larger field of Animal Studies and its sub-disciplines, and the educational mega-fields we know as "the humanities" and "the sciences"?

In Chap. 3, **Donald Broom** brings an animal welfare science perspective to bear on a range of animal welfare law questions. These include the role played by welfare law in addressing issues of sustainability and food quality, grounded in consumer expectations. The focus then shifts to significant recent developments in animal welfare, and their implications for the law. Broom argues for the importance of a duty of care based on scientific evidence, and the use of welfare-based outcome measures. Finally, he emphasises the need for international bodies, such as the International Whaling Commission and the W.T.O., to pay greater attention to scientific studies of animal welfare.

Joy Verrinder, **Nadine McGrath** and **Clive Phillips**, in Chap. 4, suggest a need for greater scrutiny of animal welfare science before it is used as a basis for practice or law. This might be because new developments in animal welfare science, including recognition of the significance of animal emotions, render existing practice and law open to challenge, or because some of the research underpinning existing law or practice reflects vested interests in animal production. In particular, they argue that greater attention needs to be given to our ethical responses to the "data" provided by animal welfare science. Scientists and lawyers need to pay greater attention to the "science of animal ethics", including through teaching the precepts of moral behaviour to veterinary and law students.

David Favre, in Chap. 5, rounds out Part I by addressing a glaring gap in international law: no international agreement is currently in place addressing the welfare and protection of animals. Reviving and updating an earlier proposal, Favre sets out the reasons for why an animal protection treaty is essential, and then provides a blueprint for establishing one. Although frankly acknowledging that prospects for the imminent adoption of such a treaty are poor, Favre argues the importance of having a well-thought out proposal ready to be put in place when the circumstances are more favourable.

Part II shifts the focus to a jurisdictional level. **Steven White**, in Chap. 6, addresses the development of animal protection law in Australia. The ethic of humaneness underpinning the very first animal welfare legislation in nineteenth century Britain continues to animate contemporary legislation, including in Queensland, Australia. The law has changed over time in important ways, most notably through the comparatively recent introduction of a statutory "duty of care", imposed on those in charge of animals, and a more detailed regulatory scheme for farm animal welfare. However, with the ethic of extending mere "humaneness" to animals becoming increasingly difficult to justify, contemporary law is left exposed as overly reliant on nineteenth century sensibilities.

David Bilchitz, in Chap. 7, reflects on animal protection in the post-apartheid South Africa. Bilchitz argues that since the introduction of parliamentary democracy in 1994 the interests of animals have been largely ignored by legislatures and courts. Through a number of case studies, Bilchitz provides evidence of an institutional unwillingness to engage with issues of animal justice. Importantly, he argues that lawyers need to do more to draw attention to these issues, in both policy advocacy and litigation, linking them with the broader commitment to compassion, humanity and justice underpinning post-apartheid South Africa.

Israeli animal protection law is explored in Chap. 8. **Yossi Wolfson** identifies a progressive strand in Israeli animal protection law which might offer some inspiration for other jurisdictions, even if the law largely facilitates exploitation of animals.

Wolfson also provides insights into Israeli animal protection law which might otherwise only be available to those with knowledge of Hebrew, bringing to the reader his practical experience as a lawyer involved some of the important Israeli animal court cases and his expert legal knowledge.

Chapter 9 canvasses the emergence of animal protection law in Brazil. **Tagore Trajano de Almeida Silva** draws parallels and identifies overlaps with the development of other areas of law, including human rights law and environmental law. A key change occurred in 1988, with a new national Constitution formally requiring protection of animals against cruelty. Trajano argues that the change effected by this provision provides a new status for animals, no longer to be regarded as mere natural resources or mere property, and instead subjects deserving of individual consideration. The ramifications of this change of status are yet to be fully explored, but the potential now exists for courts and legislators to embrace a stronger commitment to animal dignity.

Questions of regulatory governance are examined by **Jed Goodfellow** in Chap. 10. Although focussing on Australian governance of animal protection, the insights from Goodfellow's chapter are likely to resonate in many other jurisdictions. Claims of industry capture are frequently aired by animal protection advocates, in a context where agriculture departments are the usual home of institutional responsibility for animal welfare. Rather than taking such claims at face value, Goodfellow scrutinises them through the lens of regulatory capture theory. While he finds that the public interest in animal protection has indeed been subverted by conflicting departmental priorities and reward structures, his arguments bring much-needed analytical rigour to this aspect of animal law. Goodfellow then canvasses some changes in regulatory design which might address existing shortcomings.

Grounding her analysis in recent tragic examples of cetacean confinement in United States theme parks, including the story of Tilikum told in the film *Blackfish*, **Joan E. Schaffner** in Chap. 11 argues the time is ripe for improving the protection of cetaceans under U.S. law. Apart from increased public consciousness of the plight of cetaceans through films such as *Blackfish*, there has been an international campaign for cetacean rights since 2010, greater scientific understanding of the complexity of cetacean cognition, and heightened concerns about public and employee safety in theme parks confining cetaceans. Schaffner addresses the legal possibilities for improved protection of the interests of cetaceans, arguing the need for a shift away from utilitarian considerations of welfare to protection of fundamental rights.

Part II closes with a chapter by **Deborah Cao** who focuses on crimes against wildlife as illustrated by the wildlife law and ivory trade in China. She outlines the legal framework and law for wildlife protection in China and in particular highlights the inherent problems in the law that regards animals, including protected and endangered species, as resources for exploitation. Despite the legal protection at the international and domestic levels, African elephants and rhinos are facing brutal onslaught and possible extinction due to the growing trade for such animal products and the ineffective protection system for their preservation. She argues that a fundamental change of attitude and conception in wildlife protection is required, that is, wildlife and animals in general need to be protected irrespective of their species and wildlife must not be seen as resources for human exploitation to their detriment. She also proposes that wildlife victims of human harms are recognized as victims of crimes, not just as valuable resources, and the definition of crime against wildlife needs to be expanded to include harms done to them both legally and illegally.

1.3 Conclusion and Acknowledgements

Animal law is a burgeoning legal discipline, with significant growth in teaching, practice and scholarship over the past 30 years. Animal welfare science is, likewise, a phenomenon of the late twentieth century and continues to increase in importance in the new millennium. As encouraging as these comparatively recent disciplinary developments may be for advancing animal protection, this book provides a cautionary note. In some jurisdictions neither field has found much traction and, even in those where the disciplines are well-established, harm to animals continues on a large scale.

A key theme of this book is that no one jurisdiction is alone in facing the challenge of animal exploitation, and that each has something to learn from or to share with others. As animal exploitation becomes increasingly globalised, the need for a global response becomes more urgent. This response needs to be founded on a shared understanding of the challenges being confronted across jurisdictions. If we are on the cusp of a major re-evaluation of human and nonhuman animal relations, if animal rights and protection are to constitute a major social justice movement of the twenty-first century, lawyers and scientists, for their part, will need to be part of a cooperative, creative and committed push for change. This book is an intellectual contribution to this project, whether diagnosing existing limitations in law and practice across jurisdictions; reshaping our understanding of animal law as part of the wider animal studies field; understanding the origins of prevailing outdated ethical approaches to care and protection; emphasising the need for a critical approach to the use of science in grounding animal protection law; or identifying the possibilities of institutional reform at a domestic and international level.

In conclusion, the editors would like to thank the Socio-Legal Research Centre (SLRC), Griffith University and, in particular, Professor Brad Sherman, Director of the SLRC, for funding the symposium giving rise to this book, held at the Gold Coast in Queensland, Australia, over 9–10 December 2013. Funding support for the symposium was also provided by the Griffith Social and Behavioural Research College, for which we are most grateful. We extend our sincere thanks to Carol Ballard, Julia Barker and Madonna Adcock for their fantastic work in organising the symposium and ensuring the symposium ran smoothly. We are especially grateful to the symposium speakers, many who travelled from far distant shores to take part, for their enthusiastic and engaging contributions to the success of the symposium. We thank also those authors who subsequently contributed chapters, helping to broaden and deepen the jurisdictional reach of this book. This book has been a collaborative venture, and we extend our thanks to each of the contributing authors for their participation. The diversity of the contributions lays the foundations for what we hope will be a growing focus on the possibilities of global responses to the need to overcome animal cruelty. Finally, we thank our publisher, Springer, and in particular, Neil Olivier and Diana Nijenhuijzen, for their editorial and other support in bringing this collection together.

References

Kelch, Thomas G. 2011. *Globalization and animal law: Comparative law, international law, and international trade*. Alphen aan den Rijn: Wolters Kluwer.

Wagman, Bruce A., and Liebman Matthew. 2011. *A worldview of animal law*. Durham: Carolina Academic Press.

Part I
General Issues in Animal Law and Welfare Science

Chapter 2
Second Wave Animal Law and the Arrival of Animal Studies

Paul Waldau

Abstract Many societies today reflect that animal protectionists are going beyond efforts to establish fundamental protections of the kind signaled by terms like "animal rights," "animal welfare," and even "animal liberation." This Chapter uses four interrelated questions to explore this new stage of animal law: (1) how should animal law courses in law school proceed given the obvious interdisciplinary reach of the larger field of Animal Studies?; (2) what is the place in second wave animal law for answers to the root question "who, what are 'animals'?"; (3) why is the field of "Animal Studies" important to legal education and, more broadly, education as a whole?; and, (4) What roles have personal connection with and meeting of nonhuman animals played in first wave animal law, and what roles might they play in second wave animal law, the larger field of Animal Studies and its sub-disciplines, and the educational mega-fields we know as "the humanities" and "the sciences"?

2.1 Introduction

How do we go about our attempt to see the future of animal law? We can speak rather specifically, of course, about the near-term projects on which we will focus in the coming few years. But what I'm after is our need to assess the possible directions and shape of animal law *decades out into the future*, which takes an altogether different imagination than does looking only at possible projects in the coming years. To meet this need, I examine the burgeoning academic field known as "Animal Studies," but which is known variously under other names such as "Anthrozoology," and "Human-Animal Studies" (there are at least a half-dozen other options that I do not list here. For further discussions, see Waldau 2013). In order to see why the field of law and the subfield of animal law need perspectives from Animal Studies, I employ what I will refer to as *three lenses*—these include a generalization, a warning about social movements, and an imaginative image to help us see the terrain we walk as ethical creatures. *The generalization* is this pithy observation by Robert

P. Waldau (✉)
Canisius College, Buffalo, NY, USA
e-mail: pwaldau@gmail.com

D. Cao, S. White (eds.), *Animal Law and Welfare - International Perspectives*,
Ius Gentium: Comparative Perspectives on Law and Justice 53,
DOI 10.1007/978-3-319-26818-7_2

Cover (1986): "*Law is the projection of an imagined future upon reality.*" *The warning* is embodied in this observation about social movements: "*It's hard enough to start a revolution, even harder still to sustain it, and hardest of all to win it. But it is only afterwards, once we've won, that the real difficulties begin.*"[1] The third lens is a simple, three-word phrase that I take to be an imaginative image luring us into a healthy intersection with lives beyond our species—this image is "the animal invitation." This image suggests that other animals have lives that, if we notice and take them seriously, call out not only to the deep ethical instincts so characteristic and defining of human lives *but also* to our desire for communion or relationship with the larger life community. Other-than-human animals invite us to inquire about them, to take them seriously, to take enjoyment in our similarities and differences, and even to relate to them as community members. Using these lenses, this chapter tries to look at both mid- and longer-term issues. So I will speculate about the future of animal law out to, say, the year 2030 and also about longer term issues regarding the shape of animal law at the end of this twenty-first century.

2.2 Focusing with Three Lenses

Peer with me through three lenses as we attempt an honest, humble, and imaginative view of the future of animal law.

The generalization: Law is the projection of an imagined future upon reality.
The warning: It's hard enough to start a revolution, even harder still to sustain it, and hardest of all to win it. But it is only afterwards, once we've won, that the real difficulties begin.
The three-word phrase: The animal invitation.

These lenses can do more than help us see further into the future. They can also help us today as we try to recognize what is currently happening in law, animal law more specifically, and allied fields such as Animal Studies. Seeing such things better, we can begin to talk to one another about what might happen in the near term with animal law, as well as the mid-term prospects and the long-term future of both animal law and Animal Studies.

One important problem that complicates any attempt to see the future of animal law is the widespread ferment one easily discerns in humans' relationship with other animals. One indication of the pervasive changes is that the *Animal Law* casebook from Carolina Academic Press, which first appeared in 2000, went into its *fifth* edition in 2014. Another indication of change in the offing is the increasing number of schools offering an animal law course. A third indication of our society's opening up to animal issues is the rapid emergence of the field of Animal Studies. It is not, in fact, *at all hard to find other signs* that people are reevaluating humans' possible

[1] Said by Larbi Ben M'Hidi to Ali la Pointe in the 1966 documentary "The Battle of Algiers" directed by Gillo Pontecorvo. This influential film in the history of political cinema focused on the 1957 revolution in Algeria.

relationships to other living beings—these signs are all about us in media, art, business opportunities, scholarship of many kinds, religious communities, local and state government, national policy debates, and on and on.

This rich, extraordinary ferment in humans' relations with other living beings has three important backgrounds or contexts. One is that the changes in the way we think and feel about "animals" are taking place even as much else is in ferment in our culture. Another background or context for increasing concern for nonhuman animals is that the increase takes place against prevailing human-centerednesses*es* that are pervasive and often altogether unhealthy. Finally, shifts in attitudes about animal issues are going forward even though we have very serious unresolved human problems. I take it for granted that today there is a great deal of first-rate evidence for three claims: (1) that change of many kinds is happening all about us (this has been going on for centuries, of course), (2) that our society has an astonishing number of features that attest we are decidedly "human-centered," and (3) that our societies are plagued by myriad unsolved human problems.

At the end of this chapter and in both *Animal Rights* (Waldau 2011) and *Animal Studies* (Waldau 2013), I argue that we will handle our human-on-human problems *far* better if we stay open-minded to animal law prospects. To explain why this is so, and especially to make the point that animal law is *in and of itself* an extraordinarily important field *even if* it does not generate mostly *human* benefits, let me ask you to look through the following three "lenses" with the goal of seeing better the possible futures of animal law.

2.2.1 The First Lens—Cover's Generalization

Recognition that a society's *law is a projection of someone's imagined future upon reality* helps us see that a version of animal law was solidly in place long before the modern animal protection movement. It also tells us that *what we do today and tomorrow* with law in our societies will *also be a projection of a future* for those who come after us. What is it that *we* will imagine in animal law? What will we project *onto our children and their children?* Just as we live in terms of laws that our parents and grandparents enacted—and thereby projected *their* vision about human-nonhuman relationships into the midst of our lives—so, too, we will shape laws and legal concepts and thereby project our ideas about our membership in a more-than-human community onto our children and their children. This means that much is at stake in the future we choose. It also means that *even if we do nothing*, we will project onto our children's future the vision that our parents and grandparents projected onto us. There is no hiding from the fact that what we do, or fail to do, will impact our children and grandchildren.

The impact of our choices on future humans is but one reason that many active citizens interested in animal law seek changes today. Of equal or greater importance is the impact our laws will have on life beyond our species. The fact that today's animal courses were overwhelmingly demanded by students rather than proposed by visionary law school administrators or tenured professors has given today's teachers an extraordinary opportunity. Those who teach animal law courses today by and large did not benefit from a form of legal education—or general education, really—that was sensitive to nonhuman animal issues in the manner that animal law courses are today. Those who are privileged to teach animal law today have been given the chance to teach a course less dominated by the human-centered concerns that shaped so decisively their own general and legal education. The upshot is that contemporary courses can fully address the inadequacies of present laws that govern the inevitable intersection of humans and other living beings.

It should not go unnoticed that most students in animal law courses today react strongly to these inadequacies. This phenomenon prevails in spite of the fact that today's students also had, as did their present instructors, relentless emphasis on the importance and "rightness" of human exceptionalism.

As in the past, today human-centerednesses of many kinds and an astonishing variety of exceptionalism and its exclusivist strategies often arrive in our lives as a virulent political correctness, even a belligerent moral arrogance that insists that focusing on humans *alone* is morality itself. If one pierces through this fundamentalist veil, what emerges is that such insistence is often less than a privileging of all humans, but, instead, merely a privileging of one human group, such as males, a single culture, one's own nation or religious tradition, or economically important consumers.

A vast array of exceptionalist fundamentalisms, then, make raising the interests of other-than-human animals a risky enterprise. Yet law schools can be, in general, a wonderful place to develop free expression. This is one of the reasons that so many law schools have been hospitable to animal law courses proposed by students and then, with the help of faculty and administrators, realized. The result is often only a single course, true, in a curriculum that continues to be dominated by strident, human-exceptionalist rhetoric and concepts, but the upshot is extraordinary—these developments in animal law are the most developed part of the broader field Animal Studies (more on this below).

2.2.2 Peering Through a Second Lens—Larbi Ben M'Hidi's Warning

This warning humbles anyone who seeks major changes in prevailing political realities, social practices, or ethical visions—all of which *any* legal system embodies.

> *It's hard enough to start a revolution, even harder still to sustain it, and hardest of all to win it. But it is only afterwards, once we've won, that the real difficulties begin.*

Certain facts today—such as the number of courses, journals, conferences, and bar association groups—suggest strongly that "animal law" has arrived not just as a topic and/or course possibility, but as a field. The earliest pioneers in this field readily confirm the first part of Larbi Ben M'Hidi's warning— *It's hard enough to start a revolution*. They struggled for decades, and today we can see well how their insight and persistence paid off.

The revolution has thus started, and it is easy to see that legal education's embrace of animal law courses thereby reflects that a first wave of the field "animal law" has arrived on our shores. From 1977, when the first such course went forward at Seton Hall Law School, to the year 2000, difficulties in getting this revolution off the ground must have been formidable, for fewer than a dozen such courses in the United States were started. But with Harvard Law School's offering of an animal law course (called "Animal Rights" and taught by Steven Wise in 2000), the field took off nationally and internationally—within only a decade, for example, the United States saw a ten-fold increase in the number of such courses. Such an increase would be significant in any field, of course, but particularly so in the case of animal law—there are just under 200 accredited law schools in the United States, and with well over 140 schools offering an animal law course in 2014 (virtually all of which were the result of *student* petitions), a large majority of American law schools now offer this kind of course. This also means that thousands of law students are taking this course each year around the world—given lawyers' proficiency at getting their voice heard in public policy circles (compare them to, say, veterinarians who historically have not been successful at having their voices heard, let alone taken seriously, in public policy discussions), the emergence of so many animal law courses means that future policy discussions touching upon "animal issues" will likely be attended by graduates of these courses, all of which augurs more critical thinking, technical sophistication, and compassion-driven content for future policy discussions in this area.

In the spirit of critical thinking, let me call attention to some important disputes and even the polarization that attends much discussion of the animal law phenomenon—it is likely that this phenomenon will seem wonderful *if* one is interested in and favors (i) the ethics-tenor of contemporary animal law education (by which I mean the prevalence of ethics-based and justice-focused concerns as students express why they pursue and hope to expand "animal law"), or (ii) a similar concern for animal protective values and outcomes in the field of Animal Studies, or (iii) developing ever-greater sophistication when scrutinizing law for evidence of compassion. *But* if one works in one of those institutions, professions or businesses where the use of law to protect nonhuman animals is seen as a threat to humans' hegemony, then these developments no doubt create a chill.

My own sense is that all burgeoning social movements include any number of these features, if only because they are energy-driven rather than parasitic on established values, as is an economics-driven field like patent law. My observations here may be biased—I'm personally interested in "the cause" and have confidence that the future of animal law is robust. Even after I correct for such "biases," it seems "obvious" to me that the vast majority of criticisms of the animal law phenomenon are based on a failure to see the movement for what it is—the opening up of law beyond the species line.

2.3 First and Second Wave Animal Law

2.3.1 *First Wave Animal Law*

It seems to me that these initial stages of the animal law movement have opened up so many questions and issues that it is clear that first wave animal law reflects a true revolution. Caution is in order, for there are always difficulties when one attempts to assess the future of a young social movement, which contemporary animal protection is even though it is sustained by roots in ancient cultures' visions that we, as humans, are capable of rich forms of animal protection. But celebration is also in order, as it were, for this early phase of modern animal law has surmounted the challenges of the first part of Larbi Ben M'Hidi's warning— *It's hard enough to start a revolution*. But no one who celebrates this will fail to recognize Larbi Ben M'Hidi's caveat that revolutions are *even harder still to sustain*. So let me use this warning to raise some issues about what I have characterized as first wave animal law.

First wave animal law has prompted forms of animal law education that relies greatly on the following:

- use of traditional law school methods (such as casebooks, teaching of income making skills);
- traditional legal reasoning patterns in litigation;
- a foregrounding of questions about "legal rights" as if that important legal tool is the "be all and end all" of legal protections;
- preoccupation with those animals we dominate and live with, and which were commonly termed "pets" until the late twentieth century when a certain political correctness prompted emergence of the now favored term "companion animals"—there has been much discussion of research animals, too, and more recently food animals (wildlife, it seems to me, receives much attention because this category of nonhumans is the focus of species-level discussions dominated by extinction risks, but the sentience of the individual "wild" animals is less talked about, although there are *many* substantial efforts to protect and rehabilitate injured wild animals of both endangered and non-endangered species);
- preoccupation with the special cognitive abilities of nonhuman animals, all of which risks a surreptitious affirmation that the human mind, often seen as the pinnacle of cognitive abilities, is the definitive measure of the universe; and,
- forms of activism that work primarily within established public policy circles, particularly courts, legislative bodies and administrative agencies as if these circles are where we (our modern, industrialized societies) create and sustain broad social values.

These are achievements both interesting and important—in many ways, they have so far proven a good road map to starting the revolution. While there can be little doubt that these achievements will continue to play important roles in the future, below I suggest that these features of contemporary animal law education and activism are first wave "stuff," if you will, and that the future of a robust animal law field requires more and different approaches.

2.3.2 Second Wave Animal Law

Consider how each of the achievements of first wave animal law listed above might be enriched by an imaginative, humble, ethics-sensitive and interdisciplinary engagement with other living beings. The result can be a *qualitatively different* stage of animal law.

- *Use of traditional law school methods*—While animal law educators' use of traditional law school methods (such as casebooks) will very likely always remain essential, *much more* is needed. The typical form of animal law education, with a few notable exceptions, is a survey/introductory course taken as an elective by interested law students. Second wave animal law must be creative in reaching beyond this important but clearly preliminary offering. As anyone who has taught an animal law course will attest, there is *no way* that a single course can address the many different dimensions and possibilities of animal law. Some of the creativity needed to create a battery of courses that cover animal law adequately will come from liaisons that animal law creates with (i) other law courses, such as environmental law, and (ii) other academic disciplines that have informative perspectives on nonhuman animals.
- *Traditional legal reasoning patterns in litigation*—Animal law activists and litigators *should be* experts at traditional legal reasoning patterns and in litigation techniques, since these forms of expertise have often been used in support of oppression of *human* "others." First wave animal law has had success engaging legal systems' technicalities, many of which have been designed to limit non-lawyers' access to courts. Such mastery has been important because such technicalities have long been utilized by the wealthy and powerful segments of modern societies who can hire the most accomplished and politically connected lawyers for the purpose of limiting claims that might divest the powerful of their privileges. There are, in addition to standard legal reasoning patterns, important perspectives that non-lawyers have to offer animal law that help illuminate the values that drive the legal system. Both law insiders and those outside law have helped everyone see that legal systems *clearly* have birthed and nurtured some very oppressive ideas and practices, such as oppressive notions of property and business that have been historically used against not only nonhumans, but humans as well. Further, since legislative fiats can override the breakthroughs achieved in courts (even at the constitutional level via amendments authorized by a super majority), in liberal democracies one *must get to the people* who are the ultimate drivers of consumer patterns, economics, and legislative pronouncements, all of which will play key roles in the future of animal law.

- *Foregrounding questions about "legal rights"*—Because foregrounding of a "question everything" mentality about the very concept of "legal rights" opens doors and minds, this practice has had a high profile in first wave animal law and thus has been historically and psychologically important. Extending rights has been a key element of first wave animal law, but it is equally important to see that this valuable tool has its limits, for it is but one tool among many others in the legal tool box. Invoking "rights for animals" is part of the larger moral revolution, but rights-based approaches have complex features that can actually disadvantage if used insensitively—for example, David Kennedy in his 2004 book *The Dark Sides of Virtue: Reassessing International Humanitarianism* supplies important examples of how public remedies involving "explicit rights formalized and implemented by the state" do not work well in some other cultures (Kennedy 2004, p. 11). On the nonhuman front, rights can surely work well in some contexts for, say, chimpanzees and other extraordinarily complex nonhuman animals, although there are also other legal tools that can work as well. But for many nonhuman animals that do not fit so well our fascination with cognitive abilities, the tool of individual *legal* rights with public remedies might not work well at all. Instead, other tools, such as legislatively mandated prohibitions on ownership, may work far better.
- *Preoccupation with companion animals*—first wave animal law's preoccupation with those companion animals who are our family members has been a brilliant populist move, but of course animal law as a field has many tasks to accomplish beyond protecting these important lives and family members. There are obvious political realities at play here, and it cannot be overstated how important these are for first wave animal law. But the revolution will surely prove *even harder still to sustain* if it stays at this level. Discussions about protecting other nonhumans, such as those used as research tools, still demand much attention, and more recently food animals are on many active citizens' ethical radar. As animal law proceeds with its revolution, we all do well to keep in mind that the very notions of "companion animal," "research animal," and "food animals" are *human-centered* categories (that is, categories we create—they do not fairly and fully define the living beings within them). Wildlife remains a category less fully addressed by first wave animal law, in part because the early revolution gained advantages if, for a variety of reasons, it concentrated on cognitively sophisticated nonhuman animals or those with a companioning genius. Simply said, such concentrations gave first wave animal law a foothold in humans' imagination. But as everyone involved in animal law is so acutely aware, there are *many* other animals "out there," and animal law needs to engage them with creativity *and* in light of these additional animals' actual realities. Doing so will necessitate much greater entry into new realms, including environmental law in order to challenge this important field's one-dimensional focus on species-level discussions dominated by extinction risks. In short, second wave animal law needs richer approaches, for the present preoccupations are but one part of the future of animal law and do not provide detailed road maps to all future stages of this revolution.

- *Preoccupation with cognition*—people in the animal protection movement have long been aware that cognition is but one of many factors that draw humans' attention to nonhuman animals, as Peter Singer's historically important emphasis on sentience four decades ago underscored (Singer 1975). Yet there is *no doubt* that the heavy concentration on the cognitive complexities easily demonstrated in certain other social mammals (the nonhuman great apes, elephants, whales and dolphins) has opened minds to the fact that out beyond the species line are brainy, amazing nonhuman animals *who* exhibit overwhelmingly rich and compelling personalities, lives and various forms of intelligence. This is one reason some publishing houses these days now suggest that their authors avoid using the relative pronoun "it" and feel free to use "who" when referring to a nonhuman living being. There is, however, something of a two-edged sword risk that lurks nearby, for as noted above a too heavy preoccupation with cognitive complexities can amount to a surreptitious affirmation that only those nonhumans that are *cognitively* somewhat *like us* get protected. If the upshot is that those nonhuman animals that are not like us therefore count *less*, the price will be very high. Many ethicists, scientists and animal protectionists are deeply uncomfortable with this sort of cognitive hierarchy as the leading edge of our ethical obligation and legal protection—the insight that drives this discomfort is the insight at the heart of the most famous quote in the western world's animal protection movement, Bentham's penetrating observation that "[T]he question is not, Can they *reason*? nor Can they *talk*? but, Can they *suffer*?" (Bentham 1970, p. 283, n. 2).
- *Focus on activism in courts, legislatures and administrative agencies*—work within public policy circles promoted by first wave animal law has been crucial, particularly courts, legislation and administrative regulation. *But* these venues, as important as they are, are decidedly *not* the only circles of modern societies that *create and sustain* our social values. Social values are rooted in deep-seated psychological factors like inherited customs and cultural stories. That these factors plainly go far deeper and well beyond specific laws and legal systems is a notion to which every law student is introduced during the first year of law school and then throughout the multi-year curriculum. This happens as every lawyer-to-be is trained to think about fact situations or legal arrangements that "violate public policy."

The literature of the first wave of animal law characteristically involved lawyers talking to judges, lawyers and law students. This important function, clearly to be expected in an early phase of a legal revolution, is only a first step in developing the conditions for fundamental change regarding humanity's inevitable interactions with other-than-human animals. In crucial ways, such a fundamental change requires numerous additional changes *beyond the legal system*. The good news is that such changes have begun, for they can easily be found throughout the ferment already mentioned. This ferment contributes in psychologically important ways to what can happen inside legal systems—for example, the rich concerns for companion animals no doubt appeal to legislators with dogs at home and those judges whose lap dogs join them on the bench, even if under the protective covering of the judicial robe. Everyone in law is also impacted by the grass roots ferment reflected in the use of legislative tools like popular initiatives, for these help us all to see the great breadth of concern for "animal issues."

Why might it be hard to sustain first wave animal law's achievements in starting a revolution? Since the achievements of first wave animal law have been so important, the risks of remaining at this stage might seem minor, especially given that the revolution has been started. But the risks are real, though less than obvious at first—they involve *future* possibilities, which everyone senses are quite remarkable. Said most simply, even though much good growth has taken place, there is a risk of animal law in its present state remaining ingrown, bound by the legal profession's traditional approaches to education and by today's legal establishment. This establishment is tolerant in some senses (it has, for example, permitted student demand to nurture animal law's growth), but it still harbors—indeed, promotes—narrow, arrogant and ultimately harmful forms of human-centeredness that will remain the unchallenged heartbeat of modern law *unless* animal law progresses to much greater fullness and power. Said another way, the revolution which has started will not long survive without growth.

Legal education has often been inhospitable to interdisciplinary considerations, and on the issue of nonhuman animals in particular, legal systems often feature inherited views (such as the claim that only humans can have legal rights) that dominate, eclipsing the common sense proposition that *any* legal system can, if the host society wishes it to do so, offer rights-based protections to nonhuman animals. Conservative approaches that are blinkered can easily become reactionary over time, and one hallmark of such an "evolution" is the fostering of negative attitudes within established legal circles toward the emergence of animal law. The following story illustrates how one reaction turned up in an unusual place. I happened to teach an animal law course at a famous law school a number of years back. When I visited the faculty lounge during my first week of teaching, I met a number of the core faculty members who were enjoying both wine and conversation. After I joined this primate social occasion, one of the core faculty members asked me what I did. When I replied, "I teach the animal law course," his immediate response was a loud, uncompromising "*Oh, that's B.S*!" (in the interest of decorum I have substituted "B.S." for the actual word used). Immediately, another faculty member rebutted the first assertion in an equally eager tone, saying, "*Oh, no it's not!*" On this occasion, the first faculty member spoke freely, I suspect, because he knew absolutely *nothing* of the course's subject matter, how the course was run, what the student dynamics were, what the level of rigor and/or scholarship was, or what the state of contemporary scholarship in this subfield of law was.

In his facile, ignorance-driven dismissal, this highly educated, very accomplished faculty member reflected attitudes that only two decades ago dominated all law schools. But today, first wave animal law has successfully started a revolution, such that animal law courses are now available at a majority of American law schools. One sign that the revolution is progressing is that such less-than-polite responses are far less likely at law schools where the faculty is secure with legal education's wonderful tradition of open-minded inquiry. But if one is familiar with such signs that the first wave of animal law has started a revolution, no one familiar with today's law schools asserts that the long-prevailing tradition of human exceptionalism in legal education is now weak. It remains *very* powerful, appearing in both overt and far more subtle forms. Second wave animal law has the task of engaging even hardened skeptics who support the "humans only" thesis that has driven law for centuries and still today dominates in the vast majority of courses offered in the modern law school curriculum.

What drives some people's continuing, wholesale dismissal of animal law is a deeper, culturally ingrained attitude that the philosopher Mary Midgley (1984) has called "the absolute dismissal" of nonhuman animals characteristic of western culture generally. Any number of things make this kind of radical dismissal *terribly* problematic—one is that most other human cultures have *not* taken this position; another is the ethic of free inquiry I describe above, for a complete dismissal is so undetermined by known facts today that it can only be characterized as ignorance-driven. As the saying goes, "those who know nothing about a topic can speak freely." Yet only a little inquiry today readily shows that many sciences, the field of ethics, and even the formerly staid study of religion and spirituality offer a variety of rich ways to inquire about other animals' *actual* realities. In a very real way, many new (that is, non-traditional in western culture) approaches to other living beings reflect humans' elegant abilities with ethics, imagination, community and humility. In effect, such approaches attempt to notice other animals on *their* terms, not on solely on our terms. Law as traditionally taught was, and in some quarters still remains, truly autistic about this possibility because it was deeply rooted in its own past absolute dismissals of nonhuman living beings.

The dominance in modern legal systems of such dismissals prompts one to ask, how much of first wave animal law has *really* penetrated the legal education establishment? Today, law still is done virtually exclusively on *our human terms*, with only human interests playing determinative roles. One can see this in what conferences exist, what courses are offered, who is hired, and whether animal law courses are expanded or confined to a single offering. In all of this the stark dualism described below—legal persons *versus* legal things—is clearly the controlling reality that functions like a heartbeat in today's law. Through the offices of first wave animal law, a door by which a few nonhuman animals may actually exit the subordinated category of "legal things" has been outlined. Further, in the larger society and in some precincts of the legal system, first wave animal law has clearly facilitated the emergence of more open-mindedness on animal law as a viable subject. Most importantly, a powerful description has been stated in first wave animal law regarding the *radical* inadequacy of the "*us versus them*" dualism of "legal persons versus legal things" that every legal student today is still indoctrinated into by virtue of attending law school. This dualism exists outside law, of course, such that virtually all domains and institutions in our society have inherited it *and by and large still operate under it*.

Some legal theorists and philosophers assert there *now* exists sufficient legal and scientific grounds to make major changes (see, for example, Wise 2000; White 2007); further, there is some evidence that sufficient political will exists today to make fundamental changes (for example, there is information on polls below that could be used to support such an assertion). But even if there are some encouraging signs around the world, we remain in a period of education about animal law where many institutions, disciplines and even professional organizations still exhibit a stark refusal to admit, let alone seek out, *other animals' realities.* Ignorance serves the status quo, but just as the economist Paul Samuelson once observed "theory advances, funeral by funeral," so too in legal education the field of animal law will have more possibilities as new administrators and professors familiar with the accomplishments of first wave animal law come into power.

As for second wave animal law developing ways to push beyond today's preoccupation in law and in legal education with humans as the measure of all things, there is an important shortcoming in the present system that must be addressed. As with so many other major fields in education that are only now engaging nonhuman animal issues, the legal system does not itself generate much expertise, if any at all, about nonhuman animals. It is true that the litigation side of law includes the important mechanisms of expert testimony in courts, but those of you who are or have been trial lawyers know that expert testimony is at best a complicated matter—each side in our advocacy system can offer "experts" whose credentials can be manipulated by crafty trial lawyers. Judges have some criteria by which to assess such ploys, but as a practical matter triers of fact (whether they be judges or juries) are not at all sophisticated in deciding which "expert" is to be believed.

Even worse problems regarding "expertise" appear in the area of legislation because it is so intensely impacted by lobbyists. The animal issue in particular is well described by Will Rogers quip (he was echoing an idea earlier advanced by Otto von Bismarck):

> People who love sausage and respect the law should never watch either one being made.

The irony of the ugly process of making sausage, a dregs-based meat where flaws are covered over by intense spices, being compared to the all-important law making of legislatures is particularly apt for animal law issues.

2.3.3 Entering the Tower of Babel

Inputs on nonhuman animals available to litigation via expert testimony and to legislators via lobbyists and research are important mechanisms, but they are, from the standpoint of an ethic of inquiry, very weak. Law, in effect, uses a non-scientific approach to other animals even if some establishment science can enter the courtroom or legislative chamber through hired experts or lobbyists. Note, however, who it is who is invited into courtrooms—it is establishment scientists with "credentials." The incomparable Jane Goodall is thus admitted, but so are research scientists whose opinions can be unduly compromised by those who employ or retain them, as has happened again and again in public health disputes over leaded gasoline, smoking tobacco, the dangers of asbestos, and the effects of widespread use of pesticides. No one who has used the expert witness system in today's litigation world or watched legislative hearings is unfamiliar with the problem of expert opinion being distorted by some value other than the truth.

So how do we get high quality information before those who make decisions about law? One reason that lawyers and legal scholars rarely attempt to go beyond their own expertise is that "out there" (that is, out in the non-law precincts of society) lawyers and legal scholars find the vast body of claims and communications about other-than-human animals to be something like the biblical and now legendary Tower of Babel. Plainly speaking, there are many different ways of talking about other animals. One of them—the common phrase "humans and animals"—already puts into place the dualism that legal systems have been anchoring for centuries now. This is odd, of course, because we know that scientifically humans are apes, primates, mammals and vertebrates (all of which are *animal* categories). But vernacular speech and the legal system use the highly anti-scientific "humans and animals" division, all of which simply anchors human-centeredness ever more firmly in people's minds. Unless we acknowledge that we have been trained to speak sloppily about this subject, there is no simple way to attack the dualism. What is needed, then, is a straightforward and honestly pursued ethic of inquiry about our slovenly language habits and ethically problematic refusal to take other animals seriously.

Interestingly, though, our society has today developed the "ferment" discussed early in this chapter, such that contemporary thinking about some of the life out beyond our species is developing rapidly away from ignorance and toward a more informed character. Thereby, traditional out-of-hand dismissals of other-than-human animals are being powerfully challenged in *many* different ways in many different fields. So lawyers will miss much if they are too inclined to stay solely within legal precincts, for in many fields today extraordinarily rich thinking is developing about our past, present and future with other living beings.

2.3.4 An Aside—Law as a Concern of Non-*Lawyers*

For obvious reasons, there is *much* talk about law outside of legal circles. Public policy specialists, commentators on globalization, social movement activists, philosophers, theologians, NGO leaders, and so many others think that, in discussions of public policy changes and possibilities, *their own fields* belong every bit as "front and center" as do law and legal technicalities. So even if lawyers have admitted expertise at public policy discussion, they are by no means the only voice needed in public policy discourse.

So here I add two observations about law that are quite sobering, though in very different ways. Each is from a non-lawyer—the first is from an activist, the second from a theologian whose ideas have sparked a remarkable world-wide movement known as "religion and ecology." These together suggest how truly important it is that lawyers and second wave animal law pay close attention to *non-lawyer* comments on animal issues and, more specifically, on the socio-cultural background that makes changes in law possible *or not.*

Let me turn to that "theologian" whose ideas drive today's world-wide movement known as "religion and ecology."[2] His name is Thomas Berry, and his work sounds the theme that much of our western consumer culture, including today's version of "law" in industrialized countries, is human-centered *in the extreme.* Clearly, he would *not* qualify as an expert witness in the technical sense that term is used in litigation, just as he would not likely be called as an "expert" of the kind that public policy debates feature. But his insights are dramatic and impressive, nonetheless, and they led him during his lifetime to write as follows:

> … [t]he deepest cause of the present devastation is found in a mode of consciousness that has established a radical discontinuity between the human and other modes of being…. (Berry 1999, p. 4)

Berry labels human-centered thinking of this kind "anthropocentrism," citing Albert Einstein's definition of this word as meaning an "optical delusion of human consciousness" where we come to regard "humanity as the centre of existence." More specifically, Berry addresses a certain feature of American political and legal life pertinent to the risk to which first wave animal law is subject—"… from its beginning the American constitution was clearly a document framed for the advancement of the human with no significant referent to any other powers in heaven or on Earth" (Berry 2002, p. 13). What Berry suggests reveals that animal law remains at great risk.

Such comments make one pause, given the iconic status of the U. S. Constitution in so many circles. But that document's "feet of clay," as it were, are well known in light of the framers' acceptance of race-based slavery, the exclusion of women, and … well, the list goes on. The American legal system is, *as everyone knows*, far from perfect, and this alone strongly suggests the importance of American law students studying *other* legal systems to see if those other systems offer insights into how to create more openness on nonhuman animal issues. First wave animal law clearly takes risks when it unreflectively assumes that *American* jurisprudential values and habits of mind represent what jurisprudence can generally do—this is part of Americans' regrettable tendency to think American practices and values represent the highest version of human achievement. In this sense, first wave animal law could, without major adjustments, project *an imagined future upon reality* that leaves much to be desired—hence, the concern to see how we might move to second wave animal law. Second wave animal law needs to engage such parallel insights, to be enriched by them, and to nuance other fields' insights so that other-than-human animals are on the topical and ethical radar, as it were, of many other disciplines.

---

[2] The leader in this ecumenical movement is Yale University's Forum on Religion and Ecology—the website is http://fore.research.yale.edu/

2.3.5 More Risks Inside First Wave Animal Law Precincts

Second wave animal law will gain energy, insight and followers *when and if* it engages these debates outside standard law precincts. I refer to these as "extra-legal debates," and it is important that those interested in animal law contribute to them so that these debates can be sophisticated, informed and nuanced in the best tradition of first-rate legal scholarship. Equally, it is important that legal scholars, educators, judges, and law-based policy makers *listen and learn* from such extra-legal debates. Such connections across disciplines ensure that the legal profession avoids one of the pitfalls into which the veterinary profession has fallen, namely, the assumption that veterinary profession alone is entitled to lead our society in developing insights about what is often called the human-animal bond. In many ways, animal law today is a protest against the veterinary profession's failure to lead in this area. But in a way eerily similar to what has happened to the veterinary profession, law on the nonhuman animal issue, when it proceeds *solely* on its own lights, is in a quandary, for it too has reflected the dramatic inadequacy of using *legal tradition alone* to structure this inevitable and essential intersection in our human lives in a more-than-human world.

In every culture and surely in liberal democracies, leadership regarding the human intersection with other-than-human animals and the broader more-than-human world will be dispersed among many human professions, disciplines and interest groups. Law and the legal profession must be part of this leadership, but the subject of animal law no more belongs to the law profession alone, or to legal education, than the subject and last word on animal welfare and the realities of nonhuman animals belong to the veterinary profession.

Hard to start a revolution? Harder to sustain, hardest of all to win. Those who compare one legal system to another talk regularly about how changes take place in legal systems, and first wave animal law clearly reflects that important shifts have already taken place in common law legal systems (a good example is the emergence in the United States of felony-level penalties for acts of animal cruelty). Comparativists suggest, however, that *truly fundamental changes* in legal systems happen apart from the stroke of the legislature's pen and before the judge's gavel comes down:

> … there are features of the law which can only be changed at the slow rhythm at which the civilization of the country itself, the sense of justice of its citizens, its economic structure, language and social matters themselves are changed. (David and Brierley 1978, p. 18)

The operative insight here is that without pervasive social change, the possibilities of changing law *at a fundamental level* are limited—"The legislators may, indeed, with the stroke of the pen modify the actual legal rules, but these other elements and features nonetheless subsist. They are not so arbitrarily changed because they are intimately linked to our civilization and ways of thinking" (David and John 1978, p. 18). Importantly, such deep, persisting features of legal system are major factors in establishing the all-important continuity sought by legal systems. At the same time, however, they foster the inherent conservatism of law. Those citizens who are favored by special privileges under the existing order of the legal system no doubt feel strongly that privilege-threatening developments, like animal law, are not in continuity with the rest of the species-centered law that anchors these elites' privileges. This is normal, for just as Upton Sinclair once observed, "It is difficult to get a man to understand something when his salary depends upon his not understanding it" (Sinclair, 1935), it is also difficult to get many citizens and businesses to support new legal protections when continued enjoyment of their own privileges depends upon their blocking such new protections.

Said another way, a certain selfishness, perhaps even autism, prevails in law when it comes to the marginalized, whether they are human or nonhuman. To change any feature of the system, one has to work *within the system*, complying with legal procedure and other legal niceties. This applies with special force to those who would change our treatment of nonhuman animals. In the United States, for example, protection of marginalized humans can call upon the ideals of the founders and fundamental protections in the human-centered founding documents. But those who would use the legal system itself to remedy abuses of *nonhumans* must make their arguments in places where, historically, human exceptionalism has been the norm. In essence, advocates for nonhuman animal causes must always play in the oppressors' power alley, as it were. Some constitutions today call out the issue of nonhuman animal protections (for example, India and Germany), but in the United States this possibility is widely considered laughable. True, the existing system allows challenges and even amendments, but these were hard to achieve for women and those subject to racial and ethnic discrimination. Further, as is well known, even at its lower levels the existing legal system has myriad ways of defeating challenges—standing limits, how statutory language is read (as with the meaning of "necessity" when it comes to justifying harms to nonhuman animals) (Francione 1995, pp. 4–5, 13, and 21),[3] which harms and measures of damages are permitted, and of course the age-old limits imposed by lack of enforcement.

So, while working matters out in courts has some potential (the ground-breaking 2000 *Rattling the Cage* creatively calls on what the author sees as the legal system's core values) (Wise 2000), this risks playing out challenges in a system so fully geared to humans that legal fictions and other dishonesties can bar common sense, science-based certainties, and open-hearted ethics. One form of balance to strike is this—play out certain challenges in the courts but avoid the risk of creating negative precedents that will cement in place an altogether ghettoized version of animal law where there will be very little critique of the fundamental features of law as it now centers on humans.

What is interesting is that legislation is no less subject to such games and limits. Some litigation-focused advocates for nonhuman animals have been known to comment that they resort to litigation because prospects in legislatures are *worse* due to the power of special interest lobbying. This is significant because today some polls suggest that *nine out of ten* Americans believe "strongly" that "we have a moral

[3] The author makes detailed points about how games are played with the meaning of words when we want to rationalize abuse of other animals.

obligation to protect the animals in our care."[4] Consider percentage figures for *farm animals* from a 2004 study of Ohio citizens by The Ohio State University—an astonishingly large majority (92 %) of the citizens of this conservative, farming-intensive state agreed or strongly agreed that it is important that farm animals are well cared for. An almost equal number (85 %) had the opinion that the quality of life for farm animals is important even when they are used for meat. 81 % also agreed that "the well-being of farm animals is just as important as the well-being of pets," and 75 % agreed, "farm animals should be protected from feeling physical pain."

How is it, in the face of such numbers, that legislation on animal issues struggles so? Second wave animal law must grapple with this. The answer surely is in cultural values, what David and Brierley called "the civilization of the country itself, the sense of justice of its citizens, its economic structure, language and social matters." The poll numbers just quoted suggest, however, that today we may be edging ever closer to a political climate in which protecting some nonhuman animals is possible in a variety of ways, including not only use of our prized legal tools such as a legislatively-mandated grants of specific legal rights but also use of other legal tools that also create effective, fundamental protections such as legislative bans on ownership of certain nonhuman animals altogether or circumscribing personal property ownership rights with special responsibilities of care and protection. As suggested above, likely candidates for such high level protections are those other-than-human animals who companion us even as they submit to our domestication.

Of course, fundamental protections for those nonhumans whose interests are not so integrally tied to humans' interests, such as wild, free-living nonhuman animals that have never been domesticated, remain elusive. First wave animal law will transition to second wave animal law when nonhuman beings are considered candidates for legal protections by legislatures and courts *no matter what their similarities to human abilities* might be—in other words, an ethic of inquiry will not pursue first and only whether other animals are "like us" or familiar and favored by humans but, instead, whether they have demonstrable cognitive, emotional, cultural and sentience-based capacities that our ethical imaginations recognize as *in and of themselves important*. There is *no doubt* that our imagination, humility and ethics can address a wide range of nonhuman animals, and that we can do so as fully as we now address our companion animals and the wonderful nonhuman great apes, cetaceans, and elephants who are on the radar screens of first wave animal law.

Second wave animal law can also press some additional constituencies, like veterinary medicine, where most people think nonhuman animals have natural supporters. Some animals get healed through the efforts of this profession, for which animal protectionists are overwhelmingly grateful. But, statistically, most of the nonhuman animals that are subjected to veterinary oversight (as happens in food production and experimental situations) are either eaten or used as experimental subjects. This happens because veterinary medicine's establishment (though not the majority of

[4] An example is the 2006 study conducted by Lake Research Partners cited by Marc Bekoff in his 2010 *The Animal Manifesto* at page 3.

veterinarians, thankfully) is addicted to the money streams from research and industry. This has led to a very ingrown set of professional ethics by which veterinary medicine's establishment allows many people to use science and the welfare of the *human race* as kind of fig leaf to cover over the profession's complicity in harming nonhumans. The result is a veterinary establishment that is human-centered in the extreme, and one that is at best deeply impatient with, at worst extremely intolerant of, animal law courses of the kind in found contemporary law schools.

2.4 Animal Studies

It is relevant to the future of animal law that *outside of law schools today*, the number of people who study harms to nonhumans is *far greater* than the number of people who study animal law. In other words, knowledge of what is happening at the intersection of human and nonhuman lives is much greater outside legal realms than within them. In *Animal Studies*, I address just some of the fields that today are fast developing their own version of an interdisciplinary approach to the more-than-human world. The list includes only the following:

> history, cultural studies, education, natural and social sciences of many kinds, political studies, law, philosophy, critical studies, literature and other arts, comparative religion, ethics, sociology, public policy studies, social psychology, geography, anthropology, archaeology, and criminology, to name only some. (Waldau 2013, p. 21)

So what is Animal Studies, then? Here's one definition:

> Animal Studies engages the many ways that human individuals and cultures are now interacting with and exploring other-than-human animals, in the past have engaged the living beings beyond our own species, and in the future might develop ways of living in a world shared with other animals. (Waldau 2013, p. 1)

There are hallmarks or shared features found in each of the disciplines that now participate in Animal Studies—these include (i) a foregrounding of certain nonhuman animals' realities, (ii) an exploration of the diverse ways that various cultural and wisdom traditions have to date negotiated (and might in the future handle) humans' inevitable intersection with nonhuman lives, and (iii) recognition of the eminently personal nature of many such encounters.

In terms of numbers and influence, *animal law is the leading edge* of contemporary Animal Studies. Interestingly, though, many other fields participating in Animal Studies do not have any inkling of this fact or what animal law does. This is, at least in part, due to the fact that scholars and students who work on nonhuman animal issues in other-than-law fields consider what they do as unrelated to animal law. And, not surprisingly, there is what amounts to a reciprocal ignoring, as it were—in animal law, there is often no awareness *whatsoever* of the kinds of work done in other fields in the Animal Studies domain. This is due in part to the fact that the range and depth of the human-nonhuman intersection have long been *so extensive and diverse* that any single discipline is challenged to address even a portion of them, let alone *all* of them and their diverse, complicated implications. Thus, not only does it quickly become obvious that many disciplines are capable of exploring some aspect of such interactions, but upon further reflection it becomes crystal clear that to plumb adequately these extraordinary phenomena, *robust interdisciplinary commitments are needed.* This is why I argue in *Animal Studies* that the field is in fact a prospective super-field crucial to developing our own species' possibilities for a more informed, ethically richer relationship with other animals. To explore, for example, the limits of what humans can know about other animal individual and social realities, one needs multiple disciplines. To gather relevant evidence of how deeply and widely cultures have across time and place engaged other animals, one needs multiple disciplines. To discern what is happening around the world today, one must peer into not only multiple academic disciplines but also actual practices and trends in media, arts, religious institutions, numerous science-based fields, and much more.

The implications of such thorough and interdisciplinary exploring are many—for example, it soon becomes clear that the conceptual resources and theoretical frameworks found in any one discipline cannot control the discourse of all other disciplines, nor exhaust description of humans' multifaceted interactions with nonhumans. Why? Because the social constructions, discourse traditions, and much more in each discipline are characteristically narrow and agenda-driven, such that they miss or obscure keen insights from other disciplines.

To be sure, some features of our shared world are *well approached* by single disciplines, but there remain many subjects that are not well studied by any single discipline (for example, humanity's internal diversity, or our evident complexities as living beings). Even with multiple approaches made possible through an alliance of a few disciplines, or, more expansively, the resources of either *or both* of the two mega-fields of the modern educational establishment (that is, science, on the one hand, and the arts and humanities, on the other), the broad and complex issues of the human-nonhuman intersection are not well studied. Instead, we need *all* of our abilities *in concert* to approach and understand the foundational aspects of our human experience with the living beings beyond the species line.

An interdisciplinary approach is also needed for another, altogether relevant reason—the broad, sometimes chaotic ferment on nonhuman animal issues found today in one academic field after another (as well as in different groups' respective social ethics and cultural attitudes, and in institutions' changing attitudes) hints at the vast array of very diverse "animal issues" that are in play today. For this, reason what is happening in different academic fields is by no means that same subset of issues—a distinct subset of issues is handled in history, another subset in literature, and similar realities exist in geography, sociology, psychology, religious studies and more. Such "ferment" is in one important sense similar to what has happened through the demands of law students for the establishment of animal law courses—students and scholars alike in one academic field after another are discovering important animal-related issues in our past, our arts, our social worlds, our cultures, religious communities and local neighborhoods. In sum, a wide variety of people are noticing that humans over recent millennia have responded to the animal invitation in a stunning variety of ways. The second wave of animal law will be challenged by, but at the same time *emboldened* by, this diversity in countless ways.

One additional reason that *multiple* disciplines need to work together (an implication of which is that non-law disciplines have much to learn from the past, present and future of animal law) is this—in today's Animals Studies, *each* discipline must now swim upstream as it engages the prevailing human-centerednesses that dominate not only the mega-field we know as "the humanities," but also *each* of the individual humanities. The same can also be said of science-based disciplines, for in these as well human-centerednesses have skewed how science has been pursued.[5] In spite of this and, in some senses, *because of* the extent and virulence of the recurring bias which is promoted by the exceptionalist tradition, the scholars working on "animal issues" (who include historians, literature experts, art experts, and comparativists who work in religion, cultural studies, anthropology, politics, sociology and many other fields) have created an astonishing body of work that today dwarfs the literature of animal law. The wide array of disciplines and constituencies interested in animal issues, then, suggests possibilities for the next wave of animal law. The widespread and uncoordinated fascination with animal issues also reveals why, despite difficulties, it has been possible *to start a revolution, sustain it*, and perhaps even *win it*. In order to achieve the developments needed in legal education and legal systems that will sustain a mature "second wave of animal law", then, higher levels of cross-disciplinary work must be promoted by legal educators and law students, of course, but also by *all* educators and indeed the entire administration of institutionalized education—in this way, both *legal and general education* can be taken to a new level.

2.4.1 Four Questions for Legal Education

There are four interrelated questions that can help legal education arrive at second wave animal law. First, *how should animal law courses in law school proceed given the obvious interdisciplinary reach of the larger field of Animal Studies?* Surely this involves starting to talk to individuals pursuing non-legal approaches, as well as becoming more familiar with non-law-focused bodies of scholarly exploration—if the conversation starts, it will grow.

Second, *what is the place in second wave animal law for answers to basic questions like "who, what are 'animals'?" and "Who are the others about whom I should care?"* While the latter is the root question of all ethics (See, for example, Waldau 2013, pp. 196ff), both questions require that humans make a reasonable attempt to know what we can of other animals' individual and social realities. Scientifically-informed and ethically-attuned answers to the former of these two questions will lead second wave animal law to responsible, justice-focused analysis about legislative proposals and litigation outcomes (Nussbaum 2006). When that

[5] Chapter 4 of *Animal Studies* (Waldau 2013) makes this argument in much more detail.

happens, animal law will have all the more traction as it provides legally sophisticated accounts of our human answers to broad ethical questions like "who are the others about whom I might/should care?"

Third, *why is the field of Animal Studiesimportant to legal education and, more broadly, education as a whole?* While this question is answered in some detail throughout *Animal Studies*, the short answer is that without better information about other animals, all of our present academic disciplines, including animal law, are at risk of supporting ethically significant harms. In effect, without such information all forms of education, but *especially* animal law, go forward blindly. Animal Studies is home to multiple disciplines that offer considered perspectives on which living beings outside our own species might be noticed and taken seriously. Law is, when on its own lights, in a poor position for answering this crucial question because law, as so many people know, must often play to *political* realities that trump underlying legal principles. For this reason, law and its goal of justice have often been subordinated to political power and privilege. It is political realities that are, by virtue of the demographics of companion animal ownership, the force behind the paradigmatic role that companion animals have in today's version of animal law—these wonderful animals can, surely, open a door, but the opening can become a trap door if it remains only *owned* companion animals, but not those in shelters or running free in human communities or beyond, who are protected by newly enacted laws. Without help from other fields in Animal Studies, the phenomenon of protecting only owned companion animals can morph into yet another human-centered feature of the current legal system.

Finally, *what roles have personal connection with and meeting of nonhuman animals played in first wave animal law, and what roles might they play in second wave animal law, the larger field of Animal Studies and its sub-disciplines, and the educational mega-fields we know as "the humanities" and "the sciences"?* This question is a key feature of all Animal Studies—as such it provides a bridge between disciplines that contribute to Animal Studies as its own mega-field that can, if well done, enrich both our sciences and our humanities.

2.4.2 Teaching Now—Other Animals' Invitations to Human Caring

Second wave animal law's task is to move the revolution forward, and it can do this by helping a coalition of human enterprises lead our society to a consensus that involves fundamental protections, even legal rights, for some other-than-human living beings. This will indeed be a true revolution. It will move away from humans as the sole center of legal protections, to multiple centers. No doubt additional centers will come slowly, but they will happen sooner, more widely, and better *if* the second wave of animal law is allowed to build on the first wave achievements by working with other disciplines.

We can now see that a second wave is not only possible, *but is beginning to arrive.* As suggested above, many people and a great variety of modern disciplines now reflect that when humans notice and take other animals seriously, humans are responding to the deep ethical instincts that define our human lives—the upshot of such attentiveness to our similarities and differences with other animals has been held by a wide variety of human cultures to be central to human existence, such that Berry referred to the citizens of the more-than-human world as "the larger community [that] constitutes our greater self" (Berry 2006, p. 5).

To nurture development of such awareness, we must recognize that ghettoized versions of law, that is, law framed only on the traditional legal values and techniques that now dominate contemporary legal systems, will *never* be the leading edge of all our interactions with nonhuman animals. Law can no more prescribe, regulate and control how humans respond to the animal invitation than law can regulate *all* aspects of human-to-human relationships. Of course law is helpful in regulating *some* features of our human-to-nonhuman interactions, just as it is valuable in regulating *some* of our human-to-human interactions, but our cultural histories repeatedly affirm that humans' acceptance of the animal invitation must take place in local, daily acts. Above all, acceptance of the animal invitation *always* requires a multifaceted approach touching upon individual responsibility, the arts and storytelling, psychological integratedness, environmental awareness, ethico-spiritual respect for one's local world, and certainly willingness to take our species' moral capacities seriously.

Learning the skills needed to live in the larger community that is home to both humans and nonhumans alike is a skill that is beyond—better said, *before* or *prior to*—the practice of law. To the extent that law is related to morality (which, hopefully is the case often, but which historically has often not been so), law is primarily a *secondary* phenomenon, by which I mean only that law is meant to embody society's deep, morally laden insights, not override them. Law can be used to promote any society's underlying values, surely, but it "embodies" them in the secondary sense of *reflecting* them—a legal system's primary function is not to create such values out of whole cloth and then dictate these back to its host society—this is the thrust of what I mean by "law is a secondary phenomenon."

Given that the second wave of animal law will need to be intensely interdisciplinary and multifaceted, no doubt it too will have some "Tower of Babel" features, for the human intersection with the more-than-human world and its nonhuman animals will always be so rich that it will take work for those educated in different disciplines to learn how to speak carefully and productively to each other. This challenge can be met in legal circles, for law is by its nature a domain where diverse constituencies must meet and negotiate a common language. If we as a species are to notice other animals and take them seriously, law will *necessarily* be in the front ranks *and flanked on all sides by many other disciplines.* Second wave animal law, then, will be prominent as our society goes forward with the animal invitation.

2.5 Five Specific Challenges in Second Wave Animal Law

First, how do we create more environments where discussions can thrive? Those who have pioneered first wave animal law know that the educational establishment has been and is now, with only a few notable exceptions, a hostile world. Importantly, though, the emergence of first wave animal law has revealed some of the remarkable strengths and flexibility of the contemporary legal education. In particular, many law schools around the world today consistently exhibit a profound commitment to the importance of critical thinking, Socratic method, and each student's and faculty member's free expression of ideas. How can this deeply important tradition be honored, strengthened *and exported*, so to speak, to other realms of the academic world where such commitments *in the matter of nonhuman animals* are less robust? Throughout history, pioneers in social movements have been at great risk in political, educational, social and religious establishment circles as they challenged the status quo. How can the success of animal law be used to open up education generally so that the educational establishment is no longer a part of the problem and becomes, instead, a contributor to creative solutions that affirm our larger community? Those educators who choose to make animal law or Animal Studies their principal focus are today welcome in more and more institutions—as with all educators, they need institutions where they can be safe. If this occurs, those who pursue animal law, Animal Studies, and the need for educational reform will be able to teach in a variety of places in modern educational systems, and even rise to positions of leadership.

Second, the importance of political realism must be recognized. Currently, the animal law issue is receiving media attention, but by and large it is not yet a major concern for many people. The lack of civility in many political systems, as well as the political risk of going against reactionary forces in "the establishment," threaten some animal protection advocates—an example is the way Harvard Law School's Cass Sunstein was taken to task during his confirmation hearings in May 2009 after being nominated to be administrator of the American federal government's Office of Information and Regulatory Affairs. The mere fact that Sunstein had previously written about animal law issues triggered efforts to characterize him as radical unsuited for such a key position. When animal law becomes even more successful in the educational establishment (a sign that the second wave has arrived), the revolution will get "harder still" for it will then appear on the radar screens of those humans who challenge expansions of modern legal systems to be generous to *all humans*.

Third, how can a heavy focus on the sweet beings we know as "companion animals" be maintained without creating challenges for other animals? Companion animals need fundamental protections, and because they are so near and dear they have understandably been a preoccupation. But allocating resources so heavily to creating protections for this group risks setting up companion animals as *the* paradigm of what nonhuman animals are, can be, and thus *should* be. In one sense, companion animals are not particularly good representatives of the group "nonhuman animals"—they are, by definition, subordinate to humans; further, of all animals, they fit most easily and often harmlessly (though hardly always) into the politically entrenched notion of personal property. *But many other animals do not.* The emergence of a paradigm of change based on our fascination with companion animal may thereby create the risk that animal law goes forward with no critique of the property paradigm that foregrounds individuals' legal right to own virtually *any* nonhuman animal. Such a paradigm is in tension with another emphasis in first wave animal law—the attention given the charismatic, powerful, sometimes fearsome and always brainy animals we know as chimpanzees, gorillas, orangutans, bonobos, elephants, orcas, humpback whales and other cetaceans. These animals are humans' competitors in very significant ways—in this, they are very unlike companion animals. They cannot easily be made subordinate to us and thus do not reinforce our self-image as rulers of the world. Offering these and other *non-companion animals* fundamental legal protections is a major challenge raised by first wave animal law, and second wave animal law can meet this challenge only if it is not burdened by a system that takes the paradigm of all animals to be our cats and dogs.

Fourth, how do we meet the challenge of moving to interdisciplinary approaches? It will be a challenge to deepen modern legal systems' *existing* virtues even as we work to minimize, eventually to abandon, their narrow-minded commitments to human interests alone. Such a challenge can be seen better when those who manage our legal systems work closely with other fields so that a more fact-based and ethically informed understanding of nonhuman animals can emerge. One way to improve the chance of meeting this important challenge is to foster ways that any legal system's host human community can coexist with thriving nonhuman communities. For this to happen, a combination of environmental, ecological, economic and political considerations must play out *together* to achieve such a *modus vivendi.* A particularly interesting example of legal work relying on multiple disciplines appears among the *amicus curiae* ("friend of the court") briefs filed in *U.S. v. Stevens* (argued before the United States Supreme Court in late 2009 and addressing whether the prevention of cruelty to nonhuman animals falls with the American doctrine of "a compelling governmental interest")—the Brief for a Group of American Law Professors As Amicus Curiae in Support of Neither Party is filled with information about cruelty-focused studies done in a variety of disciplines. This American tradition of citing empirical evidence in legal briefs submitted to the court, pioneered in 1908 by Louis Brandeis, reflects how the insights and findings of multiple disciplines already have a major role in legal thinking.

The relevant issue is bringing interdisciplinary, rigorous, and critically thought out explorations that address nonhuman animals themselves *and* the depth and breadth of humans' inevitable intersection with other animals to bear on policy discussions and legal analyses of different issues. It is a common sense proposition consistent with the core values of science and those of human ethics that the realities of other animals *should* guide our thinking about how we might treat them. In addition, as already suggested, there are extremely important perspectives on humans' ethical abilities and obligations that have been developed outside of legal traditions.

Thus, both our science traditions and our moral traditions make it clear that at least some nonhuman animals deserve attention as important members of the larger community to which we and these other animals belong.

Finally, how can legal education train young lawyers to think and speak in ways that open minds, not close them? The last of the challenges mentioned here brings us back to how we speak—George Orwell once observed that our use of the English language "becomes ugly and inaccurate because our thoughts are foolish, but the slovenliness of our language makes it easier for us to have foolish thoughts" (Orwell 1946).[6] Orwell's motive in making these observations was simple—"the point is that the process is reversible" (Orwell 1946). We stack the deck *against fundamental change and in favor of continued human-centeredness* by using the common phrase "humans and animals." This verbal habit reinforces powerfully the legal system's dismissive legal persons/legal things dualism because it makes the division seem natural. The division is *not* a natural division, but a cultural habit masquerading as the order of nature. It should not go unrecognized in second wave animal law that this verbal habit is *anti*-scientific. In this regard alone it undermines our ability to talk about nonhuman animals as important cousins or, as Henry Beston said in *The Outermost House*, "other nations, caught with ourselves in the net of life and time, fellow prisoners of the splendour and travail of the earth" (Beston 1928/1962, p. 25). We know in our own bodies and *from our own animality* how an animal can hurt, suffer, and on and on. The linguistic habit "humans and animals" pushes us constantly away from such awareness and the reality of how many experiences we share fully with our fellow primates, mammals and even many other nonhuman animals.

It may seem radical to suggest that second wave animal law can effect such fundamental change—after all, I have quoted above legal scholars (David and John 1978) who point out that fundamental change follows a "slow rhythm." But we know from recent experiences that care with language matters, and being willing to notice problems and change them takes courage.

Second wave animal law needs to challenge the many different ways that a powerful, sometimes hidden agenda dismisses all nonhuman animals in the name of human superiority—speaking plainly and in scientifically accurate ways would be one way to begin to open up minds to the question of what future we want to create in our inevitable relationship with the other-than-human lives who are part of the Earth community.

[6] Orwell's 1946 essay "Politics and the English Language" is available at www.mtholyoke.edu/acad/intrel/orwell46.htm

2.6 The Real Difficulties Begin—Can We *See* What the Future Will Bring?

Primates' dominant sensory ability is sight, and thus we rather naturally respond "I *see* what you are saying" whenever we grasp a claim that a conversation partner is making. Sometimes we use colloquialisms like "I *hear* you" or perhaps "what she said *touched* me" to convey comprehension. We do not often use olfactory images in such positive ways, but we do in negative ways—"I could smell what he was up to" or "her story stunk."

Can we *see or otherwise sense* what will happen in animal law in the coming decades? Where will animal law be in the year 2030? What will the shape of animal law be at the end of this century?

I make some suggestions here in order to stimulate your thinking. Granted, you may not "hear" what I'm saying because you "see through" these suggestions or conclude that "you smell a rat." Such reactions will serve my purposes, for if we are to think out our educational priorities, we must try to sense what will happen—and what can happen—in this century. As the dynamics in any animal law class so well attest, when we discuss in a group either our impressions of general prospects or a more specific plan for going forward, each member of the group is exposed to a wide variety of approaches, to different reasoning processes, to new subject matter, and even a general openness that models how each of us individually might "think out into the future." We can see the twenty-first century (as I suggest in a forthcoming book entitled *The Animal Invitation*) as the century of the species line. How might this process unfold?

A primary task we *must* accomplish is keeping other animals' realities in the foreground. Today, most people easily recognize the obvious relevance of the notion of other animals' "interests." This is a very imprecise term that covers very complex issues talked about by a variety of philosophers. Despite the generality of this term and the inevitable problem of imprecision, "interests" is an appealing notion by which virtually everyone recognizes that other animals have their realities that need to be described fairly and, when appropriate, protected. The protection can come in a variety of forms, some of which are within the legal toolbox, others of which call upon our senses of ethics, humility, community, mystery, aesthetics and even spirituality. What is paramount as we use any and all of our abilities to protect other living beings (just as these are appropriate in protecting human animals) is that we go forward in a frame of mind that allows us to notice and take seriously the *realities* of the living beings we are addressing.

The relevance of formal legal systems to this necessarily multifaceted process of noticing and then protecting other animals cannot be underestimated. Legal systems may be but one of our protection schemes, but today they are both extremely flexible *and* highly systematized in ways that other schemes, such as a moral scheme anchored in a culture or by specific religious communities, are not. Accordingly, "law" is particularly relevant to future protections because it is, in effect, a large toolbox by which we organize ourselves and in which one finds an astonishing variety of protection methods.

Rights for individuals is often thought of as the paradigm tool because it carries such important possibilities *and* in our current era has psychological ramifications—the affirmation of rights means the recipient has "arrived" in a political sense. So the creation of specific legal rights today receives great attention in those cultures and the modern industrialized nations that have historically used this legal tool to liberate and protect. This is one reason some citizens in various societies have been so parsimonious with recognition of specific legal rights.

But there are many **other tools in the legal toolbox by which the interests of other living beings can be given fundamental, effective legal protections**—there are other forms of rights (such as collectively held rights), legislative protections of many kinds that do not invoke rights, and equitable measures that call upon deep traditions and insights within our legal heritage regarding the importance of fairness, justice and "doing the right thing."

Below I speculate on two distinctly different topics that will no doubt be part of animal law as grows into the year 2030 and finally toward this century's closing years. The first topic is possible changes in various substantive areas of the law. The second topic is increasing awareness of different kinds of reasoning and valuing that might dominate the law—the latter will be in competition with the human-centered, economics-driven calculations that now are so paramount. Some focus on our reasoning patterns is important because it helps us evaluate, even parse out, why a decision is being made in a court, a legislature, an administrative body, a popular vote, etc. By identifying the assumptions that control debates and decisions we see how virulent has been past human-centeredness and how exclusivist and speciesist are contemporary policies against which so much animal law is reacting today.

Criminal law has played an important role because of the anti-cruelty tradition that developed in all cultures—in the common law tradition, for example, such a compassion-driven tradition developed before and then *outside of the common law legal tradition*. Anti-cruelty was included into the law only by way of nineteenth century statutes. This tradition has recently seen conflicting trends—the vast majority of American states, for example, now offer felony-level convictions for some forms of cruel behavior toward certain nonhuman animals, but at the same time the vast majority of the nonhuman animals that American society impacts (farm animals) are now exempted from anticruelty provisions. There is some irony here, given that the original targets of anti-cruelty legislation were farm and work animals. Law here reflects Will Rogers' insight quoted above, and thus reflects that law-making mechanisms can be dominated by interests that do not reflect popular will; they can also compromise the more compassionate strands of our cultural heritage regarding the importance of humans noticing and taking seriously the animal invitation.

My personal guess is that criminal law will continue to be important because in many ways modern societies continue to send powerful signals to individuals that *nonhuman* individuals count far less—and often not at all—when human individuals' interests are at stake. So some bad apples, as it were, will take this message as license to harm nonhuman others, and thus will have to be dealt with by criminal law-based forms of anti-cruelty protections. Yet, as the century progresses, criminal law may become less important because we resolve some of the awful problems that criminal courts must now address because many individuals get the signal from our society that it is humans alone, not nonhumans, who *really* matter. By changing the cultural evaluation of animals to have more positive content, we will lessen the amount of cruelty that our dismissal of other animals now spawns.

Further, to the extent that some of the provisions of criminal law are parasitic on our tradition of owning and holding powerful, exotic animals captive (think of the horrible harms inflicted on a woman in Connecticut, U.S.A., in 2009, by a chimpanzee who was living in a human home), I also think that our societies will eventually defuse the captivity fascination by the end of the century, and thus tolerate less and less the presence in our communities of animals that belong in their own communities. So I think criminal law will be a factor through 2030, but will be *far* less so by the end of the century.

Tort law is a major contributor to ferment in animal law today because it addresses a wide range of issues. My guess is that by 2030 it will have addressed fully the problems of measures of damages that now draw so much attention. These problems are really human-on-human problems (who pays damages of how much *to the human* who owns the harmed animal). There is the enticing possibility of awarding damages to certain injured nonhumans (we are clearly capable of creating trust-like arrangements and fiduciary duties to handle such awards in fair ways), and this will, I guess, be in place by the end of the century.

But to the extent that tort law must deal with dangerous wild animals, by the end of the century that problem will likely have subsided because we will have learned that dangerous wild animals can no more thrive in our midst than we can survive in their midst. We will likely have learned to respect other animals by the end of the century in ways that prompt us to protect their communities so that they can lead lives undisturbed by human dominance. In other words, tort law by the year 2100 is likely to be less important only because today's debates are premised on (1) human domination of wild animals, and (2) the radical inadequacy of using market-based measures of damage for the companion animals that are our family members. By the end of the century we will have found a *modus vivendi* with domesticated animals and free-living animals such that many of today's debates will have become moot.

Constitutional law in places like the United States has such iconic status that many people refuse to even consider it possible that nonhuman animals could be protected at this level. But developments in other countries reveal, as noted above, that constitution-based protections can be a part of the future of animal law. More countries are likely to create constitution-level protections, although the form is likely to be very general and, in the end, rely heavily on cultural values that exist outside, and thus are not led by, law.

Property law is, as the leading casebook in this area suggests, overwhelmingly the dominant factor today in substantive animal law (Wagman et al. 2010, p. 51).

By the year 2030, we are likely to see nuanced, responsibility-focused notions of personal property ownership of animals in the United States, much as we already see such things in other legal systems. The fact that both political and legal realms in the United States have been dominated by a notion of personal property that is so closely connected with individual rights has been a boon in some areas, but a disaster in animal law matters. Recognition of how limited the debates over property rights have been in, for example, the United States will eventually come, for the property notion has been far more flexibly applied in other traditions. In effect, the peculiar American tradition of hallowing the property concept has distorted tort and other substantive law in ways that obscure the true potential of legal systems to help us manage how we handle the animal invitation.

Since animal law in its modern version was born out of the radical dissatisfaction with the law as it handles nonhuman animals, evolution of the law and ferment in social values will lead to a more realistic handling of the property concept—this alone will open up animal law possibilities of great power.

These comments are not meant in any way to deny that *some* protection of property is deeply important to our society. *Every* society has included some provision for personal property in one way or another. But reifying the property concept in the way American politics, law and history have is destructive of not only nonhuman lives but also human thriving.

With property law more realistically and compassionately conceived, the law's true potential to be a partner in our society going forward on animal protection can be seen far better. I doubt this process is particularly far along by 2030, but hopefully this process will be well under way by the end of this century.

Contract law is not one of the areas of substantive law in ferment, although the area gets much attention because the commercial realities governed by contract law reflect fully the maladies visited upon life if a one-dimensional idea of property rights dominates a society, as has clearly happened in the United States. The result has been the stark commercial realities of factory farming, which have led to lawyers making sad comments such as the following without any pangs of conscience. This story, which appears in the 2002 book entitled *Dominion: The Power of Man, the Suffering of Animals, and the Call to Mercy* by Matthew Scully, formerly the senior speechwriter to President George W. Bush, suggests the crucial role that legal systems, through contract and property law, play in subordinating nonhumans and also placing humans in the position of ignoring the animal invitation:

> Just how bereft of human feeling that entire industry has become was clear at a municipal court case heard in Warren County, New Jersey, in the fall of 2000. A poultry company … was convicted of cruelly discarding live chickens in trash cans. The conviction was appealed and overturned, partly on the grounds that [the corporate owner] … had only six employees overseeing 1.2 million laying hens, and with workers each left to tend two hundred thousand creatures it remained unproven they were aware of those particular birds dying in a trash can. The company's initial defense … asserted outright that this is exactly what the birds were anyway—trash:
>
> **[Attorney to the Judge]**: We contend, Your Honor, that clearly my client meets the requirements [of the law]. Clearly it's a commercial farm. And clearly the handling of chickens,

> and how chickens are discarded, falls into agricultural management practices of my client. And … we've litigated this issue before in this county with respect to my client and how it handles its manure. …
>
> [**Judge**]: Isn't there a big distinction between manure and live animals?
>
> [**Attorney**]: No, Your Honor. Because the Right to Farm Act protects us in the operation of our farm and all of the agricultural management practices employed by our firm. (Scully 2002, p. 285–286)

As this century progresses, it is possible that animal law advocates, now so riveted on companion animals, will use ideas such as implied contracts or *quantum meruit* in creative ways. Ideas from outside the legal system, such as social contract theory and feminist ethics of caring, might also have a part to play in what we consider "fair" and even "equitable." Such ideas can even be applied to food animals who "serve" humans in one sense. But since substantive contract law is a weak area of the law dominated by the harsh, exclusivist notions of human animals alone having personal property rights, this area is not likely to be a leader unless the background social realities change such that cultures insist on better treatment of nonhuman animals. If that occurs, the contract notion could turn out to be quite valuable.

Others areas of substantive law, such as the law of **wills and trusts**, have already incorporated important changes that will bring protections of different kinds to some animals. By and large, these are now focused on companion animals, but there is little to prevent this important set of legal rights for humans from being mobilized to protect farm animals, rescued research animals, and wildlife.

Apart from changes in substantive law areas, there are many other aspects of law that offer prospects for an ever more robust form of animal law in the future. **Philosophy of Law** and other theoretical enterprises are sometimes disfavored in first wave animal law, which has had much practical business to attend to as it tries to blunt the worst of harms that contemporary law permits (as Hegel is alleged to have said, *what the law permits, the law encourages*—this is especially true where, as in the United States, the foregrounding of private property rights trumps virtually all other values). Theoretical approaches admittedly can be mere abstraction or detached theorizing of the kind that has historically led to a pronounced anti-intellectualism in some circles. But the *broad thinking* that theory encourages *can be helpful* in social movements. Philosophers have had, as many recognize, an important role in the modern animal protection movement (for example, Peter Singer and Tom Regan).

By 2030, though, animal law will have become an important force in areas such as philosophy of law for a number of reasons. The inclusion of nonhuman animals in legal protections offers unrivaled opportunities for seeing how legal systems were formed, now operate, and *can operate in the future.* If thinking about the larger issues in animal law can avoid the pitfalls of *mere* theory, and instead be helpful with the essential practical work that animal law has attempted in its first stage, then thinking about the larger issues will certainly take its place in both education generally and legal education in particular.

Consider how even discussing the possible relevance of philosophy of law to animal law illuminates what might be called "the one class problem." First wave animal law has been content, with only a few exceptions, to accept *a single course* called "animal law" or "animal rights" within the law school curriculum. The title of the course risks being a misnomer for a variety of reasons—"animal rights" is, given what has been said about the crucial importance of non-rights legal tools to fundamental animal protections, only partly what animal law courses are about. Even the title "animal law" for the one course offered misleads, for there is *so much* to be discussed that all possible "animal law" topics can in no way fit into a single course—to attempt such a survey is as foolish as trying to survey all of the "humanities" or all of "science" in a single course.

Second wave animal law will cross many thresholds as it gets beyond the one class problem. Entire programs are possible, as the well-developed Lewis & Clark Law School (Portland, Oregon, U.S.A.) program has so remarkably demonstrated. Additional courses will also help second wave animal law get over the anti-intellectualism that prevails in some corners of first wave animal law. This field will *never* be captured by mere theory, for the motivations for attending to animal law are deeply personal for most people. But *some* theorizing has a place as this complex, incompletely formed field goes forward. Comparative studies of law in different kinds of legal systems (which vary widely) will help everyone see the distinctive—and limiting—features of each system now operating. Such comparative work requires high-level generalizations, which are a feature of any mature field of education, animal law included.

Legal ethics is another area where animal law considerations will enrich the debate dramatically. The passage above from Scully *begs* the question about whether professional ethics within law can handle fundamental, substantive questions, or will remain focused solely on the narrow version of professional ethics now taught. Non-lawyers will naturally be inclined to condemn the lawyer in the Scully passage because they are unaware that in some systems, such as that which governed the American attorney who made the comments quoted by Scully, lawyers are obligated when arguing in court to be a "zealous" advocate for the interests of their client. This duty includes citing existing law when it is legally relevant to the question before the court. These are important obligations, but animal law can help legal professional ethics pause just long enough to imagine how the attorney's assertions here would sound to new law students, ordinary citizens, and even our grandparents. Surely the system has room for the judge's response of disbelief that live chickens discarded into trash cans might be fairly compared to manure. Second wave animal law will, by the end of this century, offer inquiries and concepts that help professional ethics engage foundational questions about the distribution of justice, the marginalization of living beings, the place of substantive ethics in the delivery of legal services, and much more.

2.7 Conclusion

Law, then, has a vital role to play in animal protection going forward. Given the successes of first wave animal law, law has recently begun to fulfill this role. But animal law has many tasks—not only beginning the revolution, but sustaining it as well, then winning fundamental change such that the revolution is won. But then, as Larbi Ben M'Hidi's warning suggests, *it is only afterwards, once we've won, that the real difficulties begin.* It is at this point that we will face the challenge of living up to our claim to be a moral species. Will we see the animal invitation? Will we notice that it is an essential part of our own health for a thousand reasons?

The answer can be an emphatic "yes" as long as we recognize that animal law must mature into its future possibilities or waves, and that this maturation will come sooner if those who promote animal law become familiar with, and then work communally alongside, other domains of Animal Studies that can supply enriched understandings of other-than-human animals. With this in mind, I think those who shape animal law today and in the coming years can contribute a vision of our species' engagement with life beyond the species line that helps not only future law students, but also all human societies around our shared Earth.

References

Bentham, Jeremy. 1970. *An introduction to the principles of morals and legislation*, ed. J.H. Burns, and H. L. A. Hart. London: The Athilone Press.

Berry, Thomas. 1999. *The great work: Our way into the future*. New York: Bell Tower.

Berry, Thomas. 2002. The mystique of the Earth. *Caduceus* 59: 1.

Berry, Thomas. 2006. Loneliness and presence. In *A communion of subjects: Animals in religion, science, and ethics*, ed. Paul Waldau and Kimberley Patton, 5–10. New York: Columbia University Press.

Beston, Henry. 1928/1962. *The outermost house: A year of life on the great beach of Cape Cod.* New York: Viking Press.

Cover, Robert M. 1986. Violence and the word. *Yale Law Journal* 95: 1601.

David, René, and John E.C. Brierley. 1978. *Major legal systems in the world today: An introduction to the comparative study of law*. New York: Free Press.

Francione, Gary. 1995. *Animals, property, and the law*. Philadelphia: Temple University Press.

Kennedy, David. 2004. *The dark sides of virtue: Reassessing international humanitarianism.* Princeton: Princeton University Press.

Midgley, Mary. 1984. *Animals and why they matter*. Athens: University of Georgia Press.

Nussbaum, Martha. 2006. *Frontiers of justice: Disability, nationality, species membership.* Cambridge, MA: The Belknap Press; Harvard University Press.

Orwell, George. *Politics and the English language*. First published in the British journal *Horizon*, April 1946, and later reprinted in the 1950 collection *Shooting an elephant and other essays.* Available http://journalism.nyu.edu/publishing/archives/portfolio/books/book51.html.

Scully, Matthew. 2002. *Dominion: The power of man, the suffering of animals, and the call to mercy*. New York: St Martin's Press.

Sinclair, Upton. 1935. *I, candidate for governor: And how i got licked.* Pasadena (self-published).

Singer, Peter. 1975. *Animal liberation: A new ethics for our treatment of animals*. New York: New York Review Book/Random House.
Wagman, Bruce, A., Sonia S. Waisman, and Pamela D. Frasch. 2010. *Animal law: Cases and materials*. 4th ed. Durham: Carolina Academic Press.
Waldau, Paul. 2011. *Animal rights*. New York: Oxford University Press.
Waldau, Paul. 2013. *Animal studies: An introduction*. New York: Oxford University Press.
White, Thomas I. 2007. *In defense of dolphins: The new moral frontier*. Malden: Blackwell.
Wise, Steven W. 2000. *Rattling the cage: Toward legal rights for animals*. Cambridge, MA: Merloyd Lawrence/Perseus.

Chapter 3
International Animal Welfare Perspectives, Including Whaling and Inhumane Seal Killing as a W.T.O. Public Morality Issue

Donald M. Broom

Abstract Most people consider that we have moral obligations to other people, to animals of other species and to ensuring the sustainability of production systems. A system or procedure is sustainable if it is acceptable now and if its expected future effects are acceptable, in particular in relation to resource availability, consequences of functioning and morality of action. Animal welfare affects public acceptability of animal usage systems and hence sustainability. The concept of animal product quality now includes: the health of human consumers, the welfare of animals used, environmental impact including conservation and pollution, the efficiency of usage of world food resources, the use of genetically modified organisms, ensuring fair payment for poor producers, and preserving rural communities. Consumers may refuse to buy unacceptable products and may pressurise retail companies and governments to ensure that they are not sold. Hence there must be codes or laws and inspection using, for example, animal-based welfare-outcome indicators. EFSA reports, which are commissioned before all changes in E.U. animal welfare legislation, have pioneered precise review of animal welfare issues and rigorous analysis of risks of poor welfare and benefits to good welfare. The best general laws on animal welfare include a requirement for a duty of care for the animals. Welfare is now discussed in the International Whaling Commission. The W.T.O. has rejected a challenge to the E.U. legislation that banned seal products because of inhumane killing procedures, the first time that it has accepted animal welfare as an aspect of public morality that can be grounds for trade restriction.

D.M. Broom (✉)
Centre for Animal Welfare and Anthrozoology, Department of Veterinary Medicine, University of Cambridge, Cambridge, UK
e-mail: dmb16@cam.ac.uk

D. Cao, S. White (eds.), *Animal Law and Welfare - International Perspectives*,
Ius Gentium: Comparative Perspectives on Law and Justice 53,
DOI 10.1007/978-3-319-26818-7_3

3.1 Introduction

3.1.1 Human Obligations to Animals

Moral systems have evolved, in humans and other species, because cooperation and tolerance are successful strategies, especially in social species (Broom 2003). Most people would say that we have moral obligations to humans and animals of other species. If we use a living animal in a way that gives us some benefit, we have an obligation to that animal. It is my view that human behaviour and laws should be based on the obligations of each person to act in an acceptable way towards each other person and to each animal that is used. It is better to base strategies for living on our obligations rather than to involve the concept of rights because some so-called rights can result in harm to others.

With increasing knowledge and increasing efficacy of communication there has been a change in attitudes to people with a broadening of the range of people for whom we have concerns. We also now consider that a wide range of animals deserve moral consideration. One view of animal protection occurs because the animals are considered to have some intrinsic value. For many people, certain animals are valued because of evidence for their cognitive abilities, their awareness, mental aspects of their needs and the feelings that they can have, e.g. pain, fear and pleasure (Verrinder et al., Chap. 4, this volume). Animals vary in the extent to which they are aware of themselves (DeGrazia 1996) and of their interactions with their environment, including their ability to experience pleasurable states such as happiness and aversive states such as pain, fear and grief. The concept of sentience affects our decisions about which animals to protect. A sentient being is one that has some ability: to evaluate the actions of others in relation to itself and third parties, to remember some of its own actions and their consequences, to assess risk, to have some feelings and to have some degree of awareness (Broom 2006a, 2014).

People are more likely to want to protect those animals that are sentient. Human opinion as to which individuals are sentient has changed over time in well-educated societies to encompass, first all humans instead of just a subset of humans, and then also: (a) certain mammals that were kept as companions, (b) animals that seemed most similar to humans e.g. monkeys, (c) the larger mammals, (d) all mammals, (e) all warm blooded animals, (f) all vertebrates and (g) some invertebrates. New knowledge about brain function and welfare has tended to show that the abilities and functioning of non-human animals are more complex than had previously been assumed so there should be some re-appraisal of which animals should be protected (EFSA 2005; Broom 2007b, 2014).

One obligation to an animal that is used by humans (Broom 2003) is to avoid causing poor welfare in the animal except where to do so would lead to net benefit to that animal, e.g. vaccinating it against disease, or to other animals including humans, or to the environment, e.g. excluding from an area or killing disease transmitters or predators. Hence some aims in animal protection are associated with concerns about animal welfare. We can consider the welfare of all living animals, including humans, but the term is not applicable to inanimate objects, plants, bacteria or viruses. Every living organism is likely to be the subject of more reverence than an inanimate object because living organisms are qualitatively different from inanimate objects in complexity, potential and aesthetic quality. This can affect decisions about whether to kill the organism and whether to conserve such organisms. Animals can respond adaptively and behave using neural control so their welfare can be evaluated.

The welfare of an animal is its state as regards its attempts to cope with its environment (Broom 1986). Welfare is a characteristic of an individual animal whilst animal protection is a human activity. Welfare includes both the ease of coping, or difficulty in coping, and any failure to cope. It varies over a range from very good to very poor and can be evaluated scientifically (Broom and K.G.Johnson 2000; Fraser 2008; Broom 2008; Broom and Fraser 2015). Coping mechanisms can be physiological, behavioural, brain systems including those that lead to feelings, and responses to pathology. Most feelings, for example pain, fear, eating pleasure, sexual pleasure, are adaptive and are components of the mechanisms for attempting to cope with the environment and regulate life (Cabanac 1979; Broom 1998; Panksepp 1998). Feelings are an important part of welfare (Duncan 1981; Duncan and Petherick 1991; Dawkins 1980, 1990) but are not all of it. Health is the state of an individual as regards its attempts to cope with pathology so health is an important part of welfare but not all of it (Dawkins 2004; Broom 2006b).

3.1.2 Some Questions About Animal Welfare Law that are Currently Considered

In addition to the issue of what kinds of animals should be protected, a number of other examples of current questions about animal welfare law will now be considered. Section 2.2 concerns how animal welfare fits into the more general issues of sustainability and product quality. What do the public want and what is the role of law in providing this? Section 2.3 describes briefly what has been happening in animal welfare law, especially in the European Union (E.U.) and Sect. 2.4, how the recent rapid developments in animal welfare concepts and science can be utilized in law. International bodies have an effect on what laws are passed by governments. Two examples considered here are Sect. 2.5: whale welfare and Sect. 2.6: the World Trade Organisation and the significance for animal welfare of the recently resolved dispute about trade in seal products. Finally some conclusions are presented.

3.2 Sustainability, Product Quality and Animal Welfare

3.2.1 Sustainability

The question of whether or not a system is sustainable is important when decisions are made about whether a system for exploiting resources should be used (Aland and Madec 2009). The fact that something is profitable and there is a demand for the product is not now sufficient reason for the continuation of production. A system or procedure is sustainable if it is acceptable now and if its expected future effects are acceptable, in particular in relation to resource availability, consequences of functioning and morality of action (Broom 2001, 2010). A system might not be sustainable for several possible reasons. For animal usage systems, examples of such reasons are: (i) because it involves so much depletion of a resource that this will become unavailable to the system, (ii) because a product of the system accumulates to a degree that prevents the functioning of the system, or (iii) because members of the public find an action involved in it to be unacceptable. Where there is depletion of a resource or accumulation of a product, the level at which this is unacceptable, and hence the point at which the system is unsustainable, is usually considerably lower than that at which the production system itself fails. Other reasons for unacceptability are exemplified below. A system could be unsustainable because of harms to the perpetrator, other people, the environment, or other animals (Table 3.1).

No system or procedure is sustainable if a substantial proportion of the local or world public find aspects of it now unacceptable, or if they consider now that its expected consequences in the future are morally unacceptable. Examples of unsustainable practices are discussed by Broom (2012). Adverse effects on people or animals can be reported in the media around the world and there are now world-wide consequences of unacceptable practices in manufacturing, animal production or other human activities because of increased efficiency of communication.

Media reports of activities or events that the public find unacceptable, may result in consumers in many countries refusing to buy animal and other products from the companies or countries involved, for example dolphins caught in tuna nets, calves kept in small crates and sheep dying on an Australian ship going to Saudi Arabia (Broom 2002, 2012).

Table 3.1 Reasons for lack of sustainability of a system

1	Resource depletion	to level that is unacceptable.
		to level that prevents system function.
2	Product accumulation	to level that people detect and find unacceptable.
		to level that affects other systems in an unacceptable way.
		to level that affects the system itself, perhaps blocking its function.
3	Other effect	to level that is unacceptable.
The acts or system functioning above could have consequences that are unacceptable because of immediate or later:		
[a]	Harm to the perpetrator:	resource loss or poor welfare
[b]	Harm to other humans:	resource loss
[c]	Harm to other humans:	poor welfare
[d]	Harm to other animals:	poor welfare
[e]	Harm to environment including that of other animals.	

Modified after Broom (2010, 2012)

3.2.2 Product Quality

The idea of quality for the goods that people buy has changed in the last 10–20 years. Quality referred formerly to immediately observable aspects, i.e. for an animal food product, its visual qualities and taste. These aspects of quality are still important, and expectations about taste are tending to become more refined, but other factors are now becoming incorporated into what constitutes good quality. Consumption has consequences and a higher proportion of these are now considered. If a food causes people to become sick, the quality is considered poor. If the food tends to make you fat, the quality is considered poor by some people. If food has added nutrients, some consider the quality to be improved. In addition, a major recent change is that the ethics of the production method are taken into account. Factors considered by purchasers include: (i) the welfare of the animals used in production, (ii) any impact on the environment, including conservation of wildlife, (iii) ensuring a fair payment for producers, especially in poor countries, (iv) the preservation of rural communities so that the people there do not go to live in towns and (v) the carbon footprint of each product as factors leading to global warming are now high on the agenda of many discriminating consumers.

3.2.3 Consumer Pressure

Consumers drive legislation and retail company codes of practice for animal production (Bennett 1994; Bennett et al. 2002). Indeed there has been a change from a "push" society, driven by producers, to a "pull" society, driven by consumers (Broom 2010, 2012, 2014, Ch. 11). Legislation on animal welfare has developed in the European Union and in many countries because of pressure from voters (Broom 2002, 2009). In general, the standards of retail companies have a substantially greater effect on the welfare of farm animals than legislation. The codes of practice of food companies have international impact. For example, many pig producers in Brazil have to comply with the animal welfare standards of United Kingdom supermarkets in order to sell to them whilst egg producers in Thailand have to rear their birds according to the standards of the increasing numbers of United States food chain companies who have animal welfare standards.

If food is not safe, in that it contains damaging levels of toxins or pathogens, most consumers will never buy it however cheap it is. Individual food production companies are expected to be responsible for this aspect of food quality but the public expects their government to ensure that adequate standards and adequate checking systems exist. The discovery of dioxin-contaminated animal-feed and human food in Belgium (Bernard et al. 2002) is an example of this. Governments have fallen and companies have gone bankrupt because of known failure on food safety issues.

Consumers will refrain from purchasing animal products if they judge that the production procedures are unsustainable and thus not of good quality. The quality may be judged to be poor on the basis of negative effects of the production or the product on: human health; human diet; the acceptability of genetic modification; animal welfare; environmental effects such as pollution, conservation and carbon footprint; the efficient use of world food resources; fair trade i.e. considering poor producers; and preserving rural communities. Each of these factors is an aspect of both product quality and the sustainability of the production method (Broom 2010, 2012).

Poor welfare of animals that are used in the production system is a major reason why some animal production systems are regarded by many of the public as unacceptable. Hence these systems become unsustainable unless there is some modification to them. Animal welfare is becoming more important to members of the public as a reason for demanding change from farmers, food retail companies and governments. Members of the European Parliament receive more letters about animal welfare than about any other subject (Broom 1999a). However, most people think about animal welfare issues infrequently, unless their attention is drawn to it by media coverage. When the information is drawn to public attention, there is a point at which the welfare of the animals becomes so poor that the majority consider the system to be unacceptable. Hence animal welfare and public attitudes toward it must be considered wherever the sustainability of an animal production system is evaluated. Efforts to use systems of animal production that are acceptable in relation to preservation of biodiversity, good animal welfare, minimal carbon footprint and other aspects of sustainability are being developed (Broom et al. 2013). In order to produce laws or codes of practice, scientific evidence is needed.

3.3 Some Laws in the European Union and the Risk Analysis Approach

What do we need from animal welfare law? Most people would say that the law should prevent people from causing poor welfare in animals: i.e. pain, fear, other suffering, severe disease, distress caused by environments which do not meet the animals' needs, or distress caused by the genetic selection used in breeding. In reality, the way that a law might do this is principally by acting as a deterrent. People who disobey the law are punished and this becomes known. Whether it is explicit or implicit in a law, there will be a principle which guides the actions of those aware of the law. Laws should provide guidance, not just a mechanism to punish (Radford 2001). One key point of the U.K. Animal Welfare Law 2006 is that it refers directly to animal welfare. A second is that it refers to people having a duty of care to the animals covered by the law. The effectiveness of laws and codes depends on the attitudes of people to them and on the efficacy of enforcement.

In parallel with the FDA in the United States, in the European Union the European Food Safety Authority (EFSA) has been set up. A difference from the FDA is (i) that many aspects of sustainability are part of the work of EFSA and (ii) that the major part of its work is done by independent scientists, appointed solely on scientific expertise and not as representatives of countries or interest groups. In producing scientific reports, a significant part of their work is the assessment of risks and benefits. The subject area covered by EFSA is wide, reflecting the public concern. One Panel deals with animal disease and animal welfare. The reports that it produces have led to many changes in E.U. legislation and science-based standards in Europe and elsewhere in the world. A scientific committee producing reports on animal welfare is of value in any major country. Measures to check that there is compliance with legislation exist in the Member States of the E.U. and in other countries with regard to food content.

In order that the ethics of the production method can be properly taken into account, products must be traceable. If foods can be traced, it is less likely that toxins, other poor quality materials or pathogens will be in them. If animals can be traced, the sources of animal disease outbreaks are more likely to be discovered and places where injuries, or other causes of poor welfare, occurred are more likely to be found (Broom 2007a). Disease, housing that does not meet animal needs, and management or handling that causes injury or stress to animals are all causes of poor welfare. Legislation and industry initiatives ensuring traceability are important for good animal welfare and efficient production.

Current E.U. legislation deals with the keeping of various farm animal species, the transport of animals, stunning and killing of animals, laboratory procedures and other matters intended to prevent poor welfare in domestic and some wild animals.

The sequence of events that lead to a Directive or Regulation about animal use in the E.U. always includes the production of a scientific report by unbiased scientists. As an example of events leading to an E.U. Directive, the welfare of calves may be considered (Broom 2009). From 1960 onwards there was some public concern that close confinement in small crates and inadequate diet lead to poor welfare in calves reared for veal production. This was a focus of the book *Animal Machines* by Harrison (1964) which led to the setting up of a national committee on farm animal welfare in the U.K. In the 1970s and 1980s there were research results giving evidence for serious welfare problems in closely confined calves. In 1988 the recommendation concerning the welfare of cattle from the Council of Europe Standing Committee on the Protection of Animals Kept for Farming Purposes stated that cattle should be able to make all normal movements for grooming, exercise etc. Some European countries passed legislation banning calf-crates. The 1990 report by a group of scientists coordinated by the European Commission was followed in 1991 by Directive 91/629/EEC laying down minimum standards for the protection of calves. This allowed the use of crates of a minimum size but required a report from E.U. Scientific Veterinary Committee by 1/10/97 which Ministers would act on. In the 1990s there was further welfare research on the effects of diet, confinement, space in groups etc on calf welfare and in 1994–1995 much public pressure for action. The scientific report was requested early, hence in 1995 the "Report on the welfare of calves" was produced by the E.U. Scientific Veterinary Committee, Animal Welfare Section. In 1996 there was a proposal for legislation from European Commission staff and the scientific report was considered by Ministry staff from each member state. A revised proposal discussed was by Ministers from each Member state and the 1997 Directive 97/2/EC phasing out the use of veal crates and inadequate diets was passed.

Similarly, within the E.U., the Council Regulation (EC) 1/2005 "On the protection of animals during transport and related operations" takes up some of the recommendations of two separate reports: (i) the E.U. Scientific Committee on Animal Health and Animal Welfare Report "The welfare of animals during transport (Details for Horses, Pigs, Sheep and Cattle)" (March 2002) and (ii) the European Food Safety Authority "Report on the welfare of animals during transport" (2004) which deals with the other species. There are now many other examples of legislation based on information from scientific reports in the E.U. and elsewhere.

The risks that a toxic substance will be in a foodstuff, or that a pathogen will enter an animal and proliferate in it, or that a management procedure will result in poor welfare in a farm animal have always been considered in scientific reports like those produced by the European Food Safety Authority (EFSA) or its predecessors in the E.U. However, in recent years the methodology for assessing risk has become more sophisticated and systematic so it is now used in many reports on topics like those exemplified above. This has resulted in scientists being more rigorous in their analyses of potential problems. If a quantitative or qualitative risk analysis is carried out, it is less likely that factors that affect the harm under consideration will be missed. Also, the relative importance of the factors involved will often be estimated more accurately. This approach was stimulated by the initial failures in the late 1980s and early 1990s to properly evaluate the risks associated with the outbreak of bovine spongiform encephalopathy (BSE) in cattle and Creutzfeldt-Jacob Disease in humans who had consumed some cattle products. A reluctance, on the part of Ministry staff and politicians, to harm the cattle and cattle-feed industries led to inadequate analysis of risks to cattle and humans. The result of the government policies might have been the deaths of millions of people, although fortunately the number will be very much lower than this, and actually caused much more harm to the industries than there would have been if a proper risk analysis had been conducted at an early stage.

The sequence of procedures during the analysis of risks or benefits is: first to list factors (hazards if negative), second to calculate exposure, third to estimate uncertainty. The analysis may be quantitative, if sufficient numerical information is available, or qualitative if it is not. The inclusion of risk analysis in scientific reports and opinions produced by EFSA and other organisations has helped decision makers to take appropriate action, for example to minimise animal disease and improve animal welfare (Berthe et al. 2012). It is desirable that this approach should be continued, with suitable modification according to the limitations associated with the relevant data.

Some factors that affect animals have beneficial effects rather than leading to a greater risk of a harm. This is most obvious when the wide-ranging components of animal welfare are considered. Food, access to other resources, human contact, social interactions and many other factors can result in benefit to the individual. Any one of these factors may also stimulate the immune system of an animal and hence confer benefit by reducing the likelihood of clinical disease. Hence every scientific review of welfare in general or of a component of health, such as the occurrence and effects of a pathological condition, should consider the possible beneficial effects of factors as well as their impact on risk. It is never sufficient in such reports to merely conduct a risk analysis. Those who formulate laws, such as government ministers or the staff of the European Commission, have to take account of all factors in determining the best course of action so they are never just risk managers. Much of what they are trying to achieve is benefit, not just reduction of risk (Broom 2009, 2014; Rault 2012; Fraser et al. 2013). This is true in legislation to reduce disease and to promote good health as well as, more obviously, in the animal welfare area. Legislation can promote good welfare and it often does.

3.4 Developments in Animal Welfare Science and How These Can Be Utilised in Law

Animal welfare scientists provide objective information about the welfare of animals. Most welfare indicators will help to pinpoint the state of the animal wherever it is on the scale from very good to very poor. Some measures are most relevant to short-term problems, such as those associated with human handling or a brief period of adverse physical conditions, whereas others are more appropriate to long-term problems. Tests of avoidance and positive preference help in the design of better conditions and procedures. In all welfare assessment it is necessary to take account of individual variation in attempts to cope with adversity and in the effects which adversity has on the animal. The general methods for assessing welfare are summarised by Broom and Fraser (2015).

The central role of animal welfare scientists is to further develop their discipline and to carry out studies in which they evaluate the welfare of animals housed or treated in different ways. The collection and analysis of data by animal welfare scientists should be carried out in an objective way that is independent of any ethical view about the outcome of the research. After the results have been obtained, scientists, like any member of the public, may make judgements about what should be done.

When scientific reports on animal welfare matters are produced, these are easier for legislators and other informed persons to use and have greater effect if the primary scientific literature is quoted giving full references. However, evaluation of the quality of the scientific information is also important. The conclusions from the data reviewed should be quoted and recommendations made based on the evidence available. Where there is little scientific information conclusions and recommendations should still be made but should make this clear. In some cases, it is valid to use information from related species but in other cases it is not. For example, every social animal will be adversely affected by being tied up and prevented from showing social as well as normal maintenance behaviour so a report on a social animal that has not been studied directly in this respect could refer to studies of other species. On the other hand, a pathogen that causes infection in one species may not cause infection in another so extrapolation from species to species is not always reliable.

Some of the causes of animal welfare problems in animals kept for human use are a result of genetic selection whilst others are a consequence of housing conditions, management methods or procedures used. In order to legislate about such matters, particular practices or systems might be prohibited. However, sometimes bad management in a good system has similar effects on welfare to good management in a less good system and even the best housing and management may result in poor welfare if the genetic selection is causing the problems. Hence the best way of designing laws may be to require that very negative consequences for animals do not occur, or to ensure that specified good welfare does occur.

Measures that predict or assess welfare that might be used in laws or codes of practice may be: resource-based, management-based or welfare-outcome indicators that are animal-based. As an example of the latter, because lameness in broiler chickens and dairy cows is a problem, it is possible to monitor the number of animals that are lame as a welfare-outcome indicator. This involves using an animal-based measure. The welfare-outcome scored is the animal's ability to walk and this is done using a scientifically designed scale of walking ability. Animals on farm or arriving at a slaughterhouse can be checked and a threshold level of lameness can be used to decide whether or not their welfare complies with the law or code of practice. For dairy cows, the EFSA report and opinions on the welfare of dairy cows proposed that the threshold for a group of dairy cows on farm or at the slaughterhouse might be 10 %. In order to facilitate this approach, EFSA have produced a series of reports/opinions on animal-based welfare-outcome indicators for several farm species.

The links between factors that might affect welfare and consequences, perhaps used as measures, need to be validated (see EFSA 2012). Some are very clear, for example a bitten pig's tail is easily recognisable, or they are clearly negative, for example frequent stereotypies indicate substantial problems. However others are less clear, for example various factors may affect heterophil/lymphocyte ratio and many causes of poor welfare do not increase cortisol.

Current plans for legal action on animal welfare in the E.U. involve further provision of information about the scientific assessment of animal welfare, including animal-based welfare-outcome indicators, a centre for making such information available and general legislation to direct its use. This should be enacted in 2015 and further specific legislation will follow.

3.5 Whale Welfare

Public concern about whales was initially focused on conservation and the main business of the International Whaling Commission (I.W.C.) has been population management and conservation. However, for the general public in many countries, the welfare of the hunted animals has become more and more important. Despite this well-documented change in public attitude around the world, the I.W.C. had never included the term welfare on any of its agendas up to 2012. It seems that this occurred because of vetos by the whaling nations. This international body includes a very large number of countries that do not permit whaling but which have relatively little influence on the very small number that do permit it. In 2012, fringe meetings and sub-group meetings of the I.W.C. considered the welfare of hunted whales. Factual information was presented by the author and others. The whale welfare issues are not just about whaling but also about whales in captivity, whale-watching and other impacts of humans on whales. The whale populations in most countries are considered to be much more valuable as a source of revenue from tourists who wish to be taken whale-watching than from killing whales. However, such activities have to be carefully regulated to ensure good welfare in the whales and continued breeding with consequent conservation of the species.

Whale welfare can be assessed using many of the measures that are used for other animals (Broom 2013). Whales are sentient, good at learning and have a pain system. In relation to the whale hunt, studies of welfare should consider the effects of: the disturbance resulting from the approach of humans in boats, chasing by boats, a harpoon entering tissue, pulling on the line attached to the harpoon, tissue damage by an explosive harpoon and procedures during capture of individuals after they have been pulled to the whaling ship. After these matters had been discussed in the informal meetings, animal welfare was placed on the agenda of the I.W.C.

One of the terms used in relation to whaling, and other killing methods, such as that of wild seals and animals in abattoirs, is humane killing. The term humane in relation to animals means their treatment in such a way that their welfare is good to a certain high degree. The welfare is either above the threshold, in which case the treatment is humane, or it is not. No killing procedure can be partly humane. Humane killing implies either that the treatment of the animals in the course of the killing procedure does not cause poor welfare, or that the procedure itself results in insensibility to pain and distress within a few seconds (Broom 1999a, 1999b, 2013). With present methodologies for catching whales during whaling, the extent of poor welfare during catching and killing always appears to be substantial. Indeed, the magnitude of poor welfare is much greater than that of any legally permitted method of killing a domestic or wild animal. The whale killing procedure during whaling would be humane for very few whales. Hence whaling for profit cannot be morally justified.

3.6 The World Trade Organisation: Significance for Animal Welfare of the Dispute About Trade in Seal Products

The rules of the World Trade Organization (W.T.O.) do not specify that animal welfare is an accepted ground for restricting trade although they do specify public morality as such a ground. This situation has resulted in difficulties for countries with strict animal welfare laws as they cannot prevent imports of animal products from countries that can produce them more cheaply because they do not have strict animal welfare laws.

The European Union passed legislation banning trade in seal products on animal welfare grounds. This was the result of public pressure over many years because many of the young seals killed for their fur, principally in Canada, were not killed in a humane way (Broom 2014). The E.U. produced hardly any sealskin products so this action was not to do with competition with Canada. The E.U. ban was challenged at the World Trade Organisation by Canada and this challenge was supported by Norway. A W.T.O. Panel was constituted (DISPUTE DS400 European Communities — Measures Prohibiting the Importation and Marketing of Seal Products) and hearings occurred in Geneva during 2013. At these hearings, Namibia and Iceland spoke in support of Canada. Other countries including Russia, which has a ban similar to that of the E.U. on animal welfare grounds, and the United States, which has a ban on conservation grounds, spoke in favour of the E.U. position.

Scientific evidence presented at the W.T.O. hearings by the author, based on several published papers and accepted by the W.T.O. Panel is summarised here.

Seals are sentient beings. They can learn complex tasks, they are social and respond differentially to individuals in their groups, they are aware of the world in relation to themselves and of feelings such as pain and fear.

There is no evidence that the pain system in seals is different from other mammals. They have: pain receptors, pain pathways and neurotransmitters, pain analysis centres, pain inhibitors such as endogenous opioids, physiological responses to pain and behavioural responses to pain. In humans and other species, extreme fear is regarded as worse than most pain. Behavioural changes during pain and fear in seals include: escape attempts (often towards water), vocalisations, freezing, rearing up in a defensive posture, opening the mouth and violent body movements.

The majority of seals that are killed in the Canadian commercial seal hunt are harp seals. 98 % of them are pups between 12 days and 3 months of age. Harp seals are called whitecoats up to 12 days of age while hooded seal pups up to 13 months old are called bluebacks. Since 1987 in Canada only weaned pups: 12–19 day-old "ragged jackets" and 19–30 day-old "beaters", are hunted. They have lost some or all of their initial fur.

Climate change has caused a great reduction in sea ice cover in the southern Gulf of St. Lawrence in recent years. Hence most seal hunting now occurs outside the mouth of the Gulf. In some years, the bulk of the killing occurs within one week.

Seals are either killed by a sealer walking to the seal and clubbing it or by a person on a moving boat shooting a seal on an ice floe, that is often moving, or shooting a seal in the water. Clubbing and shooting can wound the seal or make the animal temporarily unconscious. Injured seals often enter the water and may escape the sealers. These injured seals may recover or die of their wounds some hours, days or weeks later. Herding or chasing seals will cause fear and other forms of poor welfare.

The major question is whether or not there is there a humane, acceptable method of killing seals. All of the arguments here are about welfare and not about whether or not it is acceptable to kill seals. According to generally accepted principles, for example E.U. legislation and American Veterinary Medical Association guidelines (commercial slaughter, disease control, veterinarian), humane killing implies that: (i) the treatment of the animals in the course of the killing procedure does not cause poor welfare and (ii) the stunning procedure itself results in instantaneous insensibility or, if the agent causing insensibility or death is a gas or injectable substance that is not detectable by the animal, no poor welfare occurs before insensibility and then death.

For a seal kill to be humane, whether in commercial conditions or not, the following steps must occur:

1. The animal should be stunned without causing unnecessary pain, fear or distress.
2. On almost all occasions, there should be no need to repeat application of the stunning method.
3. The animal should be immediately monitored to confirm unconsciousness.
4. A killing method (such as bleeding) should be carried out without delay so that recovery of consciousness does not occur before death.

Seals can be killed humanely, for example by a veterinarian in conditions like those in a laboratory or zoo. For seals on ice floes that have been clubbed or shot, some will be conscious, some will be unconscious and some will be dead but identification of unconsciousness and death is difficult. The main methods used by sealers are unreliable. A seal with the top of its cranium (calvarium) partly caved-in may be conscious. A seal with no cranial damage may be unconscious or dead. Without bleeding by cutting a major blood vessel, death cannot be certain. Bleeding a conscious animal causes extreme pain and skinning a conscious animal will cause extreme pain for much longer so a high magnitude of poor welfare. There is often failure of the stunning and killing method and it is not humane if the animal is not rendered instantaneously unconscious, monitored without delay, and bled before recovery of consciousness.

During a commercial seal hunt, some seals will be killed humanely. However, even if all rules are obeyed, the welfare of a substantial proportion of seals will be very poor. It is not possible to stun and kill seals in commercial conditions without the welfare of many seals being poor.

A key question is how seal killing on ice floes compares with abattoir killing. Hence it is relevant to compare the rates of miss-stuns and delays in commercial sealing and abattoirs. In a series of papers on the Canadian seal hunt, the frequency of inaccurate stun, involving the need to repeat the stunning method, ranged from 11 to 92 %. The sealers knew that they were being watched so the figures may be underestimates of the actual frequency in normal sealing conditions. In studies in abattoirs, the figures for several species were 0 %, 1.9 %, 1.6 % and 9.2 %. However, in every case in the abattoir, a repeat stun occurred within 3 s. As with the seal hunts, these figures were sometimes obtained when the abattoir staff knew that they were being watched so less efficient stunning might sometimes occur. However, the stunning procedure is much easier in an abattoir than on the ice and repeat stunning is quick and easy if legal requirements, e.g. for back-up stunner provision, are observed. Sealers often could not get to the animals quickly so the delay was a mean of 8.6–27 s and often much longer. In the Canadian seal hunt the delay between stunning and monitoring was 16–307 s, with delay to bleeding a little longer, and some were still conscious but this was not detected. In abattoirs, the check is within a few seconds and bleeding is within 20 s. The Canadian government figures for seals struck and lost are a mean of 5 %. These seals are injured and may take many hours or days to die. This does not occur in abattoirs.

Although it is possible to kill seals humanely in controlled conditions on land, commercial seal killing causes much worse welfare than any legal abattoir killing. It is not possible to improve seal killing so that most animals are humanely killed.

The W.T.O. panel found on 25 November 2013 that the E.U. Seal Regime does not violate Article 2.2 of the Technical Barriers to Trade (TBT) Agreement because it fulfils the objective of addressing E.U. public moral concerns on seal welfare to a certain extent, and no alternative measure was demonstrated to make an equivalent or greater contribution to the fulfilment of the objective. The E.U. ban on seal products was found not to violate the general Agreement on Tariffs and Trade (GATT). However, the E.U.'s exception to the seal product ban for indigenous peoples and marine resource management was found to be unacceptable by the W.T.O. This means that seal products resulting from inhumane killing methods may not be traded in the E.U. in any circumstances.

This is a very important result as it is the first time that a product ban on animal welfare grounds, as part of public morality, has been accepted by the W.T.O. The possibility is raised that other product bans on the same grounds might be accepted. At present, certain farming practices are banned in some countries because their effects on animal welfare are unacceptable to the public in those countries. Examples include the E.U. bans on keeping calves in small crates, keeping pregnant sows tethered or confined in stalls that do not allow the sow to turn around, and keeping laying hens in battery cages with a space allowance of less than 550 cm^2 per bird. However, these systems are allowed in many other countries and it has been perceived that the import of veal, pig meat and eggs from such countries could not be blocked by the E.U. because of W.T.O. rules. Following the W.T.O. ruling in the seal products case, it may be that this situation will change because the public in the E.U. consider that the very poor welfare of closely confined calves, sows and hens is unacceptable and a matter of public morality.

3.7 Conclusions

1. In relation to animal usage throughout the world, there will be increasing demand from consumers for the avoidance of adverse effects on human welfare, animal welfare and the environment.
2. A system or procedure is sustainable if it is acceptable now and if its expected future effects are acceptable, in particular in relation to resource availability, consequences of functioning and morality of action.
3. Animal welfare is one of the factors determining sustainability and is a part of product quality.
4. Animal welfare has been developing rapidly as a scientific discipline.
5. Laws should emphasise duty of care, be based on scientific evidence, and some should use welfare-outcome measures (animal-based).
6. International bodies, such as the I.W.C. and W.T.O., take notice of scientific studies of animal welfare. Whale welfare is now on the I.W.C. agenda and, with presently available killing methods, whaling for profit cannot be morally justified. The upholding by the W.T.O. of the E.U. seal product ban on animal welfare grounds as part of public morality may lead to other product bans on the same grounds.

References

Aland, A., and F. Madec (eds.). 2009. *Sustainable animal production*. Wageningen: Wageningen Academic Publishers.

Bennett, R.M. (ed.). 1994. *Valuing farm animal welfare*. Reading: University of Reading.

Bennett, R.M., J. Anderson, and R.J.P. Blaney. 2002. Moral intensity and willingness to pay concerning farm animal welfare issues and the implications for agricultural policy. *Journal of Agricultural and Environmental Ethics* 15: 187–202.

Bernard, A., F. Broeckaert, G. De Poorter, A. De Cock, C. Hermans, C. Saegerman, and G. Houins. 2002. The Belgian PCB/Dioxin incident: Analysis of the food chain contamination and health risk evaluation. *Environmental Research* 88: 1–18.

Berthe, F., P. Vannier, P. Have, J. Serratosa, E. Bastino, D.M. Broom, J. Hartung, and J.M. Sharp. 2012. The role of EFSA in assessing and promoting animal health and welfare. *EFSA Journal* 10(s1002): 19–27.

Broom, D.M. 1986. Indicators of poor welfare. *British Veterinary Journal* 142: 524–526.

Broom, D.M. 1991a. Animal welfare: Concepts and measurement. *Journal of Animal Science* 69: 4167–4175.

Broom, D.M. 1991b. Assessing welfare and suffering. *Behavioral Processes* 25: 117–123.

Broom, D.M. 1998. Welfare, stress and the evolution of feelings. *Advances in the Study of Behavior*, 27: 371–403.

Broom, D.M. 1999a. Welfare and how it is affected by regulation. In *Regulation of animal production in Europe*, ed. M. Kunisch and H. Ekkel, 51–57. Darmstadt: K.T.B.L.

Broom, D.M. 1999b. The welfare of vertebrate pests in relation to their management. In *Advances in vertebrate pest management*, ed. P.D. Cowan and C.J. Feare, 309–329. Fürth: Filander Verlag.

Broom, D.M. 2001. The use of the concept animal welfare in European conventions, regulations and directives. *Food Chain* 2001: 148–151. Uppsala: SLU Services.

Broom, D.M. 2002. Does present legislation help animal welfare? *Landbauforschung Völkenrode* 227: 63–69.

Broom, D.M. 2003. *The evolution of morality and religion*. Cambridge: Cambridge University Press.

Broom, D.M. 2006a. The evolution of morality. *Applied Animal Behaviour Science* 100: 20–28.

Broom, D.M. 2006b. Behaviour and welfare in relation to pathology. *Applied Animal Behaviour Science* 97: 71–83.

Broom, D.M. 2007a. Traceability of food and animals in relation to animal welfare. In *Proceedings of the second international conference on traceability of agricultural products*. Brasilia: Embrapa.

Broom, D.M. 2007b. Cognitive ability and sentience: Which aquatic animals should be protected? *Diseases of Aquatic Organisms* 75: 99–108.

Broom, D.M. 2008. Welfare assessment and relevant ethical decisions: Key concepts. *Annual Review of Biomedical Sciences* 10: T79–T90.

Broom, D.M. 2009. Animal welfare and legislation. In *Welfare of production animals: Assessment and management of risks*, ed. F. Smulders and B.O. Algers, 341–354. Wageningen: Wageningen Pers.

Broom, D.M. 2010. Animal welfare: An aspect of care, sustainability, and food quality required by the public. *Journal of Veterinary Medical Education* 37: 83–88.

Broom, D.M. 2012. Defining agricultural animal welfare: From a sustainability and product quality viewpoint. In *Animal welfare in animal agriculture*, ed. W.E. Pond, F.W. Bazer, and B.E. Rollin, 84–91. Boca Raton: CRC Press.

Broom, D.M. 2013. The science of animal welfare and its relevance to whales. *Animal Welfare* 22: 123–126.

Broom, D.M. 2014. *Sentience and animal welfare*. Wallingford: CABI.

Broom, D.M., and A.F. Fraser. 2015. *Domestic animal behaviour and welfare*, 5th ed. Wallingford: CABI.

Broom, D.M., and K.G. Johnson. 2000. *Stress and animal welfare*. Dordrecht: Kluwer/Springer (first published Chapman and Hall 1993).

Broom, D.M., F.A. Galindo, and E. Murgueitio. 2013. Sustainable, efficient livestock production with high biodiversity and good welfare for animals. *Proceedings of the Royal Society B* 280. 20132025.

Cabanac, M. 1979. Sensory pleasure. *Quarterly Review of Biology* 54: 1–129.

Dawkins, M.S. 1980. *Animal suffering: The science of animal welfare*. London: Chapman and Hall.

Dawkins, M.S. 1990. From an animal's point of view: Motivation, fitness and animal welfare. *Behavioral and Brain Sciences* 13: 1–61.

Dawkins, M.S. 2004. Using behaviour to assess welfare. *Animal Welfare* 13: 53–57.

DeGrazia, D. 1996. *Taking animals seriously: Mental life and moral status*. New York: Cambridge University Press.

Duncan, I.J.H. 1981. Animal rights – Animal welfare, a scientist's assessment. *Poultry Science* 60: 489–499.

Duncan, I.J.H., and J.C. Petherick. 1991. The implications of cognitive processes for animal welfare. *Journal of Animal Science* 69: 5017–5022.

EFSA. 2005. Aspects of the biology and welfare of animals used for experimental and other scientific purposes. *EFSA Journal* 292: 1–136.

EFSA. 2012. Statement on the use of animal-based measures to assess the welfare of animals. *EFSA Journal* 10(6): 2767.

Fraser, D. 2008. *Understanding animal welfare: The science in its cultural context*. Oxford: Wiley Blackwell.

Fraser, D., I.J.H. Duncan, S.A. Edwards, T. Grandin, N.G. Gregory, V. Guyonnet, P.H. Hemsworth, S.M. Huertas, J.M. Huzzey, D.J. Mellor, J.A. Mench, M. Spinka, and H.R. Whay. 2013. General principles for the welfare of animals in production systems: The underlying science and its application. *The Veterinary Journal* 198: 19–27.

Harrison, R. 1964. *Animal machines*. London: Vincent Stuart (Reprinted with commentaries 2013, Wallingford: CABI).

Panksepp, J. 1998. *Affective neuroscience. The foundation of human and animal emotion*. New York: OUP.

Radford, M. 2001. *Animal welfare law in Britain: Regulation and responsibility*. Oxford: Oxford University Press.

Rault, J.L. 2012. Friends with benefits: Social support and its relevance for farm animal welfare. *Applied Animal Behaviour Science* 136: 1–14.

Chapter 4
Science, Animal Ethics and the Law

Joy M. Verrinder, Nicki McGrath, and Clive J.C. Phillips

Abstract Within the paradigm of the desirability of an age of plenty, science has delivered a rapid expansion in animal production by changing the living standards, morphology and physiology of animals. While various philosophers have espoused more compassionate and just treatment of animals for at least twenty centuries, it is only in recent times that animal welfare research has evolved, providing evidence to inform standards and legal requirements and meet public expectations about the treatment of animals. However this science has struggled to maintain its independence. Its focus has been on the physical impacts of painful procedures, though more recently there has been a growth in studies on animals' emotions, both negative and positive. While it is important to determine how animals think, feel and experience pain and pleasure, ethics should be central to all these investigations, and the resulting decisions. There has been confusion about what ethics means and how it can contribute to decision making in relation to animal ethics issues. With little ethics research and no unified ethics competencies in animal-related professions, a narrow focus on legislated codes of practice and relativist or pluralist approaches to ethical frameworks are often used. This chapter explains how animal ethics can be addressed more purposefully as the science of morality regarding humans' treatment of animals, with a structured approach to moral development based on moral psychology.

4.1 Introduction – What Has Science Got to Offer the Ethical Debate About How to Treat Animals?

> "Science" … can deliver "an expansion of material well-being beyond anything that has yet occurred in human experience". Only "human folly or sub-human crime" could "deny all nations the inauguration and enjoyment of an age of plenty" (Winston Churchill 1946)

Science has delivered a dramatic expansion of animal production. From the battery cage to the feedlot, animals are raised for consumption in vastly greater numbers

J.M. Verrinder • N. McGrath • C.J.C. Phillips (✉)
Centre for Animal Welfare and Ethics, School of Veterinary Science, University of Queensland, Gatton, QLD 4343, Australia
e-mail: c.phillips@uq.edu.au

D. Cao, S. White (eds.), *Animal Law and Welfare - International Perspectives*, Ius Gentium: Comparative Perspectives on Law and Justice 53,
DOI 10.1007/978-3-319-26818-7_4

now compared with in 1946. Along with the green revolution, this has increased availability of high quality food to a much larger proportion of the world's population, although it is rarely considered whether equivalent or even greater benefit might have been achieved if the intensification had focused just on improving the quality and quantity of plant production. Animal intensification has included an increase in both animal numbers and productivity per animal. Growth rates in livestock, milk yields of dairy cows and egg production by hens have all increased several fold through genetic selection and higher quality diets. Alongside these changes in output, the morphology and physiology of the animals has changed: dairy cows eat more and have larger, more productive mammary glands; meat chickens grow faster and have larger breast muscle, beef cattle have increased muscle growth at the expense of internal organs; pigs grow faster and are leaner, in line with market demands. For example, meat chickens currently grow to slaughter weight of 2 kg within 35 days, about one half of the time that it took 50 years ago. To facilitate labour-efficiency and a rapid turnover of meat animals, as well as high yields from milk and egg producing stock, food-producing animals are kept in barns with automated feeding and drinking systems, at stocking densities much higher than the natural density of the animals. In the most intensive systems, for pigs and poultry, animals are kept in cages with little room to perform natural behaviours. The welfare of animals in the most intensive systems is attested by scientists to be low, whereas extensive systems have the potential to provide for good welfare if basic needs, in particular food, water and freedom from disease, are provided, because of the inherent advantages of freedom and space for natural behaviours to be performed.

The obvious impact of intensive production systems on animal welfare led governments of countries most involved in developing these systems, especially in northern Europe, to develop legal protection for the animals involved. In the early stages it was assumed that consumers would be unlikely to pay more for animal products, and it was therefore inevitable that industry would resist enforced minimum resources for animals, some of which reduced production efficiency. Government bodies, such as the Farm Animal Welfare Council in the United Kingdom, the E.U. Scientific Veterinary Committee and the Australian federal and state government Animal Welfare Advisory Committees, were established to advise governments on the best course of action in relation to animal welfare issues. Scientific evidence provides support for their advice; however, resource limitations, and perhaps a mistrust of science from other parts of the world, have restricted this activity.

Despite the prominence of animal welfare issues in the media and the legislative changes in Europe, in some countries the public have been slow to understand the extent of the cost of intensive production systems to animal welfare. Their knowledge of the systems has been, even in this digital age, extremely limited, particularly in the growing urban populations that have little contact with animals kept for food production. Regular exposés of acts of cruelty in the media inform about the worst treatment of animals, particularly in slaughterhouses, as well as generating discussion and mutual feelings of compassion (Tiplady et al. 2013). However, in reality there is a myriad of production systems giving rise to a variety of welfare impacts, some of which will be acceptable to the public, some not. The increasingly rare contact between the public and farm animals gives cause for concern that they either don't know or don't care how animal products are prepared. For example, most members of the public believe that meat chickens are produced in cages and even many animal advocates share this view (Ross and Phillips 2015).

In some countries that have a relatively long history of using intensive animal production systems, there is evidence of declining meat consumption. For example the proportion of the young people (18–24) in the U.K. claiming to be vegetarian had increased to 17 % in a recent survey (Eating Better 2013), but of perhaps even greater importance is the growing proportion of people in countries using intensive animal production systems that are eating less meat. In other countries, concerns that meat is less healthy and that meat production damages the environment predominate (Izmirli and Phillips 2011; Newmyer 2006). At the forefront of any changes in public perception of the ethics of eating meat is a growing band of activists, philosophers and scientists who recognise the significant impact of intensive systems on animal welfare, and campaigns to alert the public and governments to injustices to animals.

The involvement of philosophers in the debate about the treatment of animals is not new; the Graeco-Roman philosopher Plutarch was actively espousing the virtues of an empathetic attitude to animals in the first century C.E. (Newmyer 2006), following in the footsteps of Greek philosophers that lived half a millennium previously, such as Pythagoras. Admittedly such views were not common, and vehement opposition was provided by groups such as the Stoics, who argued that animals can neither reason nor experience emotions. Arguments in Graeco-Roman history are reminiscent of those circulating in the aftermath of the birth of intensive animal production in the modern era by philosophers such as Peter Singer and Tom Regan (Armstrong and Botzler 2003), supported by activists such as Ruth Harrison. However, today's debate is different because it is supported by a rapidly growing field of scientific endeavour: animal welfare science. Since 1980, the numbers of scientific publications on animal welfare have grown exponentially, far more than most other scientific disciplines, with only limited signs of the increase plateauing around 2010 (Fig. 4.1). Most of these publications come from Western countries (Robins and Phillips 2010), presumably because consumers there can more easily afford to pay for less intensively produced animal products, on the assumption that welfare is better (Phillips et al. 2012).

Expansion of animal production systems is now occurring in developing countries, some of which, historically, have made very little contribution to the scientific endeavour towards improved animal welfare (Robins and Phillips 2011). The rapid growth of, for example, the SE Asian economies has produced a consumer confidence that rarely considers the animal ethics of their increasing meat consumption. However, in view of the fact that some of these countries have a greater awareness of animal welfare issues through religion, it is likely that increased ethical sensitiv-

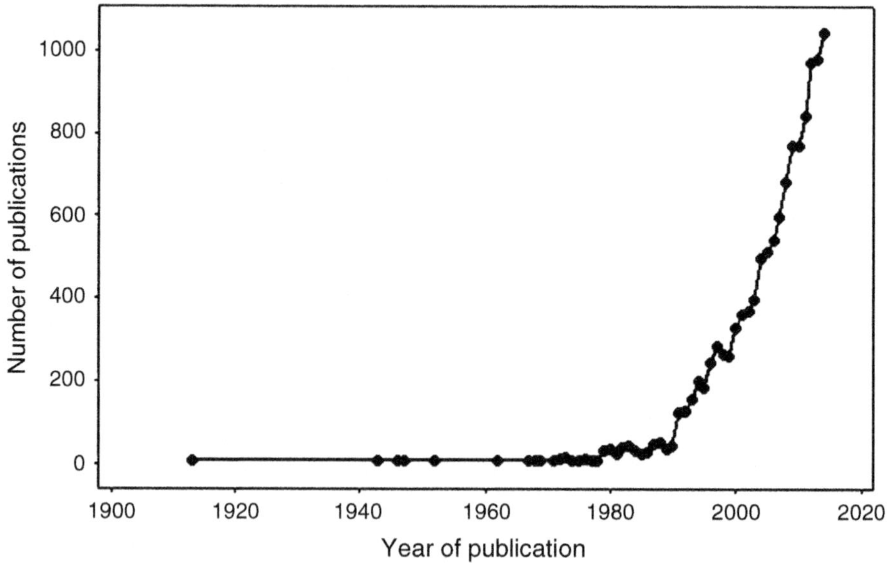

Fig. 4.1 Growth over time in scientific publications with the words "animal" and "welfare" in the title or abstract

ity to the welfare of meat animals will develop over time. The current lack of concern may primarily be due to consumers' inability to pay more for welfare-enhanced animal products (Phillips et al. 2012), and worldwide the dominance in the meat market of big multinational companies with responsibility to shareholders and little regard for public sentiment about the welfare of animals.

4.2 Research in Animal Welfare

The growth of animal welfare science was initially spawned by a cohort of scientists who were largely focused on detecting welfare problems in animal production practices of the day. They were supported by government grants, universities and philanthropists in Western countries. Legislative changes followed, mostly in Europe, which outlawed or strictly regulated the worst forms of animal suffering; for example, intensive veal calf production, and long distance journeys by livestock across Europe. This had a significant impact on the animal industries, and a second wave of scientists has emerged with a broader range of perspectives. Some used research in an attempt to refute claims of poor welfare in some industry practices, such as the transport of bobby calves (Phillips and Petherick 2014). Many had industry support for their positions or their research. Advocacy groups have also started to support research, although typically with smaller funds than provided by the powerful animal industries and their representative bodies. The entry of advocacy groups has been cautious and largely confined to more moderate organisations, the most radical believing that the use of animals in research is not justified from a deontological perspective.

The significant growth of animal welfare science has had a focus on painful procedures (Phillips 2008). Almost as many papers have been written on pain and injury in animals as all other welfare issues put together, including freedom from fear/stress, hunger/thirst, discomfort and freedom to behave normally. This may be because painful procedures are perceived to be anthropogenic in origin, whereas other serious welfare issues attract less interest because they are believed to derive from natural events, such as drought or flood, even though susceptibility of animal production systems to drought is greatly influenced by human practices, such as stocking densities. However, direct stakeholders in the livestock industries testify that nutrition during drought can have greater impact than the temporary adverse welfare from invasive practices, such as mulesing, dehorning and castration, which are mainly recognised as damaging by animal advocacy groups (Phillips et al. 2009). The question arises as to whether those directly involved have become inured, intentionally or otherwise, to the welfare impact of invasive practices, or whether animal advocates focus on these because they fail to recognise the long-term consequences of events that may be seen as outside of human control. An alternative explanation for the focus on painful procedures is that humans readily empathise with these, whereas issues such as long-term undernutrition are not commonly experienced by humans.

Animal welfare scientists follow a difficult path to providing the evidence on which to base legislative changes. In theory their publications can directly influence animal welfare, or they may be absorbed into current knowledge first and eventually find their way into accepted practice. Even though there are more funding bodies than ever before, funding that is not associated with vested interests is rare. However, the range of journals in which animal welfare scientists can publish their work has increased, with Animal Welfare, Journal of Applied Animal Welfare Science, Applied Animal Behaviour Science, Animal, Animals, Anthrozoos, and Society and Animals being the most popular. Many scientific book publishers also have significant numbers of books on animal welfare, most notably Springer and Blackwells, both of which have a series on animal welfare, and CABI, Earthscan and a variety of university presses.

Pressures placed on animal welfare scientists, along with other scientists, to generate income and publications may encourage some to sacrifice their independence or even honesty (van der Schot and Phillips 2013). However, most bias in animal welfare publications probably derives from the failure of scientists (and editors) to publish the results of studies that are ambiguous or negative. The vested interests of funding agencies may influence scientists in their reporting of results (van der Schot and Phillips 2013). This may set up funder reciprocity, in which scientists report the right message for the funder to get continued funding. Some studies may be tied to non-disclosure agreements, in which the reporting of the results by the scientist is suppressed, perhaps until a period of time after the research, or completely if the message is not one that the funder sought. In animal welfare science, there is a reluctance to believe studies that rely on variables that are hard to measure or interpret, in particular behaviour and emotions, compared to health variables such as disease prevalence and injury rate. Researchers may, for the best of reasons, record multiple variables and report only those demonstrating significant, understandable or beneficial results. Hypotheses may even be adapted after the study in order to support the results, or scientists may assume results indicate effects on animal emotions that are hard to conceptualise, let alone measure.

New techniques are becoming available to derive information on one of the most elusive but important aspects of animal welfare science, the emotional states of animals (Edgar et al. 2012). An understanding of emotional states is fundamental to determining the impact of husbandry systems on animal welfare. Emotions in both humans and animals can have both psychological and physical manifestations. Neuroscientific studies have demonstrated that the processing of emotions such as fear can be detected in particular areas of the brain (Berridge 2003; Panksepp 1998). The homologies between human brains and the brains of other animals mean that neuroscience may eventually indicate exactly what subjective states animals experience. In the meantime, the physical manifestations of emotions can be detected in behaviour (e.g. Gourkow et al. 2014) and also autonomic reactions.

Emotions are adaptive responses that offer animals the ability to solve problems they would be unable to with just simple reflexive reactions. Most are short-term and intense, creating physical responses which enable animals to avoid harm or to approach critical resources that may improve their chances of survival. Scientists have developed methods of measuring these physical responses to emotions, which in some cases appear to tell us, without any doubt, that an animal is experiencing an emotional response to stimuli. These methods include physiological measurements, such as measuring heart rate, respiration rate and electrodermal responses (Désiré et al. 2002; Reefmann et al. 2009), hormones such as cortisol, adrenaline and noradrenaline, and the immune response. However it is difficult to correctly discriminate specific emotions from general arousal through physiological measurements, and correlating physiological measurements with behavioural observations may present a more complete picture.

Behavioural measurements have long been used as measures of pain and stress in animals. The more simple tests observe the postures, locomotion, facial expressions and vocalisations of animals. However, scientists are continually designing more novel methods to detect other emotions in animals. Place preference tests, "consumer demand" tests, and operant conditioning studies to examine consumption, approach and avoidance are now used by some scientists to detect emotions in animals. Other scientists have developed more qualitative methods, such as free choice profiling, which takes a holistic look at the animal as a "behaver" rather than simply investigating individual behaviours or postures (Wemelsfelder et al. 2009).

Cognitive tests are also widely used with some animals. Methods include examining whether animals find stimuli pleasant or unpleasant, exposing animals to a range of emotions using different eliciting stimuli, including those that are novel or, on the other hand, predictable (Paul et al. 2005). One group of scientists believe that the mood state of animals can be detected within a core affective framework: emotions are measured according to predictions about what situations will generate a particular affective state and the decisions animals may make in these states (Mendl et al. 2010). Identifying emotions associated with these states can demonstrate underlying mood.

Many of the methods above are designed not only to detect negative emotions related to animal welfare, such as fear and stress, but also more positive emotions, such as excited anticipation of reward or satisfaction. A key current question in animal welfare science is examining what animals want. However, in the past, many experiments have been undertaken in which animals are deliberately harmed and the parameters measured would not have been expected to effectively record animal welfare impact. This can lead to erroneous conclusions that there was no harm involved. Considerable debate has also surrounded the value of cortisol as a physiological marker of stress in farm animals experiencing suspected welfare reduction, e.g. Caulfield (2013).

Cognitive or judgement bias testing, in which the unwillingness of animals to test ambiguous cues, is suggested to indicate negative emotions or even depression (Burman et al. 2011). In livestock, which being derived from prey animals are undemonstrative in the face of danger, the laterality of eye gaze and other behaviours can be used to infer whether the viewer perceives danger in the stimulus (Robins and Phillips 2010). The eyes of livestock are more lateralised than humans, and left eye vision, connected to the right brain hemisphere, is used more than right eye vision for responses to danger.

There are limitations to some of the methods used to assess the emotional states of animals. Assessments which look at purely physiological measurements risk attributing certain emotions to states of arousal, which could actually be produced by other underlying moods or emotions. Sampling methods, such as restraint during blood sampling, can confound results. Behavioural measurements such as facial expressions, postures or vocalisations, may often refer to our knowledge of the meaning of human responses, leading to possible inaccurate assessments. Measurements, therefore, cannot be made in a generic manner, rather they must be species-specific. Behaviours also sometimes vary widely between individuals, and also in intensity and duration. Qualitative methods may also introduce personal bias into behavioural assessment. The motivational state of animals when they first encounter stimuli can affect their individual reactions to subsequent presentations of the stimuli. Animals sometimes transition quickly between one emotional state and another, making it hard to definitively investigate and categorise discrete emotions. However, using more than one method concurrently eliminates some of these difficulties, and maintaining rigour in the approaches to the design of these studies provides scientists with the confidence that emotional states may accurately be inferred from their results. Repeating these tests and experiments then enables scientists to provide evidence that forms the backbone of standards for animal welfare assessments.

Formulating animal welfare science into standards that can be used to improve practice is fraught with difficulty, as animal systems are highly variable and welfare outcomes relatively under-researched. Hence, most standards are based largely on expert opinion or what stakeholders consider to be best practice. Most are written as a set of compulsory or prohibited practices to provide for a prescribed minimum quality of life for the animals concerned. These need to be reliably related to welfare, as well as being verifiable, and regularly verified by qualified personnel. Sometimes *desirable* practices are advocated but not required to be implemented, e.g. the guidelines in Australian Standards and Guidelines for the Welfare of Animals which attempt to distinguish between things that must be done and those that should be done (in an ideal world), in support of animal welfare. In addition, consumer groups and retailers establish optional standards that provide an opportunity to demonstrate that an individual farmer is performing to a higher welfare standard than the average farmer. The choice of what to measure, relating to inputs to the system, or outputs related more closely to the animal product, is still keenly debated, with the former, e.g. space availability, being easier to measure but often less closely related to welfare than the latter, which in this case of different stocking densities might be behaviour and physiological responses to overcrowding.

So while science has delivered a dramatic expansion in animal production through intensification, animal welfare science is still endeavouring to cater for the change, identifying appropriate measurements to inform standards and government policy. In both these areas, the ethical questions: "How should we treat animals?" and "How can we ensure we do what we should?", have largely been ignored.

4.3 What Science Can Do for Animal Ethics and Animal Law?

Mepham (2008, p. 5) distinguishes between two big questions we face as human beings: "'What is all this, and how does it work?' – the answers to which are sought by science; and 'What should we do?' – the answers to which are sought in ethics". "That humans have ethical duties to animals is an assumption that underlies the study of animal welfare. There would not be much point in studying how animals fare in livestock production systems, for example, if we did not think that humans had any duty to look after animals in their care" (Sandøe et al. 1997, p. 1). However it is the degree to which we recognise, value and act to protect animals' needs and interests that requires ethical consideration. In essence, animal welfare science is the "meat in the ethical sandwich" (Main et al. 2005). While it is important to determine how animals think, feel and experience pain and pleasure, ethics is required to decide what we should do about it, how it should impact on our behaviour toward animals. Ethics is also important to inform animal law questions: How should society be required to treat animals?

However, confusion about what ethics, and more specifically animal ethics, actually means and therefore whether it can be reliably developed to inform decision-making and action, provides both evidence of the lack of education in this area and a reason for this education not occurring in the first place. Some people hold the view that "ethics is first and foremost, a realm of theory and arcane formulations" (Longstaff 1994). Some regard ethics or morality as a personal matter for individuals, so ethical judgements are 'subjective' opinion and not 'fact', and thus not subject to rational discussion and adjudication (Longstaff 1994). Some also argue that ethics is relative to the society one happens to live in, while others claim that there is a common social ethic across cultures (Rollin 2006a), and others that ethics takes a universal point of view (Singer 1993b).

4.4 What Is Ethics?

The words ethics and morality are often used interchangeably (Preston 2001; Rollin 2006a, b; Tannenbaum 1995) and are in this chapter. There are two central issues which create the ethical aspect of living. "One fundamental fact which shapes and necessitates the ethical life [is] our interconnectedness" (Preston 2001). Every action we take (or don't take) has an impact on other living beings. The second issue is that all living sentient beings have needs and interests, needs being defined here as what is essential for survival, including both physical and emotional needs, and interests defined either as capabilities for suffering or experiencing pleasure (Singer 1991) or as "growth and flourishing" i.e. "being engaged in the struggle for survival and prosperity", or "life and well-being" (Steiner 2005). "Often we have competing interests, beliefs, temperaments and traditions, which constrain the decision that a group settles on" (Churchland 2011). Our capacity to ignore others' interests or prioritise our own, perhaps because of greater knowledge of our own and because any decision will directly affect our own happiness, makes equal consideration of the effect of our decisions on others' interests more difficult. Addressing this bias requires considerable sensitivity, skill and will. This should be the central domain of ethics education.

Ethics programs commonly include explanations of the range of ethical frameworks. Briefly, two main frameworks of ethics are, first, utilitarianism, based on our need to balance interests to create the greatest well-being and least suffering of all capable of these, and second, deontology, founded on defining our duty to the fundamental principles which protect each individual's interests. A third prominent perspective seeks to develop the moral virtues, e.g. courage, perseverance and compassion in each person to achieve happiness, which has been interpreted from Aristotle's writings as "a certain sort of activity of the soul in accord with complete virtue" … achieved "through some sort of learning and attention"(Aristotle 1999). Other perspectives include the focus on caring (Gilligan 1982).

Considerable debate has occurred about which of these ethical frameworks is best as a basis for moral behaviour and whether ethics is personal choice (relativism) or universally applicable (universalism) to everyone. This ignores the complementarity of the various frameworks and the interdependence of all life and the common interests of sentient beings in life and well-being (Hare 1995; Preston 2001). Singer argues: "If ethics is a jigsaw puzzle, then we are now at the stage where we have laid out all the pieces, and are beginning to see the outline of the picture. For ethics is not a meaningless series of different things to different people in different times and places. Rather, against a background of historically and culturally diverse approaches to the question of how we ought to live, the degree of convergence is striking. Human nature has its constants and there are only a limited number of ways in which human beings can live together and flourish" (Singer 1993a, b).

Taking a standard definition of science as "a systematically organized body of knowledge on a particular subject" and "the intellectual and practical activity encompassing the systematic study of the structure and behaviour of the physical and natural world through observation and experiment" (Oxford Dictionary 2014), ethics can be described as the science of morality (Harris 2010); and animal ethics as the science of morality regarding humans' treatment of animals. However, just as science works within accepted paradigms influenced by social factors (Kuhn 1970), which have often excluded or misrepresented the needs and interests of animals, the science of morality has often excluded or misrepresented animals with a persistent anthropocentric prejudice (Steiner 2005). Only more recently has there been a growing interest in the area of animal ethics in relevant professional courses. A 2010 survey found no clear description of ethics competencies within the regulations for veterinary training in Europe (Magalhães-Sant'Ana et al. 2010). That more needs to be done proactively to address ethics issues was highlighted when, in summing up the 2009 American Veterinary Medical Association Animal Welfare Symposium entitled "Swimming with the Tide: Animal Welfare in Veterinary Medical Education and Research", it was acknowledged: "It is unclear to what degree, if at all, ethics is incorporated in many curricula in veterinary schools around the world and yet ethical consideration is key to practising as a good clinician" (Morton 2010).

In ethics courses for many professions, the ethics of role, i.e. professional ethics, is often a focus. "The word 'profession' means 'to testify on behalf of' or 'to stand for something' (May 1989) and is thus more akin to a vocation with "an intrinsic commitment to the public good" and "a strong emphasis on community service" (Preston 2001, p. 173). Professional education can encompass micro or macro concerns. Micro concerns may encourage the virtues of a good professional in terms of their qualities, e.g. honesty, trustworthiness, often aligned with a professional association's Code of Ethics. Macro concerns, however, seek to address the wider picture of the moral duties of the profession to address ethical concerns in society.

In the science professions, animal ethics is often seen as synonymous with the narrow definition of 'reduction, replacement and refinement' in the legislated codes of practice regarding the use of animals in research and teaching. Animal ethics committees undergo ethics training to oversee individual research and teaching projects involving animals within the given paradigm of animals being used for scientific purposes, and animal scientists are trained in how to apply the code. Apart from this narrow perspective, veterinary and animal science education has often taken relativist or pluralist perspectives regarding animal ethics issues. The first veterinary ethics text in 1995 argued that while there is ethical truth, ethics is intensely personal i.e. each of us must decide for ourselves what we think is right and wrong (Tannenbaum 1995). "The Ethics of Animal Use", a textbook published in 2008 for veterinary and animal science courses, uses a pluralistic approach outlining five prominent competing ethical positions selected because they have direct and obvious implications for the ongoing debate on animal use, and three views about what makes a good animal life. These conceptual tools are applied to a range of animal ethics issues "to facilitate mutual understanding and respectful dialogue" (Sandøe and Christiansen 2008). The five ethical positions are:

- Contractarian view – one shows consideration for other rational self-interested persons who have entered into an agreement or contract with you where both parties stand to gain from the contract. According to this theory, animals cannot enter into a contract with people, consequently people can do what they like with animals.
- Utilitarian view – the interests of every living being affected by a decision deserve equal consideration, interests being the capacity for suffering and/or enjoyment
- Animal rights view – it is unacceptable to treat a sentient being merely as a means to achieve a goal. While other animals were not originally included in Kant's view of the "intrinsic worth" of human beings, many have since argued that this principle should be extended to animals.
- The relational view – one should not view animals in an abstract and uniform way. Animals differ from a moral point of view in the relationships they have with human beings. Therefore pet dogs and other companion animals have a special status as individuals.
- Respect for nature view – Protection of the species is more important than the individual

A third approach has been the rational study and examination of the prevailing consensus social ethics that we learn from parents, friends, teachers, churches, movies, books, internet, attempting to justify and seeking inconsistencies in the principles of the social ethic, and drawing out principles that have been previously ignored or been unnoticed, unrecognised and uncorrected (Rollin 2006a, b).

The above approaches assume variability in individuals' or societal perspectives in deciding how animals should be treated. We support a more structured developmental approach to moral behaviour toward animals, based on the psychology of moral development and supported by neurological science.

4.4.1 The Psychological Perspective of Moral Development

In the 1950s, behaviourism dominated psychology and the dominant view of moral development was socialisation. "Moral development was a matter of learning the norms of one's culture, of accepting them and internalising them, and of behaving in conformity with them" (Rest 1994, p. 2). Kohlberg theorised that it was the individual who determines right and wrong, interpreting situations, deriving psychological and moral meaning from social events, and making moral judgements. He argued that sometimes conformity to social norms was morally wrong, e.g. Adolf Eichmann, administrator of the Nazi concentration camps, and non-conformity is morally right, e.g. Martin Luther King defying legal authorities to overcome racism (Kohlberg 1984).

4.4.2 Stages of Moral Judgement

Kohlberg emphasised moral judgement as the most interesting process of moral development. Like Piaget, he concentrated on cognition and conducted extensive longitudinal interviews to identify sequential stages of moral development (Kohlberg 1984). His three levels were originally based on Dewey's philosophy of impulsive, group forming and reflective stages of moral development (Gibbs et al. 2007). However throughout his life, Kohlberg continued to elaborate on and revise his ideas (Kohlberg 1984, p. xiii). There are similarities with Rawls' Theory of Justice stages of morality, i.e. morality of authority, association and principles, based on ever-widening and more complex levels of cooperation from child/parent, to school/neighbourhood and finally to the wider world of complex institutions, which necessitate a move to more principled thinking to balance the needs of unfamiliar parties, beyond the subjective caring for family or groups one knows (Rawls 1971).

The following is a summary (Hedgepeth 2005) of Lawrence Kohlberg's six-stage hierarchy of moral development:

Level I: Preconventional

Stage 1: Egocentric Orientation. Individual is motivated by obedience to authority figures and avoidance of punishment. Doesn't consider interests of others or see more than one point of view.

Stage 2: Instrumental Orientation. Self-interested and exchange-oriented: "You scratch my back, and I'll scratch yours."

Level II: Conventional

Stage 3: Interpersonal Conformity. Individual does good deeds to gain approval and meet expectations of own social group: "Do unto others as you would have them do unto you."

Stage 4: Social Order Orientation. Rule and law-oriented. Conforms to maintain status quo (social or religious). Fears a breakdown of the system if enough people do wrong.

Level III: Postconventional

Stage 5: Social Contract Orientation. Individual respects others' rights and is aware that people hold a variety of opinions and values. Recognizes some universal rights like life and liberty. Realizes that law and morality sometimes conflict.

Stage 6: Universal Principles Orientation. Follows self-chosen universal principles of justice, such as equality and dignity of all human beings. When laws violate these, individual follows the principles.

While Kohlberg developed these stages based on scientific analysis of interviews, Rest developed a written questionnaire, the Defining Issues Test (DIT), to facilitate quantitative analysis and be less labour intensive. Based on Kohlberg's six hierarchical stages of moral reasoning, it identifies respondents' use of three schemas for moral judgement when provided with a variety of moral dilemmas and options for action to resolve the dilemmas:

- Schema 1 Personal Interest (PI) – recognition of authority and reciprocal relationships which result in reward or punishment
- Schema 2 Maintaining Norms (MN) – abiding by existing expectations in rules and regulations set by governments or professional groups with uniform categorical application society-wide, even though the laws may not benefit all participants in an equitable way.
- Schema 3 Post-conventional (here described as Universal Principles [UP]) – a broader less partisan approach than Kohlberg's justice orientation with an emphasis on primacy of all moral ideals which are constructive, sharable and not self-serving at the expense of others, i.e. must be fully reciprocal by benefiting all participants in an equitable way. (Rest et al. 1999a)

In Rest's assessment tool the moral dilemmas concern decisions which affect how people are treated. We have adapted this tool to assess veterinary and animal science professionals' moral judgment about the treatment of animals (Verrinder and Phillips 2014a). We presume, as did Rest, that development occurs through adoption of higher level schemas, though an individual may use various stages of reasoning and may be in transition between schema (Rest et al. 1999a, b). The most commonly used score has been the Principled Score (P Score) based on the number of times the subject picks a Stage 5 or 6 item. However, the subject can also be scored on PI and MN judgment. Studies using the VetDIT suggest that Australian veterinary students use more UP reasoning, similar MN reasoning and less PI reasoning on animal ethics issues, than on human ethics issues, and reason similarly on human ethics issues to U.S. students of a similar educational level (Verrinder and Phillips 2014a). However, this is not unique to veterinary students as medical and arts students have shown similarly higher levels of UP reasoning on animal ethics issues, compared with human ethics issues (Verrinder et al. 2015a, b).

While other moral judgement measures exist, Rest's Defining Issues Test (DIT) has been the most prominent (Gibbs et al. 2007) and frequently used (Carlin et al. 2011) by researchers. It has convergent-divergent validity (p. 21), i.e. it correlates in the 0.6–0.7 range with other measures of moral thinking, i.e. Kohlberg's Moral Judgement Interview (MJI) and the Comprehension of Moral Judgements test (Rest 1994). Kohlberg's claim of the universality of moral development across cultures has been tested with the results of a 2007 database of 75 moral judgement studies using an alternative dilemma-free measure (Sociomoral Reflection Measure) in 23 countries. This was group administered to children as young as 8 or 9, used collectively with the results from Snarey's database of 45 MJI studies conducted in 27 countries. The study led to the conclusion (as did the DIT) that: "Kohlberg was in principle correct regarding universality of basic moral judgement development, moral values, and related social perspective taking processes across cultures" (Gibbs et al. 2007, p. 491).

More recently, developments in neurological science tend to support this universality, showing that morality is grounded in our biology (Churchland 2011). The origin of morality or ethics seems to be survival and well-being and sociality is an important contributor. "Caring is a ground floor function of nervous systems. …The circuitry for self-maintenance and the avoidance of pain is the source of the most basic values – the values of being alive, and of well-being" (Churchland 2011, p. 30). The crucial steps to caring about others that is typical of mammals "depend on the neural-body mechanisms that 'maternalise' the female mammalian brain, which in turn depend on the neuropeptides, oxytocin (OXT) and arginine vasopressin (AVP), along with other hormones. … Once in place, the modification that yields caring for others that are offspring could be further modified, perhaps in quite minor ways, to yield caring for others that are not offspring (Churchland 2011, pp. 31–32). These peptides are not unique to humans, are at least 700 million years old and predate mammals, in terrestrial animals and amphibians (Porges and Carter 2011). Both the insula which appears essential for the nastiness of pain, and the anterior cingulate cortex (ACC) which dominates the motivational (do something) aspect of pain, not only respond to physical pain, but also to social pain, triggered by separation, exclusion, or disapproval, and to pain resulting from errors and poor predictions (Eisenberger and Lieberman 2004). Psychopaths who lack the moral emotions of remorse, guilt and shame have shown different paralimbic regions of the brain both in size and functionality (Churchland 2011, p. 40). Morality therefore seems to be "a natural phenomenon – constrained by the forces of natural selection, rooted in neurobiology, shaped by the local ecology, and modified by cultural developments"(Churchland 2011, p. 191).

4.5 Can Moral Judgement Be Developed?

The DIT has been used extensively to assess moral judgement in a range of educational and professional contexts. General conclusions from a systematic review of 172 DIT studies published since 1980, which were specifically designed to investigate moral judgement development in undergraduate college students in the USA, found that:

- intentionally or unintentionally moral development is an outcome of higher education, at least as measured by the DIT
- students tend to decrease their preference for conventional reasoning (MN) and increase their preference for post-conventional (UP) reasoning during college

- participation in higher education makes a substantial contribution to development in moral judgement beyond that attributable to age alone
- development of moral reasoning is affected by collegiate context e.g. the environment of liberal arts colleges tends to be more conducive to fostering the development of moral reasoning than other types of universities/colleges
- research on the development of moral judgement within different academic disciplines has yielded results which are inconclusive. Several researchers have developed disciplinary-specific tests based on the DIT in journalism, management, education and environmental studies. However whether these adapted instruments will prove to be valid measures of moral judgement will depend on the results of future research (King and Mayhew 2002).

A large body of research shows that grappling with challenging moral issues in discussions or in activities that require the resolution of conflicting opinions contributes to the increasing maturity of an individual's moral judgement, especially when the instructor draws attention to important distinctions, assumptions, and contradictions (Pascarella and Terenzini 2005). However, an ethics or other moral philosophy course that does no more than critique one theory after another may lead students to believe that all ethical perspectives are seriously flawed and that, therefore, all ethical questions are matters of personal taste and opinion (Colby 2008). Perry's empirical studies of college students' progression (Perry 1999) reveals that many students move from an initial "dualistic" stage of seeing the world in polar terms of right vs. wrong and good vs. bad, to more relativistic positions where knowledge and values are seen as contextual and relative, with very few reaching the most advanced level of commitment where moral identity and responsibilities are affirmed. "Student moral relativism" as Ricks (1999) has called it, includes elements of cultural relativism (moral standards are relative to culture), ethical subjectivism ("right" means "right for me"), moral scepticism (nothing can ever be proven in ethics, since people will still disagree), moral nihilism (there are no truths in ethics), and an overriding concern for moral tolerance and respect for others' views (even though if this is the case, i.e. that there is no basis for arguing one position over another, there are no grounds for justifying the claim that we ought to be tolerant) (Colby 2008). "College students' relativism ought to be cause for concern among educators, because beliefs such as 'everyone is entitled to his own opinion and there is no way to evaluate the validity of opinions' prevent students from engaging fully in discussions of ethical issues, learning to articulate and effectively justify their views, and adopting new perspectives when presented with high quality evidence and arguments" (Colby 2008, p. 399). In essence "the stakes drop out of ethical deliberation" and students are less likely to take it seriously (Trosset 1998).

Interventions designed to promote moral judgement development vary widely in content and focus, e.g. ethics courses, courses with an ethics component, social diversity courses, participation in service learning or community service programs, outdoor education programs, and character development programs (King and Mayhew 2002). Virtually all these approaches are effective in promoting moral judgement (King and Mayhew 2002). A further longitudinal DIT study (Mayhew et al. 2010) has provided evidence that links the quality of teaching to developmental gains in moral reasoning. Regardless of gender, race, verbal ability, motivation, political orientation, entering levels of moral reasoning, or course-taking behaviours, the practices that faculty enact to encourage students to reason about moral issues from the broader societal perspectives that serve the public good are important for moral development. Mayhew et al. conclude that "what appears to be important for facilitating the development of moral reasoning are learning environments and practices that provide opportunities for students to consider issues of fairness from broadened, less-egocentric perspectives" (Mayhew et al. 2010, p. 383).

Opportunities for direct involvement in addressing ethics issues have been shown to be important. A study which controlled for class size, instructor, class exercises and text used across two sections of an ethics class (Boss 1994) found that an ethics course which included discussion of moral dilemmas and moral development was effective only for the class whose members engaged in twenty hours of community service work "that involved working directly with people in need." Similarly a course that put students into direct contact with examples of social injustice and opportunities to reflect on and discuss field experiences produced higher post test results in a DIT than those with no service components (Gorman et al. 1994).

Moral judgement can be enhanced through deliberate psychological education, and the more direct and targeted approach of teaching the component skills of moral judgement, for example, skills of logic, role taking and justice operations (McNeel 1994). "Just as students are not likely to develop skills in higher level mathematical and science reasoning without direct teaching and modelling, it is it is unlikely that students will develop skills in higher level moral reasoning without direct teaching and modelling" (Penn 1990). In a study by McNeel (1994) direct teaching, including dilemma discussions, achieved 80 % of the average effect size associated with 4 years of liberal arts college education in just three and a half months. A 15 h veterinary ethics course which included four 2 h small-group case study discussions produced an increase in female, but decrease in male, DIT scores (Self et al. 1995). A 3 h small-group intensive workshop which included moral judgment theory and ethical decision making modelling and practice applied to animal ethics issues produced an increase in moral judgment scores on the VetDIT, whilst similar content in a lecture format did not (Verrinder and Phillips 2015).

4.6 Does Moral Judgement Development Produce Moral Action?

Blasi (1983) points out, "morality requires by definition the investment of knowledge in action." Thoma suggests "the degree to which we can understand moral action is the acid test for the whole research endeavour" (Thoma 1994). Blasi defines this "responsible actualisation of what one knows to be right and true" as integrity. "Reasoning is cheap and painless, action and integrity are not. Integrity requires the development of the whole person" (Blasi 1983).

Research literature suggests that moral judgement development is related to moral action, but the magnitude of the relationship is small (Thoma 1994). Specifically the reviews that provide a statistical summary indicate that the strength of association between moral judgement and actions is estimated at between 10 and 15 % (Thoma 1994). This is not to say that moral judgement is not important. Even if other factors influence moral choices to a similar (or even greater degree), moral reasoning is critical because it produces the moral meaning that an intended action has for the individual (Bredemeier and Shields 1994). Nevertheless, development of moral judgement is often insufficient.

4.7 Other Components of Moral Behaviour

Through review of the morality literature and research, Rest formulated three other major psychological determinants of moral behaviour which, together with moral judgement, he defined as the Four Component Model of Moral Behaviour:

- moral sensitivity – interpreting the situation
- moral judgement – judging which action is morally right/wrong
- moral motivation – prioritising moral values relative to other values
- moral character – having courage, persisting, overcoming distractions, implementing skills (Rest 1994).

All four are determinants of moral action. These components continue to develop across an individual's life span. They are distinct from each other and moral failing can result from a deficiency in any one of them (Bebeau 2008). Kohlberg also saw moral judgement as only part of the psychology of morality and in his later work described a judgement-action model very similar to Rest's model. He also argued that there was an increasing likelihood of subjects at each higher stage [of moral reasoning] judging that they are responsible for carrying out the right action and are therefore more likely to perform that moral behaviour (Kohlberg 1984).

While many programs focus on developing one aspect of the Four Component Model, through dilemma discussion (moral judgement), character education (moral character), sensitivity training for improved communication (moral sensitivity) or the Communitarian Approach (moral motivation) to shift focus from self to community, it is rare to find education programs that encompass all four (Bebeau et al. 1999, p. 24). In the dental profession, strategies such as reflective essays, tests, action plans, dialogues, observations of interactions, and examining case studies have been used to develop all four components (Bebeau 2002). We have investigated the capacity of veterinary students to address animal ethics issues through a questionnaire on moral sensitivity, moral motivation and moral action. This study suggested that most veterinary students are sensitive to animal ethics issues, do experience moral distress, believe that veterinarians should address the wider social issues of animal protection, but have little or no experience in addressing animal ethics issues (Verrinder and Phillips 2014b). Further research suggests teaching and assessment strategies can enhance the capacity to express ethical sensitivity (Verrinder and Phillips 2015, unpublished).

Research studies suggest a need for improved moral development to address ethical conflicts in both animal science and law professions. Veterinary practitioners regularly experience stressful ethical dilemmas related to animal treatment, with most reporting one or two ethical dilemmas weekly, and one in three reporting three to five dilemmas per week. Respondents identified as highly stressful convenience euthanasia of a healthy animal, the client wishing to continue treatment despite compromised animal welfare/quality of life, and financial limitations of the client restricting treatment options, with females rating the first two more stressful than males. Other dilemmas reported were "all euthanasias" and senior staff recommending treatments that the respondent did not agree with. A total of 78 % of respondents reported that they felt they had inadequate training in ethics during their veterinary degree (Batchelor and McKeegan 2012). Two studies by Self (f) suggest that the experience of veterinary medical education seems to inhibit the increase in moral reasoning of veterinary students that would be expected in young adults of this age, while another identified growth. Paul and Podberscek (2000) found that veterinary students in the later years of their course rated dogs, cats and cows as having lower levels of sentience, and while female students maintained relatively high levels of empathy (a component of moral sensitivity) throughout their three final clinical years of study, male students in their final year showed lower levels than male students in their first clinical year.

Studies have consistently found law students report unusually high levels of stress, psychiatric symptoms, substance abuse, anxiety, depression, and internal conflict soon after beginning law school (Hartwell 2004). "Academic law school education tends to limit students to what is called conventional moral thinking and hinders them from attaining post-conventional moral thinking." … Law students "enter law school and immediately discover that they have virtually no real autonomy nor any realistic moral choice in representing the interests of others" (Hartwell 2004, p. 117). Veterinarians face similar pressure for moral conformity through owners' legal entitlement to treat animals used for different purposes differently, with broad exemptions for cruel treatment of animals, regardless of similar sentience levels.

What are the implications of this for animal welfare science and animal law? Animal welfare science and animal law have developed from concerns about how animals are being treated, each based on the assumption that humans have an ethical obligation to animals. However neither can actually address these concerns without ethics. Both require knowledge and skills in moral sensitivity, judgement, motivation and character. Whereas most college students increase their moral reasoning during their courses (Rest et al. 1997), particularly if undertaking liberal arts programs (McNeel 1994), studies suggest that attending veterinary (Self et al. 1996) and law (Hartwell 2004) schools arrests students' moral development. If animal ethics is taught scientifically to develop and assess the four components of moral behaviour, it is likely that both animal science and law professionals will be better equipped to address welfare issues.

4.8 Conclusions

Reflecting on the increase of animal use and abuse, it appears that science and law professionals do not have the necessary knowledge and skills to address the complex area of moral decision-making and moral action concerning the treatment of animals. Animal welfare science has faced difficulties in providing sufficient and impartial information to inform government and industry policy. Animal welfare science should be informative, independent and objective in its approach to welfare issues that are most likely to harm animals. The focus has been on painful procedures, yet the scope of animal welfare problems is much larger than this, and includes both man-made and natural issues. There is a need for scientific rigour in animal welfare studies, avoiding the misleading effects of bias, selective publishing and vested interests being allowed to influence the reporting of findings. This rigour will be particularly important as scientists start to unravel the impact of harmful practices on animal emotions, using a newly developed battery of tests designed for that purpose. Reluctance by scientists to accept the body of proof of these effects on animal emotions will set back the advancement of animal welfare several decades. A failure to allow scientists to provide the information on emotional responses will result in the public attitude, which frequently relies on limited understanding of the problems, prevailing. Although this often provides animals with the benefit of the doubt, a misunderstanding of animal responses is likely to lead to lack of progress in developing standards.

While animal welfare science is developing knowledge and skills to identify how human behaviour impacts on animals, simultaneously animal ethics needs to develop moral sensitivity, moral judgement, moral motivation and moral character in both science and law professionals to enable them to confidently prevent and address animal ethics issues. Accepting that ethical frameworks are complementary rather than competitive, and applying a scientific approach to moral development will avoid the confusion and disenchantment of relativism and pluralism often associated with ethics. It will enable analysis of the paradigms within which science sits, to expose, for example, the human folly of "the inauguration and enjoyment of an age of plenty" at the expense of other animals, and human integrity. Science, animal ethics and the law are inextricably linked. With a greater emphasis on moral development, what we *can* do will be moderated and enhanced by what we *ought* to do.

References

Aristotle. 1999. *Aristotle: Niomachean ethics*, ed. T. Irwin. Indianapolis: Hackett Publishing Company Inc.

Armstrong, S., and R. Botzler. 2003. *The animal ethics reader*. London: Routledge.

Batchelor, C.E., and D.E. McKeegan. 2012. Survey of the frequency and perceived stressfulness of ethical dilemmas encountered in UK veterinary practice. *Veterinary Record* 170(1): 19–22.

Bebeau, Muriel J. 2002. The defining issues test and the four component model: Contributions to professional education. *Journal of Moral Education* 31(3): 271–295.

Bebeau, M.J. 2008. Promoting ethical development and professionalism: Insights from educational research in the professions. *University of St. Thomas Law Journal* 5(2): 366–403.

Bebeau, Muriel J., James R. Rest, and Darcia Narvaez. 1999. Beyond the promise: A perspective on research in moral education. *Educational Researcher* 28(4): 18.

Berridge, K.C. 2003. Pleasures of the brain. *Brain and Cognition* 52(1): 106–128.

Blasi, A. 1983. Moral cognition and moral action: A theoretical perspective. *Developmental Review* 3(2): 178–210.

Boss, J.A. 1994. The effect of community service work on the moral development of college ethics students. *Journal of Moral Education* 23: 183–198.

Bredemeier, B., and D. Shields. 1994. Applied ethics and moral reasoning in sport. In *Moral development in the professions: Psychology and applied ethics*, ed. J. Rest and D. Narvaez, 173–187. Hillsdale: Lawrence Erlbaum Associates.

Burman, O., R. McGowan, M. Mendl, Y. Norling, E. Paul, T. Rehn, and L. Keeling. 2011. Using judgement bias to measure positive affective state in dogs. *Applied Animal Behaviour Science* 132(3): 160–168.

Carlin, N., C. Rozmus, J. Spike, I. Willcockson, W. Seifert Jr., C. Chappell, H. Pei-Hsuan, T. Cole, C. Flaitz, J. Engebretson, R. Lunstroth, C. Amos, and B. Boutwell. 2011. The health professional ethics rubric: Practical assessment in ethics education for health professional schools. *Journal of Academic Ethics* 9: 277–290.

Caulfield, M. 2013. *Science and sense, the case for abolishing sow stalls*. Occasional Report. Sydney: Voiceless.

Churchland, P.S. 2011. *Braintrust: What neuroscience tells us about morality*. Princeton: Princeton University Press.

Colby, A. 2008. Fostering the moral and civic development of college students. In *Handbook of moral and character education*, ed. L.P. Nucci and D. Narvaez, 368–385. New York: Routledge.

Désiré, L., A. Boissy, and I. Veissier. 2002. Emotions in farm animals: A new approach to animal welfare in applied ethology. *Behavioural Processes* 60(2): 165–180.

Eating Better. 2013. *New survey shows support for eating better messages* 2013 [cited 21 November, 2013]. Available from http://www.eating-better.org/blog/23/New-survey-shows-support-for-Eating-Better-messages.html.

Edgar, J.L., C.J. Nicol, C.C.A. Clark, and E.S. Paul. 2012. Measuring empathic responses in animals. *Applied Animal Behaviour Science* 138(3–4): 182.

Eisenberg, N.I., and M.D. Lieberman. 2004. Why rejection hurts: A common neural alarm system for physical and social pain. *Trends in Cognitive Sciences* 8(7): 294–300.

Gibbs, J.C., K.S. Basinger, R.L. Grime, and J.R. Snarey. 2007. Moral judgment development across cultures: Revisiting Kohlberg's universality claims. *Developmental Review* 27(4): 443–500.

Gilligan, C. 1982. *In a different voice: Psychological theory and women's development*. Cambridge, MA: Harvard University Press.

Gorman, M., J. Duffy, and M. Heffernan. 1994. Service experience and the moral development of college students. *Religious Education* 89: 422–431.

Gourkow, N., A. LaVoy, G.A. Dean, and C.J.C. Phillips. 2014. Associations of behaviour with secretory immunoglobulin A and cortisol in domestic cats during their first week in an animal shelter. *Applied Animal Behaviour Science* 150: 55–64.

Hare, R.M. 1995. Universal prescriptivism. In *A companion to ethics*, ed. P. Singer, 451–463. Oxford: Blackwell Publishers Ltd.

Harris, Sam. 2010. *The moral landscape: How science can determine human values*. London: Bantam.

Hartwell, Steven. 2004. Moral growth or moral angst? A clinical approach. *Clinical Law Review* 11(1): 115.

Hedgepeth, E. 2005. Kohlberg's hierarchy of moral development. *School Administrator* 62(4): 38.

Izmirli, S., and C.J.C. Phillips. 2011. The relationship between student consumption of animal products and attitudes to animals in Europe and Asia. *British Food Journal* 113(3): 436–450.

King, P.M., and M.J. Mayhew. 2002. Moral judgement development in higher education: Insights from the defining issues test. *Journal of Moral Education* 31(3): 247–270.

Kohlberg, Lawrence. 1984. *The psychology of moral development: The nature and validity of moral stages*, vol. 2. San Francisco: Harper & Row.

Kuhn, T.S. 1970. *The structure of scientific revolutions*. 2nd enlarged ed. Chicago/London: University of Chicago Press.

Longstaff, S. 1994. What is ethics education and training? In *Ethics for the public sector: Education and training*, ed. N. Preston, 138–160. Annandale: The Federation Press.

Magalhães-Sant'Ana, M., I.A.S. Olsson, P. Sandoe, and K. Millar. 2010. How ethics is taught by European veterinary faculties: A review of published literature and web resources. In *Global food security: Ethical and legal challenges*, ed. Carlos Maria Romeo Casabona, San-Epifanio Leire Escajedo, and E.-C.A. Wageningen, 441–446. Wageningen: Wageningen Academic Publishers.

Main, D.C.J., P. Thornton, and K. Kerr. 2005. Teaching animal welfare science, ethics, and law to veterinary students in the United Kingdom. *Journal of Veterinary Medical Education* 32(4): 505–508.

May, W. 1989. *Vocation, career and profession*, vol. 1, no. 1. Australian Institute of Ethics and the Professions Monograph Services, St Lucia, Qld.

Mayhew, M.J., T.A. Seifert, and E.T. Pascarella. 2010. A multi-institutional assessment of moral reasoning development among first-year students. *The Review of Higher Education* 33(3): 357–390.

McNeel, S. 1994. College teaching and student moral development. In *Moral development in the professions*, ed. J. Rest and D. Narvaez, 27–49. Hillsdale: Lawrence Erlbaum Associates.

Mendl, M., J. Brooks, C. Basse, O. Burman, E. Paul, E. Blackwell, and R. Casey. 2010. Dogs showing separation-related behaviour exhibit a 'Pessimistic' cognitive bias. *Current Biology* 20: R839–R840.

Mepham, Ben. 2008. *Bioethics: An introduction for the biosciences*, 2nd ed. New York: Oxford University Press.

Morton, David B. 2010. A commentary on the animal welfare symposium, with possible actions. *Journal of Veterinary Medical Education* 37(1): 107–113.

Newmyer, S.T. 2006. *Animals, rights and reason in Plutarch and modern ethics*. New York: Routledge.

Oxford Dictionary. 2014. *Science*. http://www.oxforddictionaries.com/definition/english/science. Accessed 14 Oct 2014.

Panksepp, J. 1998. *Affective neuroscience: The foundations of human and animal emotions*. New York: Oxford University Press.

Pascarella, E.T., and P.T. Terenzini. 2005. *How college affects students: A third decade of research*, vol. 2. San Francisco: Jossey-Bass.

Paul, E.S., and A.L. Podberscek. 2000. Veterinary education and students' attitudes towards animal welfare. *Veterinary Record* 146(10): 269–272.

Paul, E.S., E.J. Harding, and M. Mendl. 2005. Measuring emotional processes in animals: The utility of a cognitive approach. *Neuroscience and Biobehavioral Reviews* 29: 469–491.

Penn Jr., W.Y. 1990. Teaching ethics: A direct approach. *Journal of Moral Education* 19(2): 124–138.

Perry Jr., William G. 1999. *Forms of ethical and intellectual development in the college years: A scheme, Jossey-Bass higher and adult education series*. San Francisco: Jossey-Bass Publishers.

Phillips, C.J.C. 2008. Pain as an animal welfare issue. In *Proceedings of the animal welfare science summit*, co-ordinators D.J. Mellor, P.M. Thornber, A.C.D. Bayvel and S. Kahn. Scientific Assessment and Management of Animal Pain, 38–46. Organisation Internationale Epizootie (O.I.E.) Technical Report No. 10. Paris: O.I.E.

Phillips, C.J.C., and J.C. Petherick. 2014. The ethics of a co-regulatory model for farm animal welfare research. *Journal of Agricultural & Environmental Ethics*. doi:10.1007/s10806-014-9524-9

Phillips, C.J.C., J. Wojciechowska, J. Meng, and N. Cross. 2009. Perceptions of the importance of different welfare issues in livestock production. *Animal: An International Journal of Animal Bioscience* 3(8): 1152–1166.

Phillips, C.J.C., S. Izmirli, S.J. Aldavood, M. Alonso, B.I. Choe, A.J. Hanlon, A. Handziska, G. Illmann, L. Keeling, M. Kennedy, G.H. Lee, V. Lund, C. Mejdell, V.R. Pelagic, and T. Rehn. 2012. Students' attitudes to animal welfare and rights in Europe and Asia. *Animal Welfare* 21: 87–100.

Porges, S., and C. Carter. 2011. Neurobiology and evolution: Mechanisms, mediators and adaptive consequences of care-giving. In *Self-interest and beyond: Toward a new understanding of human care-giving*, ed. S. Brown, R. Brown, and L. Penner, 53–71. New York: Oxford University Press.

Preston, N. 2001. *Understanding ethics*, 2nd ed. Sydney: The Federation Press.

Rawls, J. 1971. *A theory of justice*. Cambridge, MA: Harvard University Press.

Reefmann, N., F. Butikofer Kaszas, B. Wechsler, and L. Gygax. 2009. Physiological expression of emotional reactions in sheep. *Physiology & Behavior* 98(1–2): 235–241.

Rest, J. 1994. Background: Theory and research. In *Moral development in the professions: Psychology and applied ethics*, ed. J. Rest and D. Narvaez, 1–26. Hove: Lawrence Erlbaum Associates.

Rest, James, Stephen Thoma, and Lynne Edwards. 1997. Designing and validating a measure of moral judgment: Stage preference and stage consistency approaches. *Journal of Educational Psychology* 89(1): 5–28.

Rest, J.R., D. Narvaez, M.J. Bebeau, and S.J. Thoma. 1999a. *Postconventional moral thinking: A Neo-Kohlbergian approach*. Mahwah: Lawrence Erlbaum Associates.

Rest, James R., Darcia Narvaez, Stephen J. Thoma, and Muriel J. Bebeau. 1999b. DIT2: Devising and testing a revised instrument of moral judgment. *Journal of Educational Psychology* 91(4): 644–659.

Ricks, V. 1999. *How not to teach moral relativism*. Paper presented at the Teach-In on Moral Relativism. Stanford: Stanford University.

Robins, A., and C.J.C. Phillips. 2010. Lateralized visual processing in domestic cattle herds responding to novel and familiar stressors. *Laterality: Asymmetries of Body, Brain and Cognition* 15: 514–534.

Robins, A., and C.J.C. Phillips. 2011. International approaches to the welfare of meat chickens. *World's Poultry Science Journal* 67(2): 351–369.

Rollin, B.E. 2006a. *An introduction to veterinary medical ethics*, 2nd ed. Iowa: Blackwell Publishing.

Rollin, B.E. 2006b. *Science and ethics*. Cambridge: Cambridge University Press.

Ross, T., and C.J.C. Phillips. 2015. Relationships between knowledge of chicken production systems and advocacy by animal protection workers. *Society and Animals* (in press).

Sandoe, P., and Stine B. Christiansen. 2008. *The ethics of animal use*. Oxford: Blackwell Publishing.

Sandøe, Peter, Roger Crisp, and Nils Holtug. 1997. Ethics. In *Animal welfare*, ed. M.C. Appleby and B.O. Hughes, 3–17. Wallingford: CAB International.

Self, D.J., A.B. Pierce, and J.A. Shadduck. 1995. Description and evaluation of a course in veterinary ethics. *Journal of American Veterinary Association* 207(12): 1550–1553.

Self, D.J., M. Olivarez, D.C. Baldwin, and J.A. Shadduck. 1996. Clarifying the relationship of veterinary medical education and moral development. *Journal of the American Veterinary Medical Association* 209(12): 2002–2004.

Singer, P. 1991. *Animal liberation*. Hammersmith, London: Thorsons.

Singer, P. 1993a. *A companion guide to ethics*. Oxford: Blackwell Publishers Ltd.

Singer, P. 1993b. *Practical ethics*, 2nd ed. Cambridge: Cambridge University Press.

Steiner, G. 2005. *Anthropocentrism and its discontents: The moral status of animals in the history of western philosophy*. Pittsburgh: University of Pittsburgh Press.

Tannenbaum, Jerrold. 1995. *Veterinary ethics: Animal welfare, client relations, competition and collegiality*, 2nd ed. St Louis: Mosby.

Thoma, Stephen. 1994. Moral judgements and moral action. In *Moral development in the professions: Psychology and applied ethics*, ed. J. Rest and D. Narvaez, 199–211. Hillsdale: Lawrence Erlbaum Associates.

Tiplady, Catherine M., Deborah-Anne B. Walsh, and Clive J.C. Phillips. 2013. Public response to media coverage of animal cruelty. *Journal of Agricultural and Environmental Ethics* 26(4): 869–885.

Trosset, C. 1998. Obstacles to open discussion and critical thinking. *Change* 30(5): 44.

van der Schot, A.A., and Clive Phillips. 2013. Publication bias in animal welfare scientific literature. *Journal of Agricultural and Environmental Ethics* 26(5): 945–958.

Verrinder, J.M., and C.J.C. Phillips. 2014a. Development of a moral judgment measure for veterinary education. *Journal of Veterinary Medical Education* 41(3): 258–264.

Verrinder, J.M., and C.J.C. Phillips. 2014b. Identifying veterinary students' capacity for moral behaviour concerning animal ethics issues. *Journal of Veterinary Medical Education* 41(4): 358–370.

Verrinder, J.M., R. Ostini, and C.J.C. Phillips. 2015a. Differences in moral judgment on animal and human ethics issues between university students on animal-related, human medical and ethics programs. *PLOS One* (Submitted; this is the first unpublished quote by Verrinder et al.).

Verrinder, J.M., R. Ostini, and C.J.C. Phillips. 2015b. Developing and assessing ethical sensitivity to address animal ethics issues. *Animal Welfare* (currently being submitted; this is the second unpublished quote by Verrinder et al.).

Wemelsfelder, Françoise, Ian Nevison, and Alistair B. Lawrence. 2009. The effect of perceived environmental background on qualitative assessments of pig behaviour. *Animal Behaviour* 78(2): 477–484.

Winston Churchill's Iron Curtain Speech. 1946. http://www.historyguide.org/europe/churchill.html. Accessed 21 Dec 2015.

Chapter 5
An International Treaty for Animal Welfare

David Favre

Abstract Currently there is no international agreement that ensures the welfare and protection of animals. Nor is there any international standard that regulates and defines the acceptable treatment of animals. This lack of international consensus leads to the current disparate treatment of animals around the world, echoing the need for an international framework addressing the issue. This Article discusses a proposed umbrella treaty, the International Convention for the Protection of Animals (ICPA). This umbrella treaty would enable animal welfare issues to gain international recognition and protection by setting the general guidelines and polices regarding the treatment and use of animals. This Article argues that this is the best way to successfully pursue international protection by reconciling the conflicting goals of making a treaty enticing to as many countries as possible, without eliminating enforcement mechanisms. This Article also suggests four companion protocols that would further delineate specific animal welfare standards and requirements. With the present economic climate, it may be difficult to convince countries to pass such a treaty. However, the ICPA could make it possible to begin the process of enacting groundbreaking international animal protection.

5.1 Introduction

Many critical environmental issues today cannot be resolved without international agreements. As part of the struggle to further environmental and human rights issues, proponents of these causes utilize treaties, conventions, and declarations to

 This article is an updated version of an article of the same name published in a law review of Lewis and Clark Law School in 2012 (18 Animal Law 237 (2012)). The initial article had considerable additional footnoting and references not found in this chapter. A full copy of the proposed treaty is available https://www.animallaw.info/treaty/international-convention-protection-animals.

D. Favre (✉)
Animal Legal & Historical Center, Michigan State University College of Law, East Lansing, MI, USA
e-mail: favre@law.msu.edu; http://www.animallaw.info

D. Cao, S. White (eds.), *Animal Law and Welfare - International Perspectives*, Ius Gentium: Comparative Perspectives on Law and Justice 53,
DOI 10.1007/978-3-319-26818-7_5

find standards and procedures to address these issues on a global basis. Until now, the issue of the care and well-being of the nonhuman inhabitants of this planet has had no equivalent international focal point. Because of human need, greed, ignorance, and vanity, untold numbers of animals throughout the world suffer and die on a daily basis. While thousands of humans work daily to reduce this toll, a more visible, efficient, and universal standard is needed. The immediate elimination of animal pain and suffering is not likely. Therefore, a mechanism is needed under which progress for animal welfare can be realized as fast as moral persuasion, technology, economic development, and political support will accommodate it. The adoption of the International Convention for the Protection of Animals (ICPA) can be one important legal and political tool in the process of legal progress for animals.

Some countries have adopted serious laws for dealing with animal welfare issues. Others have welfare laws but inadequate enforcement resources or political will to carry out their laws. Some countries have neither enacted meaningful laws, nor expressed a legitimate interest in dealing with animal issues. At present, there is no international animal welfare standard by which to judge the legislative efforts within one country. Also, at the moment, every battle has to be fought over and over again in each country as those trying to help animals seek to justify new laws and restrictions. With the ICPA, there will be an accepted standard, which is immediately accessible to all the nation states, organizations, and individuals of the world. The adoption of the ICPA will give more credibility and weight to all of the local efforts on behalf of animals. With the ICPA, local groups could stop using their finite resources to advocate for what the standards ought to be, and instead could use them to further the implementation of the standards that have been accepted within the Convention.

The term "animal rights" is part of the legal conversation in only a few countries (see Kelch 2011, p. 271, p. 304). "Animal welfare," and its supporting concepts, is the best available and most acceptable term in most countries (see Garner 2006, p. 161, p. 163 and pp. 169–170). Certainly, among diplomats, the term "animal rights" will only cause confusion and suspicion. Species protection is the only language culturally acceptable in many countries. At this point in time, seeking to draft a treaty for the legal rights of animals would not be useful. First, most governments would not give it serious consideration. Second, it would be disrespectful to seek legal rights for animals when so many humans around the world still struggle for their rights.

This Article will first develop a quick background on the nature of treaties in Part 2 and reasons for seeking an animal welfare treaty in Part 3. It next focuses on how to structure a specific umbrella treaty in Part 4. Finally, Part 5 considers the prospects for the adoption of the treaty.

5.2 The Nature and Role of Treaties

Historically, the cornerstone of international law has been the treaty. Sovereign states controlled the world, its people, commerce, information, and military might. Creation of rules by treaty became the international rules of conduct between nations. Up until 1990 or so, the countries of the world were the primary deciders of most international issues, but with the breakup of the Soviet Union and the coming of the technology revolution of the 1990s, the sovereign states have had an increasingly modest role in international issues (see Friedman 1999, pp. 200–202). The growth of large international corporations and the communication capacity of the Internet have diminished the role of nation states in international affairs. This Article will discuss this topic more thoroughly below. For the moment, the focus will be on understanding what a treaty is and how it could be helpful for the well-being of animals.

A treaty is a formal written agreement between nation states, either bi-lateral or multi-lateral, which commits the state to engage, or not engage, in certain actions.[1] In the absence of a treaty, under the internationally accepted concept of sovereignty, the leaders of a country may choose any course of action internally or externally that they may wish. The leaders of a country may change their minds at will, but most find it in their best interests to restrict their freedom of action by becoming members of a treaty under which other countries also agree to restrict their freedom. Thus, while a country may believe it would be cheaper to dump chemical waste in other countries rather than treat the waste at home, most nation states have made the collective decision that this is a dangerous and unfair path to take. As a result, most countries have agreed to limit their freedom of action by signing the Basel Convention on the Control of Transboundary Movement of Hazardous Waste and their Disposal, which prohibits the dumping of certain waste in other countries.[2]

So long as a country is a member of a treaty, it is expected to conform to the treaty. Even if a new president is elected, a country nevertheless continues to be bound by all previously signed treaties—unless it formally withdraws from the treaty. For this reason, the creation of a treaty, and the decision to become a "member state" of a treaty, is recognized as a significant national event requiring a formal process. This is referred to as "the process of ratification." While treaties do not dictate how a nation state decides whether to become a member of a treaty, they do have a process by which a nation state formally acknowledges the decision of ratification to other countries.

The United States, for example, ratified the treaty on global warming in October 1992, becoming a member state. However, when the Parties to the United Nations Framework Convention on Climate Change developed the Kyoto Protocol

[1] Vienna Convention on the Law of Treaties pt. I, art. 2(a) (23 May 1969).

[2] Basel Convention on the Control of Transboundary Movements of Hazardous Wastes and their Disposal art. 4 (5 May 1992), 28 I.L.M. 649.

(Protocol)[3] establishing emissions limitations for greenhouse gases so as to limit global warming, unlike the 192 nation states, which are Parties, the U.S. declined to ratify this Protocol. The U.S. therefore is not bound by international law to restrict greenhouse gases. (Of course the U.S. could decide at any point to adopt any level of restriction that it might choose under the power of sovereignty.) As the U.S. is not a party to the Protocol, the U.S.'s national policy for economic development is not limited by the restrictions of the Protocol. An example of a significant multi-lateral treaty that the U.S. promoted internationally is the Convention on International Trade in Endangered Species and Wild Fauna and Flora (CITES).[4] The obligations of the U.S. under this treaty—primarily to issue permits and restrict importation and exportation of listed, protected species—are implemented through the U.S. Endangered Species Act (ESA).[5] To import or export a specimen of a species listed on a CITES appendix may require private parties to obtain government issued permits.

It is important to understand that some international instruments are not treaties. Declarations are not treaties. Declarations may be public statements about something, but they do not bind States to actually do or stop doing anything. An example of this is the Universal Declaration on Animal Welfare (Declaration).[6] This short declaration primarily states that countries should acknowledge that animal welfare is an important public policy issue and more needs to be done for the benefit of animals around the world. While a number of countries have acknowledged support of the document, it is not yet a formally accepted declaration. Whenever it may become an international declaration, it will not be considered a treaty. A review of the language of the Declaration makes it clear that any country associating itself with the Declaration is not actually committing itself to doing or refraining from doing anything in particular. Likewise, declarations announced at economic summits do not have status as treaties, but instead are simple, feel-good statements about the current thoughts of countries that are represented at the meeting.

The final point to make is that treaties are agreements only between sovereign nations. Within the U.S., the state of California, even though its population and economic activity might suggest it is as important as many countries, does not have the power to enter into a treaty. Neither may Apple, Inc. or Boeing make treaties; nor may the Humane Society of the U.S. or other non-profit organizations. In today's world, Bill Gates, who oversees a foundation with over $40 billion in assets, may have as much economic power, information, and ties to others with political power as any number of countries. Apple, Inc. has as much cash on hand as any number of

---

[3] Kyoto Protocol to the United Nations Framework Convention on Climate Change (11 December 1997).

[4] Convention on International Trade in Endangered Species of Wild Fauna and Flora (3 March 1973), 27 U.S.T. 1087, 993 U.N.T.S. 243.

[5] 16 U.S.C. §§ 1531–1544 (2010); 50 C.F.R. § 23.1.

[6] See Universal Declaration on Animal Welfare (Declaration), http://s3.amazonaws.com/media.animalsmatter.org/files/resource_files/original/Latest%20draft%20UDAW%20Text%20-%202011.pdf?1314177486 (accessed 8 April 2012) (copy of proposed Declaration, amended in 2011).

countries. But money and power alone do not allow these entities to make treaties. With all these other sources of power in today's world, the importance of a treaty is more limited than it was in the past. However, a treaty can still be a powerful statement and motivator for action.

5.3 The Need for an Animal Welfare Treaty

Many individuals and organizations around the world feel that the use of animals and the conditions under which many are kept are unacceptable and they thus seek to enhance the protection and status of animals. They believe that animals are subjected to considerable pain, suffering, and death, much of which is not necessary or appropriate. The fact patterns under which the discussions arise are very diverse. How should pets be treated? If commercial animals are being transported, what conditions are appropriate for them? Should animals be used in research, testing, or educational settings? Which wild animals are acceptable for exhibiting and what standards should there be for the conditions of exhibits? These are not easy questions, but a horse is a horse regardless of what country it lives in, and it is not appropriate that it can receive high care in some places and no concern in others. For the wellbeing of the animals we need to seek a more universal view about how to treat animals.

The animals that would gain the most from an animal welfare treaty are agricultural animals. While within the European Union (E.U.), and Europe generally, some level of protection for animal welfare is provided, for most of the world there are very few legal limitations on the use of agricultural animals. Within the U.S., this is a particular problem. When introduced to commercial animals and issues of welfare, most students assume that some law must exist to provide a minimum level of protection for farm animals such as chickens. This is not the case. Most state anti-cruelty laws have a specific exemption for agricultural practices.[7] For example, the Michigan anti-cruelty law has the following provision:

> This section does not prohibit the lawful killing or other use of an animal, including, but not limited to, the following:
>
> (a) Fishing.
> (f) Farming or a generally accepted animal husbandry or farming practice involving livestock.

The language above would extend to hogs, cattle, and other animals, as well as chickens. A key issue in this area is what constitutes "traditional agricultural practices," as most statutory exemptions use this or an equivalent phrase. While apparently everyone can agree that starvation is not a "generally accepted farming

[7] See e.g. Mich. Comp. Laws § 750.50b(7) (2001); Mo. Rev. Stat. Ann § 578.007(8) (West 2009); Wash. Rev. Code Ann § 16.52.185 (West 2009).

practice," it is unclear what the word "traditional" might mean: those practices used by the average producer or the high-volume producer; 100-year-old practices or farming practices that are only 20 years old. State prosecutors seem entirely disinterested in pursuing with a jury trial what is or is not included in such exemptions. Therefore, state cruelty laws in the U.S. do not really provide welfare protection for commercial animals.

Even when the legislature seeks to act on behalf of agricultural animals, it does not always result in a beneficial outcome. In 1996, New Jersey took the progressive step of directing the state agricultural agency to adopt humane standards for the raising of agricultural animals.[8] However, the regulations adopted were nearly identical to many of the existing industrial practices.[9]

In the U.S. we have one national law that touches upon animal welfare for a limited number of issues, the federal Animal Welfare Act.[10] Under this law, the outcome for agricultural animals is no better at the federal level than at the state level. The Act specifically says that agricultural animals are exempt from its provisions.

5.3.1 The Lack of Existing International Law

Considering existing international treaty law about animals requires dividing animals into the categories of wildlife and domestic. While among the European States there are a number of treaties that deal with domestic animal issues, there are no global treaties dealing with these issues. Wildlife on the other hand, with their natural tendency to move about regardless of national borders, are the subject matter of a number of treaties. The primary focus of these treaties is the preservation of endangered species or the management of commercial valuable species so that they will not become endangered. There are a number of treaties dealing with migratory birds and fish and the International Convention for the Regulation of Whaling deals with whales.[11] The welfare of individual wildlife is not normally the focus of these treaties.

One exception to the lack of concern for the welfare of wildlife is a provision within the Convention on International Trade in Endangered Species and Wild Fauna and Flora (CITES). Article III, paragraph 2(c) and Article IV, paragraph 2(c) states the requirements for granting an export permit of live wildlife protected by the treaty: "[T]he State of export is satisfied that any living specimen will be so prepared and shipped as to minimize the risk of injury, damage to health or cruel treatment." Satisfaction of this requirement is normally met by simply agreeing to

[8] N.J. Stat. Ann. § 4:22–16.1 (West 1998).

[9] *N.J. Socy. for the Prevention of Cruelty to Animals v. N.J. Dept. of Agric.*, 955 A.2d 886, 907–16 (N.J. 2008).

[10] 7 U.S.C. §§ 2131–2159.

[11] International Convention for the Regulation of Whaling art. IV (2 December 1946).

abide by the shipping standards adopted by International Air Transport Association (IATA).[12] The countries that are Parties to CITES have not explicitly defined what constitutes cruel treatment. Also, attempts to extend this welfare concern to the capture and holding of wildlife during the domestic portions of travel, rather than just the international component have failed.

More representative of present treaties is the Convention on Biological Diversity (CBD), which seeks to promote the protection of the environment and the wildlife contained therein, but says nothing about the welfare of individual animals.[13] Thus it is clear that while the environmental perspective of the importance of wildlife as part of ecosystems is well accepted, the conditions of life and death of individual animals at the hands of humans around the world is not yet a focus of legal drafting.

5.3.2 The International Movement of Animals

Animals move internationally, slipping from one jurisdiction and set of laws to another; from full protection in one country to little or no protection in another. For all practical purposes the slaughter of horses in the U.S. for food has been eliminated; now horses are exported to be slaughtered in Mexico, where the conditions of the slaughterhouse are not as regulated as in the U.S. This outcome might make some of the people in the U.S. feel better, but it is not a better outcome for the horses that are shipped to Mexico. The laws of the U.S. for horse protection are frustrated since other countries do not have equal laws of protection and the horses can be easily transported to other countries.

Whales and sharks move globally on their own volition. Some countries, particularly in Asia, use the fins of a shark to make soup that is in high consumer demand as a prestige food. This demand results in the cutting off of the fin of live sharks resulting ultimately in pain, suffering, and death. Other countries have judged that the consumption of the soup does not justify the pain, suffering, and death of the sharks and prohibit the practice. Depending on where the shark is swimming, they will either be protected or subject to a cruel death.

Another international issue is the live shipment of livestock from countries like Australia to the Middle East, because of a preference for local slaughter rather than frozen meat. Even though Australians may have good laws for the welfare of livestock within their borders, conditions on the high seas and treatment at the import country are not under the same welfare standards. It is not clear whether this use of agricultural animals should be allowed at all, let alone what might constitute acceptable welfare conditions.

---

[12] Transport of Live Specimens, CITES Resolution Conf. 10.21, (1997) (rev. CoP14) (available at https://cites.org/eng/res/all/10/E10-21R14.pdf, accessed 24 August 2015).

[13] Convention on Biological Diversity art. 8 (21 May 1992), http://www.cbd.int/convention/articles/?a=cbd-08

An animal welfare treaty would be one method of seeking a broader global consensus on how a particular species of animal should be used, and, if a use is acceptable, what level of welfare must be provided during the use.

5.3.3 Fostering Discussion and Consensus

Within a particular country, it is often the case that citizens, citizen groups, and nongovernmental organizations have a concern about a variety of animal welfare issues, but it is very difficult to get the attention of the national government to address these issues. The creation of an international discussion about the possibility of an animal welfare treaty will provide space for conversations on the national level. The adoption of such a treaty will necessarily impose some level of obligation on countries to modify existing laws or adopt new laws. This will provide political leverage for individuals and organizations to raise internal issues with their governments and foster a public discussion about what the laws ought to be.

Additionally, within the U.S., the existence of a treaty would bring federal jurisdiction to what is otherwise now primarily a state jurisdictional issue. At the moment, under the U.S. Constitution, the only jurisdiction for animal welfare issues arises indirectly out of the power of the federal government to control international and interstate commerce. However, the federal government is fully empowered to engage in the treaty process, and, once that is done, the federal government has jurisdiction over the subject matter of the treaty for purposes of implementation. For example, CITES gives the federal government the power and duty to implement domestic legislation for the protection of endangered species. Likewise, the discussion of the International Convention for the Protection of Animals (ICPA) would raise important animal issues in a national, federal debate. Whether this is positive or negative for the animals of the U.S. cannot be known at this point in time.

5.3.4 Reducing Negative Economic Incentives

A critical reason for a treaty is the reality of global corporations that operate internationally and that seek to make a profit off of animals and animal products. These corporations have concern for the welfare of the animal in only three circumstances: the buyers demand it, it has a negative effect on the profit of the corporation, or the law requires it. Under the rules of capitalism, usually the least cost producer of a product will receive increased market share and higher profits. To the extent that a single country seeks to enhance the welfare of agricultural animals, corporations may seek to avoid the cost of increased animal welfare measures by raising the animals in another country. Even if a corporation seeks to do a better job of caring for its animals, it risks losing market share—and potentially going out of business—because other corporations can operate more cheaply without welfare measures.

Unrestrained economic competition will always impose the most inhumane conditions on the animals within the system, as they tend to be the cheapest management practices. If there were worldwide standards within commerce that assured a cruelty-free life for animals, then the forces of capitalism would accept this as a base and seek the least cost products within that set of rules.

Even if a country could reach political consensus on enhanced protection for farm animals and adopt protective legislation, it may not be able to affect the conditions that exist outside the borders of that country. As an analogous example, consider that while American laws protect American children from child labor abuses, the U.S. is not in a position to prohibit the importation of items made by children in other countries that violate our labor standards. The result is that products overseas made with cheaper labor compete with products made in the U.S. There is no reason to think that the same pattern would not come into play if the U.S. laws for animal welfare increased product cost in this country while other countries did not change their law.

State sovereignty is a fundamental concept that permits these different levels of protection, whether of child labor or animal welfare. Under this long-standing international legal concept, it is accepted that each country has the right to decide what level of protection to give its citizens, animals, or environment. No one country has the right to enforce its standards on another country. Under this principle, a country may set up all sorts of barriers to trade coming into their country. The counterpoint to that perspective is found in free traders who believe that maximum wealth production on a global basis will occur when there are no barriers to the movement of goods between countries.

This conflict of principles between sovereign protectionism and free trade is front and center in the World Trade Organization (W.T.O.) and associated treaties.[14] At the moment, this regime may be the most important legal framework regarding animals as it controls international trade of live animals and animal products, such as meat and skins. One hundred and sixty-one countries are members of this treaty (April 2015). With Russia's accession to the W.T.O. in 2012, Iran and Iraq remain as major countries not part of the treaty. Unfortunately, "animal welfare" is not a phrase found in the treaty.

A full discussion of the workings of the W.T.O. is not possible in this article. Consider these issues: whether the E.U. may ban the importation of seal skins because of the methods of death carried out in Canada; and whether the U.S. may restrict the importation of tuna from Mexico or the Philippines if dolphins have been killed in the process of catching the tuna. In both of these examples, one country is asserting its standards against the sovereign power of another country to decide when and how to kill animals. Promotion of free trade under the W.T.O. umbrella is advanced by the general principle that it is illegal to restrict the importation of a product based on its method of production. Thus, the manufacturing of soccer balls

---

[14] See World Trade Org., *W.T.O. ... Why it Matters: A Guide for Officials, Legislators, Civil Society and All those Interested in International Trade and Global Governance* 17–18, http://www.wto.org/english/thewto_e/minist_e/min01_e/wto_matters_e.pdf, accessed 24 August 2015.

with child labor would not be a justification that the U.S. could use to ban the importation of soccer balls manufactured in another country. Nor would the creation of industrial pollution in a country justify the imposition of trade bans by another. Under this general principle, if New Zealand passes a law requiring that only eggs from free-range chickens be sold, this provision might be a breach of W.T.O. obligations for eggs from caged chickens that Australia seeks to import into New Zealand. However—and it is a big however—under the specific treaty establishing this principle, the General Agreement on Tariffs and Trade (GATT), there are exceptions and arguments that can be made that animal welfare concerns may qualify to justify limitations of trade between countries.

In 2014 a hearing panel within the W.T.O. released an opinion about a dispute between Canada and the European Union (E.U.). The E.U. had adopted a ban on the importation of seal skins to which the Canadians strenuously objected. They claimed it constituted a violation of the responsibilities of the E.U. under W.T.O. However, the panel accepted that the moral concern for the welfare of the seals was acceptable justification for the import ban stating that the ban was acceptable as 'necessary to protect public morals.'[15] This seems to open a new door of animal welfare considerations, but it is too soon to suggest the potential scope of this new path.

Another exception to the provisions of the W.T.O. is when a nation state carries out the requirements of another multilateral treaty. If the ICPA is subsequently adopted—prohibiting the sale of eggs from caged hens—countries adopting that ban in domestic law would not be in violation of W.T.O. responsibilities. Any subsequently adopted animal welfare treaty will supersede the general restrictions of the GATT, thus allowing the use of trade import restrictions to enforce international obligations.

5.3.5 Default Standards of Organization for Protection of Animal Health Are Not Sufficient

There is one international body that is seeking to create international standards for animal welfare issues: the World Organization for Protection of Animal Health, known as O.I.E.[16] O.I.E. is an international organization created by an agreement between 28 countries in 1924 to deal with issues of animal health, particularly those in international trade where the risk of disease can transfer from one country to another. Presently there are 178 countries that are members of this international organization.

[15] All the documents can be located at: http://www.wto.org/english/tratop_e/dispu_e/cases_e/ds400_e.htm

[16] See World Org. for Animal Health, The O.I.E.'s Achievements in Animal Welfare, http://www.oie.int/animal-welfare/animal-welfare-key-themes/ (stating that animal welfare was identified as an O.I.E. priority in O.I.E. Strategic Plan 2001–2005).

The human delegates to the body tend to be from the national veterinary administration or the agricultural administration of a country. They have adopted a detailed and science-based process for the adoption of sanitary standards. These standards are not immediately binding on the Parties, but are recommendations for them to follow. These standards are significantly enhanced by the fact that they are recognized by the W.T.O. system and may be the basis for international trade restrictions.

Since 2001, the organization has also decided to take upon itself the development of animal welfare standards. While the issues of animal health are strongly based upon scientific information, issues of animal welfare are public policy decisions, adopted with scientific input, but not by scientist. Therefore, this organization is not the best place to address the difficult questions of the quality of life for animals versus the economic and social consequences of imposing limitations on the use of animals. Rather the role of science is to inform the decision makers of present facts and possible future consequences of alternative causes of action.

A review of the adopted O.I.E. standards on live animal transport will reveal their significant limitations. The standards include no numbers, no prohibitions, no required inspections, and no limitations on operations. Rather, the standards read like a checklist of issues that should be considered if one is going to engage in live animal transport. While this list is useful for policy makers, it is not an actual standard that limits or prohibits practices that are harmful to animal welfare. Therefore, the existence of this organization, and it's proposed standards does not diminish the need for an international animal welfare treaty that directly seeks to deal with public policy issues.

5.4 How to Build an Animal Welfare Treaty

Any treaty dealing with animal welfare should utilize an umbrella treaty approach. The first step is to draft a framework treaty with the intent to further refine and resolve specific welfare issues in subsequent protocols. A model treaty has already been drafted and currently the most crucial issue facing its adoption is the absence of a country to sponsor the treaty. Necessary provisions for an umbrella treaty are discussed in relation to the chosen language in the International Convention for the Protection of Animals (ICPA) and the companion protocols.[17]

[17] Full draft available at https://www.animallaw.info/treaty/international-convention-protection-animals

5.4.1 The Nature of Umbrella Treaties

When faced with large and complex problems that impact all countries, it is often difficult to obtain consensus on specific answers to problems and concerns even when there is general agreement that a problem exists and ought to be addressed. During the early 1990s the global community began to realize that the negative impacts on the planet arising out of the number and consumption patterns of humans were causing serious global problems. There was considerable political energy to move toward treaty agreements that address a number of these broad issues. The United Nations (U.N.) body, United Nations Environment Programme (UNEP), oversaw the Rio Summit in Rio de Janeiro, Brazil in June 1992. At this summit, two umbrella treaties were adopted: the Convention on Biological Diversity (CBD) and the U.N. Framework Convention on Climate Change (UNFCCC).

For both treaties, the vast majority of states agreed that (a) increasing greenhouse emissions represented a risk of global warming, and (b) that present human patterns of consumption and land use were destroying the natural ecosystems around the world. However, there were serious disagreements about who was responsible for the problems, who would accept responsibility for solving the problems, and even how they would be solved. This Article is not the place to relive those debates. Rather, it is useful to consider the structures of these adopted treaties and how each was intended to lead to future agreements about difficult issues.

Again, an underlying policy point is the right of a sovereign state to control its own destiny. Most nation states are very reluctant to enter into a treaty whereby they could be restricted in their future actions and policy by a majority vote of other Parties. The concept of majority rule, as accepted within the United States (U.S.) political system, is normally rejected as a basis for treaty decision making. A treaty that created an assembly that could, by majority vote, impose restrictions on energy development or require the setting aside of 10 % of a country's land as national parks, would not be acceptable as it could give rise to violations of the countries' right of self-direction. But, a country might well agree to become part of an organization that would seek to solve problems and make recommendations in future legal documents that would contain specific restrictions. These future legal documents that would arise out of the discussions of the primary treaty body can be referred to as protocols of an umbrella treaty.

A protocol, which can be thought of as a mini-treaty, must be ratified by Parties just like a full treaty. Therefore, each country would have the sole right to decide whether or not to be part of the protocol. While the CBD has generated two subsequent protocols,[18] the classic example of a protocol is the Kyoto Protocol, under the UNFCCC. A future goal that was reflected in the language of the initial umbrella

[18] The Cartagena Protocol on Biosafety to the Convention on Biological Diversity (29 January 2000), http://bch.cbd.int/protocol/ (accessed 8 April 2012) (international agreement which aims to ensure the safe handling, transport and use of living modified organisms (LMOs)); *see generally* The Nagoya Protocol on Access and Benefit-sharing to the Convention on Biological Diversity (29 October 2010), http://www.cbd.int/abs/ (accessed 24 August 2015) (international agreement which

treaty was the adoption of specific emissions limitations by the developed countries of the world. After an additional 5 years of negotiations, at a formal meeting in Kyoto, the Protocol was agreed upon. Most developed countries formally ratified the Protocol, agreeing to comply with the limitations. However, the U.S., which had been a significant part of the negotiation process, declined to ratify the agreement. The Protocol went into effect without the U.S. being bound by it even though it was a member of the umbrella treaty. The U.S. exercised its sovereign right and decided not to be a part of the Protocol.

The same reasons that supported the adoption of the environmental umbrella treaties in the 1990s support the approach of the umbrella treaty for the ICPA this decade. International rules for animal welfare are a new topic on the global scale, and many countries would be concerned about what this might mean in the future. Although a number of countries might agree that animal welfare is a problem that should be addressed, there is no consensus on what the outcome should look like. The creation of a formal dialogue process is a positive step forward. However, there may be some broad points of policy that could be agreed upon and, wherever possible, these should be sought out and reduced to a written provision. The ICPA stipulates that any state who joins the treaty must also become a member of at least one protocol. Draft protocols exist for the following topics:

Companion Animal Protocol
Protocol for the Care of Exhibited Wildlife
Protocol for the Taking of Wild Animals
Protocol for the International Transportation of Animals

Whether this approach would survive the actual negotiations is not known, but as the topic is a new one for international relations (e.g. who at the U.S. State Department would take the lead for such a treaty negotiation is unknown), it is also important to set out some level of detail so the direction and scope of the treaty can be better understood.

5.4.2 Development of Draft Treaty

During the 1980s this author attended a number of the international meetings of the Party States to the Convention on International Trade in Endangered Species (CITES) as a representative of the Animal Legal Defense Fund. While in attendance I met a delegate, named Bill Clark. We both decided at some point that an animal welfare treaty was needed as it became clear CITES was not interested in such a topic. A committee was formed, and a draft produced. Two public hearings were held: one in Geneva, Switzerland and one in London, England. Once a polished draft evolved, it became clear that an international sponsor was required to give the

---

aims at sharing the benefits arising from the utilization of genetic resources in a fair and equitable way).

treaty international attention. We asked to be put on the agenda of a Board meeting of World Society for the Protection of Animals (WSPA), attended a meeting in Paris, and presented the treaty. They were not interested. In the early 1990s we were joined by another person from the United Kingdom who had higher visibility and political connections. With this person's efforts, an approach was made to a particular country as to their willingness to be a state sponsor of the treaty; but after some careful deliberations, the country declined to go forward with the treaty. So, here it sits without an international sponsor that has the resources to start the process. It is prepackaged and ready to go.

5.4.3 Suggested Language

In drafting the language of the proposed treaty, a number of concerns had to be balanced against each other, as the treaty must: attract the most countries possible; provide provisions that would actually promote animal welfare, sooner rather than later; allow for some enforcement mechanism that has consequences but will not prevent countries from joining the treaty; provide for future development by creating a community of individuals seeking consensus of policy; provide for information gathering and sharing about present practices; and support transparency that allows non-government actors to be included, even if they do not have a vote at the end of the process.

Generally, treaties have clusters of expected provisions: words of general policy, definitions of key words, substantive provisions, procedural provisions for ongoing concerns, and provisions that will define the starting point of the treaty. The words of general policy are in the Preamble of the treaty. They simply set the stage for the treaty, usually suggesting the need that gave rise to the treaty. As a rule, preamble language does not create any obligations for nation states. In the draft, ICPA Article 1's Fundamental Principles also do not create specific obligations but can be used to interpret the obligations found elsewhere in the treaty.

ARTICLE 1 – Fundamental Principles

1. Humans and animals co-exist within an interdependent ecosystem. Humans and animals share an evolutionary heritage. Humans, as moral beings, have an obligation to act responsibly toward animals.
2. Life has intrinsic value. No animal should be killed unnecessarily or be subjected to cruel acts or to unnecessary suffering.
3. When humans have control over specific animals they have a positive obligation to provide these animals with an environment and care appropriate for the species.

Article 2 contains some key definitions. Definitions often represent political compromise for the scope of the treaty. Thus "animal" is defined as "any non-human mammal, bird, reptile, amphibian or fish and any other organisms which may be included specifically within a particular protocol." Obviously the biological definition of animal would be much broader, but this is more likely the broadest definition that global concerns would support. While lobsters and octopuses are very interesting, they are unlikely to support a broader definition of the term animal at the international level. But if other specific species outside this broader definition gain political support for consideration, inclusion it could be achieved within the context of a particular protocol.

Articles 4 through 9 contain policy positions for particular categories of use by humans. This approach represents a compromise. Many individuals may believe that a rabbit should be treated the same regardless of the category of use that a particular rabbit might find itself: as a pet, food animal, scientific research subject, or game animal. But the laws around the world have been crafted with a focus on the categories of use. It is not possible to think that this perspective could be changed on a global basis. Also, the reality is that the political process will not accept equality of all rabbits, as the value of the human use will be weighed differently between rabbit categories, thus resulting in different outcomes for the particular rabbits. The language of these sections seeks the reduction of cruelty, but admittedly within the context of the use in question.

The substantive provisions of the Treaty are found in Articles 10 through 13. Each party agrees to designate an administrative agency to deal with treaty issues. While this may seem very minor, it is imperative to the goals of the treaty. If there is no authorized agency within a government, then there is no focus for the implementation of the treaty. No one will care if the treaty is fulfilled. Agencies propose budgets to implement programs and that will be necessary to realize treaty goals. The other obligation represents a political judgment about how to best structure the treaty and its protocols. Understanding that, in all likelihood, not every country will have equal political support to engage all of the possible topics and that some commitment to change is desirable at the beginning of the treaty, a nation state does not have to join all the protocols, just a minimum of one. It is not desirable to have a process where states would join the master treaty, not have any obligations under a protocol, but have the power to cause difficulties for other nation states. Agreeing to be part of a protocol will require some level of national implementation, depending on the topic.

As treaties can create stand-alone organizations, something must be said about the internal operation of a treaty. One key issue is whether or not to have an administrative staff organization that exists permanently, not just during the meetings of the Parties. Today most global treaties have a Secretariat to carry out much of the communication and program development for the treaty regime. These provisions are set out in Articles 14 through 25. These provisions are within the norm of existing treaties and follow some of the provisions of CITES. These articles also contain a number of procedural points for the operation of the treaty that will not be described in this Article, but should not be controversial.

Articles 26 through 31 contain the provisions necessary to bring the treaty into legal existence. To be bound by the provisions of a treaty or a protocol, a country must ratify or accept the treaty using whatever internal process the nation state has adopted. Then the nation state gives notification to the world of its acceptance of obligations by depositing a formal document with the depository government. The government that will perform that task is subject to negotiations and is not stated in this draft. Also, a remaining issue to be negotiated is how many nation states must agree to acceptance, and thus ratification, of the treaty before it becomes international law. Article 29 suggests ten nation states as the threshold of becoming an effective treaty.

As with the umbrella ICPA, in each protocol there are initial operating principles and definitions that give focus for provisions of a particular protocol only. In the context of the animal welfare issues being discussed in the U.S. today, the topics, scope, and restrictions of the proposed protocol will seem modest. And while to those active in animal welfare issues, U.S. law often seems inadequate, the U.S. is ahead of the law and public policy discussion of most of the world. An exception is the EU, which may well be ahead of the U.S. in a number of areas. To draw in a majority of countries around the globe, the provisions of an international treaty must represent modest steps forward. They must help solve issues that exist within other countries. For example, pets are often seen as family members within the U.S. This view, however, is not shared throughout the world. While in the U.S. the issues around veterinary malpractice and how to measure damages for harm to a pet are of major importance, in most of the developing world the mere establishment of a veterinary service available for pet animals would be a significant step forward. Any suggestion that pets should not be property is a non-starter from a global perspective. And because a key goal of the treaty is to build an international community with a focus on animal issues, the provisions of the treaty and its protocols can only be as progressive on animal welfare issues as is politically feasible at a particular point in time. The idea of an ongoing community is to allow the politically acceptable consensus to grow more protective over time.

As might be expected within the Companion Animal Protocol, the first articles imposing obligations on the states create the positive duties of pet ownership and to prohibit the negative effect of cruelty.

ARTICLE 4 – Care

1. The keeper of a companion animal shall be responsible for the animal's health and welfare.
2. The keeper of a companion animal shall provide the care and attention required by the species and breed of the animal, and in particular shall:

   (a) give it suitable and sufficient food and water
   (b) provide it with adequate shelter from adverse environmental conditions;
   (c) provide it with adequate opportunities for exercise; and
   (d) take all reasonable measures to prevent escape.

3. The keeper of a companion animal shall not:

   (a) subject the animal to cruel acts or conditions;
   (b) restrain an animal so as to preclude it from obtaining adequate food, water and shelter;
   (c) train an animal in a way which is detrimental to its health or cause unnecessary pain or suffering;
   (d) use an animal by forcing it to exceed its natural capacities or strengths.

4. Annex A to this Protocol shall contain species specific requirements for the appropriate care and welfare of companion animals.

This Article uses a tool that is important to the structure of the various protocols, an annex to the protocol. Treaties by their nature are short. Additionally, once a treaty or a protocol is adopted, it is extremely hard to amend. Therefore, there needs to be a mechanism to allow some levels of change as time goes by without changing the general nature of the treaty. Thus, Article 4 has four topics of prohibitions: cruel acts, nature of restraints, training, and physical use (overburdening). If these categories are considered in regard to all the possible species that are pets around the world, the Article would turn into an encyclopedia. Instead, the Article says that the details about a particular species can be set out in Annex A to this protocol. During the initial negotiations, it is expected that some particulars will populate this Annex, but the key is that over time the Parties to this Protocol can modify and add to the Annex at conferences of the parties without having to go through the full protocol amendment process. This allows for incremental change over time.

Article 5 deals with the issue of inappropriate species to keep as pets. It is presumed that a fair consensus can be obtained to prohibit most endangered or dangerous species as pets: lions, tigers, chimpanzees, wolves, bears, unicorns and venomous snakes. But again, an annex is used as the listing mechanism so it can be added to over time.

Article 9 might be considered radical in the context of the U.S. because the cutting off of animal parts to have pets conform to "standards" is an accepted practice by many. Article 9 states:

1. Surgical operations for the purpose of modifying the appearance of a companion animal or for other non-curative purposes shall be prohibited and, in particular:

   (a) the docking of tails;
   (b) the cropping of ears;
   (c) devocalizsation; and
   (d) declawing and defanging.

Other operations or procedures, which the Contracting Parties believe inappropriate, may be listed as an Annex to the Protocol (unless necessary for the well-being of the particular pet).

While these prohibitions might be acceptable to Parties within the category of pets, the cutting off of body parts for commercial animals (sheep tail docking) is a much more difficult issue as it goes to perceived economic interests, thus making it much more difficult to initially address.

Article 13 requires each nation state to domestically implement the provisions of the treaty.

ARTICLE 13 – Domestic Implementation

1. Each Contracting Party shall adopt such domestic measures as are necessary to implement the provisions of this Protocol. Each Contracting Party shall designate an appropriate authority to carry out the responsibilities of this Protocol. A Party may adopt stricter, more protective, domestic legislation.
2. Each Contracting Party should provide a process for the confiscation and forfeiture of animals being held in violation of the Protocol and being treated in a manner in violation of this Protocol. The disposition of a confiscated animal will always be in a manner suitable to the needs and interest of the companion

A treaty does not create law binding on the individuals within a nation state, but only the nation state itself. The passage of domestic law is expected after the ratification of the treaty by a nation state. Admittedly, the present draft does not set a deadline for adoption of the domestic law, and such a deadline (within 2 years, perhaps) would be helpful.

As with all international agreements, the key issue is enforcement of the provisions of the treaty. Regrettably, many international agreements lack a compelling process to make countries comply with their obligations.

There is no international police force that can compel compliance with treaty obligations. Again, the concept of sovereignty limits the willingness of countries to sign treaties subjecting themselves to the will of others. Therefore, the implementation provisions of the protocols are modest and, perhaps, politically acceptable. Article 15 (2) states: "Each Contracting Party may prohibit companion animal trade with any State which refuses to comply with the provisions of this Protocol and its Annexes." This at least allows unilateral enforcement efforts by nation states with the political will to confront other nation states, whether or not a party to the treaty. If utilized, this provision would also be an incentive to join the treaty since a nation state would have to comply with the limitations of the treaty even if it is not a member.

5.5 Prospects for Adoption of a Treaty

While private parties and non-profit organizations may initiate consideration of the treaty by holding international meetings, they do not have the capacity to call for a treaty negotiations process; a nation state must take up that task. As the existence of the language in the Appendix proves, writing possible treaty language is the easy part. The difficulty is in finding a country willing to initiate the process. For a country to do so requires a level of political concern within the nation state, such that the expenditure of human and financial resources is justified. This initial hurdle has not yet been overcome, but it is possible a country can be found who is willing to take a leadership role. There is no expectation that it would be the U.S. It should be understood that there are any number of politically powerful voices within the U.S. who would not want a treaty that might interfere with the existing commercial use of animals. The lobbying by these groups would be a simple message to elected officials: "Don't go there, just leave the issue alone, the status quo is what we are paying you to keep."

Historically there have been two sources of treaty initiation outside of nation states themselves. A division of the United Nations might initiate a process that the nation states become a part of, or large international non-government organizations could lobby governments to take up the issue. Nongovernmental organizations might be willing to do the initial drafting, spend the money for holding a hearing and discussion meeting, and invite government participation. The hope would be that a sufficient level of interest could be generated, such that one or a group of governments would take a leadership position and call for formal nation state-to-nation state negotiations.

It is also possible, and perhaps a prerequisite to action, that economic users of animals, faced with a great diversity of animal welfare standards around the globe, will decide to support a treaty in order to obtain uniform global standards. Some corporations might well have the policy position that "we would enhance animal welfare for the animals under our care, if our economic competition would be required to meet the same standards." Therefore a global treaty with protective animal welfare standards would raise the economic playing field to a more comfortable, ethical level.

But if we assume that one country can start the process, then the decision of other states to join the negotiations is slightly different. Other countries would come to the negotiations out of self-interest. They would not want a global treaty to proceed without being a part of the process, allowing them to help shape the provisions of the treaty and protect or assert their self-interests. There would be a fear of a change of the welfare rules that would put them at a disadvantage. Now this presents the second pressure point. The call for negotiations should not occur unless there is some level of belief that the results of the negotiations will not simply make present animal welfare treatment the global standard.

The risk is that present commercial users of animals would capture the process through government representatives and establish such a low bar of welfare for animals that there would be no enhancement over existing negative conditions. Indeed the risk is that the treaty could be transformed into something enhancing the economic, and therefore political, power of the global and national corporations that control so many millions of animals. Considerable judgment must be made about when and under what conditions negotiations of a treaty might reduce or control the risk of a loss for the animals.

If a treaty were negotiated, would the U.S. adopt it? This is doubtful. The U.S. Senate has chosen a path of non-participation with any number of important international agreements over the past several decades. There is political gridlock and an arrogance that the U.S. is so powerful by itself that it does not have to compromise and join other states in international agreements. Perhaps public support can change this view.

At present, the largest hindrance to the consideration and adoption of an animal welfare treaty is the negative conditions of much of the global economy. Animal issues cannot rise to a level of international concern unless the economics of the world recover, allowing the middle class of most countries to become comfortable and political issues other than economic survival to become important again. If a country is not taking care of the economic needs of its citizens (which include education and health care), then there is very little political space for the welfare of animals. At the moment, the stars are not aligned for the adoption of an animal welfare treaty. There is hope, however, that the alignment might occur in the future; those interested in animal welfare issues must be ready to take advantage of a political opening whenever it might occur.

References

Friedman, Thomas L. 1999. *The Lexus and the olive tree*. New York: Farrar, Straus & Giroux.

Garner, Robert. 2006. Animal welfare: A political defense. *Journal of Animal Law & Ethics* 1: 161–170.

Kelch, Thomas G. 2011. *Globalization and animal law: Comparative law, international law, and international trade*. Alphen aan den Rijn: Wolters Kluwer.

Part II
Developments in Animal Protection in Different Jurisdictions

Chapter 6
Animal Protection Law in Australia: Bound by History

Steven White

Abstract This chapter examines changing attitudes to the treatment of animals in nineteenth century Britain, tracing the effects of change through to the first British animal protection legislation in 1822, and beyond to the law of the Australian colonies. By the early part of the twentieth century a number of key facets of animal protection regulation were established: the adoption of the generic 'no unnecessary suffering' standard in assessing the extent of cruelty to animals allowed; the use of exemptions from the generic prohibition against cruelty; the imposition of duties to provide for the needs of an animal; and the establishment of one of the key institutional actors in the animal protection field, the RSPCA. As well, the Australian colonies faithfully reproduced an understanding of domesticated animals as personal property. Analysis of the development of animal protection law in Queensland provides a "representative sample" of the adoption of animal protection law in the States and Territories more broadly. In Queensland, as in other similar jurisdictions, the question remains whether the present day animal protection regulatory framework amounts, in essence, to a nineteenth century answer to twenty-first century concerns.

6.1 Introduction

This chapter examines the evolution of animal protection law in Australia. The United Kingdom (U.K.) antecedents of Australian animal protection regulation will be explored first, followed by a description of the way in which regulation unfolded across newly colonised Australia. Drawing on the established literature addressing the history of U.K. legislative developments, this chapter summarises the historical context from which contemporary Australian regulation of animal protection emerged. An understanding of the historical foundations of the prevailing regulatory regime is important for at least two key reasons. First, the broad model of regulation established by around the end of the nineteenth century in Australia is one

S. White (✉)
Griffith Law School, Griffith University, Brisbane, QLD, Australia
e-mail: steven.white@griffith.edu.au

D. Cao, S. White (eds.), *Animal Law and Welfare - International Perspectives*,
Ius Gentium: Comparative Perspectives on Law and Justice 53,
DOI 10.1007/978-3-319-26818-7_6

which has persisted to the present day – a prohibition against cruelty under threat of state sanction and an exemption from prosecution for a range of practices concerning animals which might otherwise be legally cruel. Second, the emergence of animal protection law in the nineteenth century was driven by changing attitudes to the significance of animal suffering. During the nineteenth century animals were increasingly recognised as sentient creatures capable of experiencing pain and suffering and therefore deserving of humane protection. The legal application of an ethic of humaneness in the protection of animals has proven to be durable, continuing to inform contemporary regulation.

Section 6.2 of this chapter provides a brief account of the attitudinal changes towards animals occurring in the eighteenth and nineteenth centuries in the U.K. These changes preceded the passage of the first Act in the U.K. proscribing cruelty to animals, legislation later replicated in the Australian colonies. Radford (2001, p. 3) points out that "[c]ruelty to animals pervaded eighteenth-century England; the majority of the population simply disregarded their suffering, but a significant proportion positively revelled in it." However, it was also during the latter part of the eighteenth century that public consciousness of animal cruelty increased, stimulated by the public nature of cruelty in an increasingly urbanised landscape and by social reform movements occurring across a number of policy fronts, domestically and abroad.

Section 6.3 summarises the key early legislative developments in the U.K. Proposed animal protection legislation was presented to the U.K. Parliament as early as 1800, but it was not until 1822 that the first bill was passed. The passage of this legislation was quickly followed by the establishment of a key and enduring animal welfare institution, the Royal Society for the Prevention of Cruelty to Animals (RSPCA), tasked with an educative and enforcement role. The RSPCA was founded by committed individuals active across a range of humanist projects, including the abolition of slavery, improved working conditions for factory workers and the protection of children. It was created at a time when private individuals enforced much criminal law. The role of inspectors of the RSPCA in enforcing the legislation was, from the outset, akin to that of a private police force. This remains the case today, despite the State assuming the near-exclusive responsibility for the enforcement of criminal law.

Section 6.4 charts the importation of animal protection law into Australia. After an initial and unapologetic "copying" of U.K. legislation, adapted for domestic purposes, a distinctively Australian approach subsequently emerged. This approach was reflected, for example, in the inclusion of animal welfare protections in police legislation rather than in stand-alone anti-cruelty statutes. After federation, the animal protection project was largely undertaken in an uncoordinated way by the States (and later Territories), with a putative lack of constitutional power sidelining the Commonwealth. Much later – under the pressure of international scrutiny – a more coordinated approach to regulation developed with the introduction of codes of practice for the welfare of farm animals. From the introduction of the first code, in 1980, the Commonwealth assumed an increasingly significant role in leading policy reform and seeking the implementation of an agenda of national consistency in animal protection regulation. More recently, the Commonwealth has abandoned the field of domestic animal welfare policy, drawing into doubt the viability of the consistency project. Significantly, along with the adoption of UK-inspired animal protection legislation, the Australian colonies embedded an underlying legal categorisation of domesticated animals as personal property.

Section 6.5 concludes this chapter by narrowing the focus to a particular Australian jurisdiction. A brief overview of the development of animal protection in Queensland provides a "representative sample" of the adoption of animal protection law in the States and Territories. Such a focus is illustrative at a broad level of regulatory structure, while at the same time acknowledging that the history of other jurisdictions will be marked by their own particular idiosyncrasies.

6.2 The Emergence of an Animal Protection Sensibility in the United Kingdom

In one sense, the story of the emergence of animal protection law in nineteenth century England[1] is the story of industrialisation, urbanisation and the growth of a civic-minded, metropolitan middle class.[2] As Armstrong and Botzler (2008, pp. 4–5) put it:

> [i]n England, the growth of towns and the emergence of an industrial order in which animals became increasingly marginal to production were significant factors in the development of concern for animals' rights. The reformist ideas were expressed by either well-to-do townspeople or by educated country clergy. The professional middle classes were unsympathetic to the warlike traditions of the aristocracy, which had valued hunting because it simulated warfare, and cock-fighting and bear-baiting because they represented private combat.[3]

As some animals, especially horses, were becoming increasingly marginalised "workers" in the face of expanding industrial production, other animals were

[1] While this chapter focusses on the emergence of animal protection law in the U.K., as the coloniser of Australia, similar developments were occurring in the newly independent United States (Favre and Tsang 1993).

[2] Turner (1980, p. 25) suggests that 'urbanisation and industrialisation in some way helped to generate the new concern for beasts. But this is merely an observation, not an explanation'.

[3] Ryder (2000, p. 147) acknowledges the significance of urbanisation but also stresses that this should not be accepted in an unqualified way. He points out that "some of the worst cruelties towards non-humans in Victorian Britain were inflicted by urban dwellers in the pursuit of objects that were neither agricultural nor sporting: vivisection, the fashion industry and the daily abuse of horses are three major examples. Furthermore, although the animal protection movement in America was largely an urban phenomenon, in Britain this was not so true; many of the movement's leaders, for instance, were country clergymen or landowners (admittedly often with business in London) … indeed the industrial middle class played little part in the British animal welfare crusade."

assuming an increasingly significant role in the households of some families (Armstrong and Botzler 2008, p. 4):

> pet keeping had been fashionable among the well-to-do as well as among religious orders in the Middle Ages, but it was in the sixteenth and seventeenth centuries that pets seem to have established themselves as a normal feature of the middle-class household … Gradually, the idea that tamed animals were property was developed. Pets were distinguished by being allowed into the house and by going to church with their human companions, by being given individual personal names, and by never being eaten. The spread of pet-keeping created the psychological foundation for the view that some animals were entitled to moral consideration.

Hilda Kean (1998, p. 28) argues that the increased visibility of animals in an urban setting was particularly important in shifting sentiment about the treatment of animals:

> Clearly what animals you saw depended on where you lived. In the country there were, of course, animals on farms, birds and wild animals in the fields and domestic pets in the home. However by the start of the nineteenth century Britain was already an urban country. The cities would be the places where animals were increasingly seen and where their treatment would be most hotly debated.

London was a focal point for the use of and trade in animals, with a large live market for farm animals on the outskirts of the city, numerous touring animal menageries, permanent exhibitions of animals, animals used in public "entertainment" such as bull-baiting and animals used for transportation of people and goods (Kean 1998).[4] Changing sentiment towards animals could not be explained by simplistic notions of a loss of connection with the countryside, or by the idea that "urban isolation from animal farming had nourished emotional attitudes which were hard to reconcile with the exploitation of animals by which most people lived" (Kean 1998). An analysis which values animals in the countryside as "real" animals and those in the towns as sentimental echoes of a lost way of life (Kean 1998, p. 30):

> fails to recognize the abundance of animals living in cities in the early nineteenth century and their economic, as well as cultural, importance for the inhabitants. It also fails to acknowledge the importance of the role of sight in developing the relationship between seeing ill-treatment and creating change. The "farm" animals that lived and worked in London would also be the first type of animals to benefit from legislation.[5]

Kean (1998, p. 24) acknowledges that those "supporting humane treatment for animals adhered to no one political or ideological set of beliefs." Increasingly,

---

[4] Bull-baiting "involved tying a bull to a stake and setting one or more dogs upon it, the object being for the dogs to get hold of, and hang on to, the bull's nose" (Radford 2001, p. 18). Bull–baiting and bear-baiting date back to the Middle Ages. The former arose as practise of butchering. In the medieval period 'baiting bulls with dogs was believed to improve the quality of the meat, and for this reason most medieval towns enforced by-laws stipulating that bulls should be baited before slaughter' (Griffin 2005, p. 42). These regulations had fallen into disuse by the eighteenth century, but the practise remained as a form of entertainment. Bear-baiting had always been an entertainment, "patronised by royalty, nobility, and civic elites" (Griffin 2005, p. 42).

[5] Siobhan O'Sullivan (2011) has utilised a "visibility" argument to underpin a political science analysis of prevailing regulation of animal protection.

though, "the way in which people treated animals became a distinguishing feature of being humane and of membership of a new middle class and respectable working class" (Kean 1998, p. 24). Debates on animal protection occurring in Parliament at the turn of the nineteenth century were not "simply about the development of changing attitudes towards animals." Such debates partly reflected significant challenges to the status quo represented by more sympathetic religious responses to the plight of animals, the emergence of a rights discourse for the disenfranchised following the French Revolution and the publication of Thomas Paine's *Rights of Man* and, significantly, the work of utilitarian Jeremy Bentham.[6] The ideas expressed at this time contributed to a climate in which the concept of animal protection could be taken seriously. Although Paine did not address the plight of animals, "the call for society to recognize that all human beings had the same moral and civil rights, on the basis of the nature of each person and regardless of their social status, caused at least some to consider applying a similar principle to other species" (Radford 2001, p. 24). Radford acknowledges that arguments for the extension of rights to animals were not "common or widely accepted" in eighteenth century Britain, "but there are examples from this period of the language of rights being applied to animals."

Concerns about the treatment of animals were part of a wider movement for humanitarian reform, with a focus on the pain caused by cruel practices. Through the eighteenth century a range of activities were targets of reform, including "blood sports, public executions, treatment of the insane, punishment by flogging in the armed services, corporal punishment of children, and sport that caused serious injury" (Munro 2001, p. 11). Animal protection took its place as part of a "general movement which gathered pace during the nineteenth century, often promoted by the same people, which was concerned to improve conditions across a range of areas" (Radford 2001, p. 48). Child protection and animal protection campaigns emerged at around the same time, with humanist activists also agitating for improved employment conditions, prisons and treatment of the poor. A notable feature of reform in these disparate areas was the similarity of the legislation which emerged addressing these societal problems. For example, legislation addressing the mistreatment of children was labelled as "prevention of cruelty to children", and made it an offence "wilfully to ill-treat, neglect, abandon, or expose a child in any manner likely to cause it suffering or injury to health" (Radford 2001, p. 50). As will be

[6] For more detailed accounts of the significance of each of these developments see Radford (2001). Radford (2001, p. 59) concludes that the "impetus for the introduction of animal protection legislation during the first part of the nineteenth century was founded on much more than mere sentimentality. Greater understanding of animal physiology, a reassessment of man's place in the world, the development of a secular morality, the increasing influence of middle-class values, concern for social discipline and stability, a political and legislative system which was responsive, the individual campaigners to carry the cause forward, and the endorsement of the higher ranks of society were all factors in legislative protection becoming a reality. It represented a confrontation between the old, pre-industrial, paternal, rural community, dominated by local customs, identity and administration, and the new, urbanized society governed from London." See also Ryder (2000) who cites a range of prominent artistic, philosophical, and cultural figures in eighteenth century Britain extolling the need for better treatment of animals (such as Samuel Johnson, Percy Shelley, Jeremy Bentham, William Blake and Robert Burns).

shown in Sect. 6.3, animal protection legislation passed earlier in the nineteenth century employed strikingly similar language.

The motivations of those seeking legal change in support of animal protection were diverse and, as Jamieson (1991a, p. 21) suggests, "dissecting the motives of the early nineteenth century protagonists and antagonists of animal protection legislation … violates both the integrity and the confusion of their ideals." However, there seems little doubt that change in the understanding of the moral significance of animals, and of their vulnerability to pain, was important. This new sensibility went beyond a concern to protect property or to control the behaviour of the masses (Kean 1998, p. 31):

> The rise of organized political discussion and then of societies to protect animals became a distinctive part of the creation of new political and moral sensibilities. Changes in the law were invoked not just to defend property nor to regulate the behaviour of the rabble and seditious agitators; they also had the effect of giving protection to those unable to speak for themselves. The role of advocate and protector was being established to invoke the cause of those literally without human speech, dumb animals.

With shifts in popular attitudes to the plight of some animals, fuelled by increasing urbanisation and an increasingly pluralistic and vibrant intellectual debate, the possibility of political change was fostered, reflected in attempts to legislate for animal protection beginning in 1800.

6.3 Early Animal Protection Legislation in the United Kingdom

Early attempts at legislative reform of animal protection in the U.K. targeted the prohibition of bull-baiting, once widely considered a respectable past-time, but by 1800 a popular recreational activity for lower class Britons (Kean 1998). Bills introduced in the House of Commons in 1800 and 1802 respectively were partly motivated by concerns about cruelty to the animals involved, but even more by concerns about the undesirable public order consequences of bull-baiting, including drinking, gambling and interference with work. Both Bills were voted down. A key opposition argument was that the Bills reflected a censorious approach to working class past-times; they betrayed a "double standard" where bull-baiting would be prohibited, but other recreational activities such as fishing, hunting and shooting, and horse racing would continue.

The next reform effort occurred in 1809, when Thomas Erskine introduced a Bill into the House of Lords prohibiting malicious and wanton cruelty to animals. As Radford points out, this Bill differed from the bull-baiting Bills in a number of important respects. First, it was not confined to one class of animal or activity, but extended to any circumstance involving cruelty to horses, donkeys, oxen, sheep and pigs (although it was subsequently amended to be confined to "beasts of burden"). In this respect, the Bill clearly anticipates contemporary regulation. Second, the controversial matter of defining "cruelty" was avoided, with the Parliament leaving it to the courts to judge whether or not an act or omission would constitute cruelty. Third, the justifications put forward for the Bill shifted sharply towards a focus on the protection of animals, even if concern about public order remained relevant. Erskine emphasised the daily cruelty meted out to animals, recognised the shared sentiency of humans and animals and acknowledged the direct moral duty of protection owed by humans to animals. Fourth, Erskine argued that mere education is not enough; State intervention was required to establish and promote the obligations owed by humans to animals. Again, this sensibility underpins contemporary regulation, and is consistent with Ryder's argument that "by the end of the eighteenth century in England the basic principles of the modern animal protection position were established. These are that nonhumans, like humans, can suffer pain, and that pain entitles them to legal as well as moral rights" (Ryder 2000, p. 72).

Although Erskine's Bill passed the House of Lords, it was comfortably defeated in the House of Commons. One of the arguments made against the Bill – in a context where "most prosecutions were brought by private individuals rather than the State" – was that the generality of the legislation meant it would be manipulated by fanatics and by over-zealous enforcers seeking to impose their sensibilities on members of the working class (Radford 2001, p. 37). The lack of definitional clarity would also mean that the legislation would be applied in an inconsistent way by different judges. In part, these represent a concern not to "open the floodgates" in an entirely new area of legal liability, and to avoid the injustice that may go along with an inconsistent application of the law. The leading opponent of the legislation, William Windham MP, argued that the proper treatment of animals was a matter of "manners and morals", and not the province of Parliament. The attempt to legislate for animal protection "raised the fundamental question of whether it was either legitimate or desirable for the State to define and impose common standards of public morality" (Radford 2001, p. 54). This concern resonated with debates across a range of areas, in which the sanctity of the classic liberal private-public divide was asserted. Windham also repeated his earlier criticism of the bull-baiting Bills – by not specifically including field sports such as hunting and shooting as cruel, legislators would be imposing a double standard on the working class (Radford 2001). Erskine tried again to pass a similar Bill a year later, but in the face of even stronger opposition withdrew the proposed legislation.

Richard Martin, MP for Galway, introduced a new Bill into the House of Commons in 1820 prohibiting ill-treatment against horses and other animals by third parties. The Bill passed the Commons, but not the House of Lords. Martin persisted, and 2 years later, even in the face of opposition by the Attorney-General, succeeded in shepherding through the first animal protection statute in the U.K. (Radford 2001). Martin's Act[7] received the Royal Assent on 21 June 1822. The Act made it "an offence for any person or persons wantonly and cruelly to beat, abuse,

[7] An Act to prevent the cruel and improper Treatment of Cattle 3 Geo IV, c 71.

ill-treat any horse, mare, gelding, mule, ox, cow, heifer, steer, sheep or other cattle."[8] Prosecutions could be initiated on complaint to a Magistrate, with fines of up to five pounds applying to offenders. The Act also contained a provision allowing a fine to be imposed on those bringing vexatious or frivolous claims.

Emboldened by his 1822 success, Martin introduced a range of further Bills over the next 4 years seeking to extend the scope of protection provided by the initial statute (including extending protection to other animals such as cats, dogs and monkeys). None of these further Bills succeeded. However, Martin also contributed to another major development in animal protection regulation – he was one of the founders of the Society for the Prevention of Cruelty to Animals (SCPA). Although Martin was an assiduous enforcer of the legislation, at a time when private individuals rather than the State largely initiated prosecutions, "an individual, working alone, could only achieve so much; for greater effectiveness, a formal organization charged with protecting animals was required" (Radford 2001, p. 40). The SPCA was founded in 1824 in a London coffee shop at a meeting comprising Parliamentarians (including Martin and Wilbur Wilberforce), clergy, and newspaper editors and correspondents. From the outset the SPCA resolved to improve the treatment of animals through education and enforcement. An inspector was appointed, and 149 prosecutions were successfully run in the first year of operations. As Harrison (1973, p. 793) notes:

> The R.S.P.C.A. is particularly interesting among nineteenth century pressure-groups because it did not rely solely on the police to enforce the law. In its protection of animals, [a] major achievement was to create agencies which specialized in enforcing the law – a sort of private police force with a strictly limited area of concern.

After running into serious financial difficulties through the 1820s the society came close to being dissolved, before recovering and prospering over the next decade. In 1837 "Royal patronage followed" and "Queen Victoria gave permission to add the royal R in 1840" (RSPCA UK n.d.). Two aspects of the creation of the RSPCA UK stand out in the context of present day regulation of animal protection. The first is that the organisation assumed the role of prosecution in an age where the State was not the primary enforcer of penal legislation.[9] Although that age has

---

[8] Although often-cited as the first animal protection statute, Radford (2001, p. 39) notes "an Irish law of 1635 which prohibited the pulling of wool off sheep and the attaching of ploughs to horses' tails which was enacted at least in part because of the cruelty caused by these practices." He also points to legislation passed in Massachusetts Bay Colony in 1641 which provided that No man shall exercise any Tirrany or Crueltie towards any bruite Creature which are usuallie kept for man's use. Anti-cruelty legislation was also passed in Maine the year before Martin's Act was passed." Tim Bonyhady (2000, pp. 49–50) suggests that a restriction imposed on the taking of birds on Norfolk Island in 1790 by commandant Robert Ross was "not just concerned to ensure a continued supply of food … Nor was he just mimicking existing metropolitan or colonial practice. His laws included what was probably the world's first prohibition of cruelty to animals."

[9] Ritvo (1987, p. 145) suggests that "[b]y maintaining what amounted to a private police force, the society defined itself as a quasi-governmental institution and its mission as the surveillance and control of the dangerous perpetrators of cruelty to animals. This whole-hearted endorsement of interventionist law enforcement seemed particularly powerful and daring in 1824, when the regular

passed, the RSPCA – a non-government, charitable body – continues to be a key prosecution agency, including in various States and Territories in Australia. Second, the main regulatory tools adopted by the RSPCA in the early nineteenth century have remained largely unchanged. Officers of the RSPCA in most Australian jurisdictions have been conferred with the power to enforce animal protection legislation (including by bringing prosecutions). Further, humane education remains central to the work of the RSPCA, and has been formally incorporated into legislation through devices such as Animal Welfare Directions – instructions provided by inspectors to animal owners to address the welfare needs of their animal – as a means of forestalling prosecution or avoiding the need for prosecution at all.[10]

In 1835 Martin's Act was repealed and replaced by legislation championed by Joseph Pease MP. Pease's Act[11] was more expansive than Martin's Act, with "torture" included as prohibited conduct, and bulls, dogs and other domestic animals were included within the category of protected animals (Radford 2001). Other changes included the extension of the prohibition on animal fighting and baiting from London to the country at large; the imposition of a duty to properly feed an animal in cases of impounding; and provisions specifying that all horses and cattle should be killed within 3 days of arriving at an abattoir, with the animals to be sufficiently fed daily before slaughter, and not put into use.[12]

By specifically prohibiting the so-called "blood sports", including bull-baiting, badger-baiting, dog-fighting and, a few years later, cock-fighting, Pease's Act was consistent with the humanitarian reform efforts initiated at the turn of the century, and canvassed earlier in this chapter. Griffin (2005, p. 233) acknowledges that this legislation "has long been regarded as evidence of increasing humanitarian concern amongst social elites." However, she cautions that the "practical significance of these laws" should not be overestimated. Bull-baiting, she argues, was in steep decline prior to 1800. Further, to the extent the legislation focussed on "blood sports" it was unnecessary, since local authorities had long had the power to control the use of public spaces. The working-class did not possess the large, open spaces required for these sports, so that they were necessarily conducted on public land. Despite this, local authorities rarely exercised the powers available to them under relevant ordinances to shut down these practices. Again, this was largely because the practices were already in decline. Griffin (2005, p. 234) argues that:

> the 1835 Cruelty to Animals Act was arguably important as an ideological statement, rather than as a political act. Its significance lay not in the new powers it introduced, but in the cultural message, concerning a civilised and enlightened ruling class, that it embodied … the decline of bull-baiting looks very much like an example of nineteenth-century social progress, and we are used to looking to social elites to explain progressive cultural shifts of

police force were still new and widely resented as invaders of the privacy of citizens." See also Anderson (2012).

[10] See Sect. 6.5 below.

[11] 5 & 6 Will IV, c 59; *An Act to consolidate and amend the several Laws relating to the cruel and improper Treatment of Animals, and the Mischiefs arising from the driving of Cattle.*

[12] Ibid.

> this kind … however … it is necessary to turn away from social elites, to the more nebulous and unquantifiable domain of plebeian cultural tastes if we are to understand the disappearance of bull-baiting.

While Griffin provides an important reminder of the need to avoid over-simplification in rationalising legislative developments in animal protection, her analysis risks the opposite – over-simplifying the law. In particular, regardless of the declining significance of some of the practices targeted by the legislation, a generic prohibition on cruelty was also established, one which continues to shape the contours of animal protection to the present day, not only in the U.K., but also in former colonies such as Australia.

6.4 Animal Protection Law in Colonial Australia

Animal cruelty was a subject of public concern at an early stage of Australian colonisation. Referring to a range of newspaper articles in 1804 and 1805, Jamieson (1991b, p. 239) states that a "genuine concern for animal welfare beyond merely protecting their value as property is clearly evident in early nineteenth century colonial Australia. *The Sydney Gazette* of that period frequently admonishes cruelty to animals." Despite this, it was not until 1837 in Van Dieman's Land (Tasmania), that the first colonial animal protection legislation was passed.[13] Although Pease's Act significantly extended the scope of protection provided in Martin's Act, as discussed above, the latter was more closely emulated in the first colonial legislation in 1837 (Jamieson 1991a, p. 22):

> the legislation for the prevention of cruelty to animals introduced in the sittings was only a pale reflection of its more extensive English counterpart of two years previously. While also extending the provisions of 'Martin's Act' of 1822 to encompass cruelty to any "domestic animals", it made no further provision, as did its English equivalent, regarding the keeping of pits for animals fights, the provision of sufficient food for impounded animals or the slaughter of horses.[14]

Specific animal protection legislation was next introduced in the colony of New South Wales in 1850.[15] The 1850 Act was comparatively detailed. It established a

[13] 8 William IV, No. 3; *An Act for the Better Prevention of Cruelty to Animals* http://www.austlii.edu.au/au/legis/tas/num_act/aaftbpocta8win3611/. The Act is brief, making it an offence to "wantonly torture or cruelly beat ill-treat or abuse any Horse Bull Ox Cow Calf Mule Ass Sheep Pig Goat Dog or other Domestic Animal." For a detailed history of the development of animal protection law in Tasmania see Petrow (2012).

[14] The available evidence suggests that the legislation was enforced through prosecutions at least from 1838. Jamieson (1991a, p. 22) states that "the *Hobart Town Courier* notes the imposition of fines for convictions of cruelty given against Thomas Dowling (14 September 1838) and Richard Hume (12 October 1838). Records at Richmond Gaol record 6 days solitary confinement for cruelty given to Charles M in September 1838."

[15] 14 Vic, No 40; *An Act for the More Effectual Prevention of Cruelty to Animals* http://www.austlii.edu.au/au/legis/nsw/num_act/ctaa1850n40240.pdf. The key provision provided that if "any person

basic cruelty offence, but also contained provisions specifically addressing cattle driving, bull-baiting and animal fighting. It adopted a wide-ranging definition of "animal".[16] The legislation was also notable for the inclusion of a provision making it an offence to "convey or carry or cause to be conveyed or carried in or upon any vehicle any animal in such a manner or position as to subject such animal to Unnecessary pain or suffering." This provision marked the first use of the no "unnecessary pain or suffering" standard in Australia, on which was adopted in subsequent colonial legislation, and which remains a staple of contemporary regulation of animal protection in the States and Territories.[17]

In New South Wales, and from 1859 in the new colony of Queensland, regulation of animal protection occurred through a specific prevention of cruelty statute. The 1850 statute governed animal protection in New South Wales until a new Act was passed in 1901 and in Queensland until, similarly, a new Act was passed in 1901. Tasmania persisted with an anti-cruelty statute, although a prohibition on animal fighting was included in police legislation in 1865.[18] In the other colonies, animal protection was effected exclusively through police legislation, including in Western Australia in 1849,[19] South Australia in 1863[20] and Victoria in 1864.[21] The inclusion of animal protection provisions in police offences legislation in the Australian colonies marked a different approach to that of the U.K. At one level, this is only significant in a formal sense. Jamieson (1991b, pp. 241–242) acknowledges a shared concern on the part of all jurisdictions to enforce social order through a focus on lower class leisure activities rather than "the more gentlemanly activities of the hunt." Accordingly "the growth of animal cruelty legislation in the colonies as a means of social control [was] reinforced by the early inclusion of such provisions almost exclusively in colonial police offences laws." However, Jamieson suggests that the exertion of social control was more significant in the colonies: "the state … was inevitably a stronger, more intrusive, legitimately interventionist instrument than Victoria's Britain … not having to contend against the traditional restraints of established church, military services, and landed aristocracy." Consequently, as a matter of substance, it was "not surprising that the police legislation adopted in the colonies was far more intrusive of individual liberty than its English counterpart."

---

shall … cruelly beat ill treat over-drive abuse or torture or cause or procure to be cruelly beaten ill treated over-driven abused or tortured any animal every such offender shall for every such offence forfeit and pay a penalty not exceeding five pounds": s 1.

[16] The definition referred to "any horse mare gelding bull ox cow heifer steer calf mule ass sheep lamb hog pig sow or goat or any dog cat or other domestic animal."

[17] The qualified nature of this standard has been a particular focus of contemporary debate: see, e.g. Sankoff (2013); c.f. Radford (2001).

[18] *Police Act 1865* (Tas) s 83 http://www.austlii.edu.au/au/legis/tas/num_act/tpa186529vn10163/

[19] 12 Vic No 20; An Ordinance for Regulating the Police in Western Australia, s XX http://www.austlii.edu.au/au/legis/wa/num_act/p12vn20131/

[20] 26 & 27 Vic No 10, *Police Act 1863* (SA) s 76 http://www.austlii.edu.au/au/legis/sa/num_act/pa10o26a27v1863168/

[21] 27 Vic No 225, *Police Offences Statute 1864* s 18 http://www.austlii.edu.au/au/legis/vic/hist_act/tpos1864248/

Apart from the adoption of a "no unnecessary suffering" standard and a residual enforcement role for police, colonial legislation through the mid-nineteenth century phase also established another key tenet of animal protection law which remains in place today. This involved extending obligations beyond a negative duty not to impose cruelty on an animal, to encompass a positive obligation to provide food and water to an animal.[22]

The next wave of animal protection law reform occurred around the turn of twentieth century. Three features of reform during this period are particularly noteworthy: first, the introduction of important structural change to animal protection legislation; second, the emergence of the RSPCA as a key enforcement institution; and third, an alliance formed between the animal protection movement, concerned with domesticated animals, and animal preservationists, concerned with protection of wild animals. As to the first of these, a significant development at this time was the introduction of exemptions from the general duty not to be cruel. The first step was taken in Victoria in 1881, with other jurisdictions following after the turn of the century. Exemptions were provided in three key areas: the control of wild animals, use of animals in experiments, and farming. Although now modified through the adoption of a code-based process of exemption, the justifications offered for exemptions have changed little over time. In relation to farming (Jamieson 1991a, p. 30):

> [the] historical concentration of animal protection legislation on domesticated animals had early fostered its perception by the rural community as mere urban meddling, the fear that such legislation "would seriously affect country districts" having in no way diminished by the beginning of the twentieth century. Agitation in South Australia in 1906 for the inclusion of a specific provision exempting from the operation of the statute the dehorning of cattle, heightened by the fining of "many people" in Victoria for engaging in this practice, led to the inclusion in the *Prevention of Cruelty to Animals Act* 1908 (SA) of a provision exempting the operation where "performed with a minimum of suffering to the animal operated upon." This early exemption in favour of the rural community was extended in Western Australia in 1912 to include "the castration, spaying, ear-splitting, ear-marking, or branding of any animal, or the tailing of any lamb." Similar exemptions were adopted in Queensland and Tasmania in 1925 and in New South Wales in 1928.

Second, once anticruelty legislation was in place in the colonies, institutional support followed through the establishment of the RSPCA. This occurred more slowly than had been the case in the U.K. after the passage of Martin's Act. Distinct societies were formed in each colony, beginning with Victoria in 1871, and progressing through Tasmania in 1872, New South Wales in 1873, South Australia in 1875 and Western Australia in 1892 (RSPCA Australia n.d.). In Queensland "the Queensland Society for the Prevention of Cruelty formed in Brisbane [in 1883],

[22] For Jamieson (1991a, p. 26) "little new development is to be found in the cruelty provisions adopted beyond the precedent earlier established by the New South Wales legislation of 1850. Nevertheless, that early legislation having generally recognized the offence of cruelty and made specific provision in respect of animal fights and the carriage of animals, both Victoria (1854) and South Australia (1863) did make further provision as to the omission to supply an animal with food and water."

following an unsuccessful attempt in 1876" (Emmerson 1993, p. 7).[23] It was not until 1923 that "the Societies were given the Royal Warrant, becoming known as the Royal Societies for the Prevention of Cruelty to Animals".[24] Munro (2001, p. 13) suggests that "[l]ike its Anglo-American counterparts, the Australian RSPCA consisted predominantly of middle-class urbanites, although in the Australian case, the RSPCA attracted affluent people from rural areas as well."

A third important strand in the development of animal protection in Australia during the period around the turn of the twentieth century was the alliance which emerged between animal protectionists and animal preservationists. The early decades of colonisation were marked by the wide-scale destruction of native wild animals and, at the same time, introduction of wild animals from the U.K., in part reflecting attempts to achieve a "Britainisation" of the landscape. Native wild animals were a food source for colonialists until domesticated farm animals were established. Later, a range of factors led to the destruction of vast numbers of wild native animals including land clearing and habitat loss; an assessment of native animals as pests; and the commodification of native animals through the production of furs, pelts and other by-products for export.[25] Non-indigenous animals were routinely afforded more protection against hunting than native animals (such as through extended "close seasons" for introduced animals, or periods during which hunting of specified animals was illegal, compared with native animals). Towards the end of the nineteenth century, however, an animal preservation movement had emerged which sought better protection of all wild animals and, especially, native wild animals. This led to an alliance between animal protectionists and animal preservationists, despite the fact that "then as now, both movements had different purposes and ideologies; the preservationists focussed their campaigns on saving species and habitats, and they thought little of animal protectionists' concern for preventing cruelty to individual animals" (Munro 2001, p. 14).

Much of the early wild animal protection legislation took the form of bird protection legislation, in part reflecting an antiplumage campaign which focussed on the indirect cruelty to birds. Munro (2001, pp. 13–14) states that the "antiplumage campaign became a synonym for anticruelty and united both protectionists and preservationists … these two movements were irrevocably entwined and, in the popular imagination, synonymous."[26] For animal preservationists the invocation of anticruelty was an important way of transforming an understanding of native animals from "vermin" to "cute creatures" worthy of protection, as occurred with a sustained

[23] In 1890 the Constitution of the Queensland Society was amended to incorporate protection of children and the name shortened to the "Society for the Prevention of Cruelty". The Society was an active contributor to the successful campaign for child protection legislation in the 1890s. However "[o]ver the twentieth century, the Government gradually assumed a greater role in the protection of children and old people and the Society's involvement declined until it finally ceased around 1970": (RSPCA Queensland).

[24] It was only much later, in 1981, that RSPCA Australia, a national organisation, was created (RSPCA Australia n.d.).

[25] For a detailed examination of this period of wild animal protection law see White (2013).

[26] Citing doctoral research by MacCulloch (1993).

campaign to protect the koala.[27] On the other hand, the animal protectionist movement benefitted through a broadening of focus, since "the animal lobby had run out of steam by 1914 … due to the movement's transformation from a social-reform lobby to a group of pet enthusiasts who emphasised the genteel promotion of kindness to animals, especially cats and dogs" (Munro 2001, p. 15).

A final but important aspect of nineteenth century Australian animal protection law is the wholesale importation of the personal property status of animals at common law. This status reflected an evolution from the medieval era, when animals were commonly put on criminal trial in the ecclesiastical courts of Britain and Europe for a wide range of offences, including murder of children and crop damage (Evans 1987).[28] However as Ritvo (1987, p. 2) points out:

> By the nineteenth century British authorities has stopped sentencing animals to suffer and die for their crimes … Nineteenth-century English law viewed animals simply as the property of human owners, only trivially different from less mobile goods. It followed that they were no longer held morally accountable for their actions ….

An influential philosophical justification for the English common law understanding of animals as a form of personal property can be found in familiar Lockean discourse. Animals, in their original state, belong to all of mankind as part of the commons. As with nature generally though, animals are given to humans for their use. When a person removes an animal out of the state of nature, applying their labour in doing so, they make the animal their property. In exercising their labour in relation to animals, humans alter their legal status, the animals becoming goods (Francione 1995).

Significantly, around 75 years after Locke, Blackstone addressed the status of animals in his *Commentaries on the Laws of England*. In Book II of his *Commentaries*, directly influenced by Locke,[29] Blackstone suggested that "[i]n such as are of a nature tame and domestic (as horses, kine, sheep, poultry, and the like), a man may have as absolute a property as in any inanimate beings" (Radford 2001, p. 100). The common law, Blackstone suggested, only protects domesticated animals to the extent that they are of some monetary worth. For example, typically it was only farm or working animals that were capable of being the subject of theft.

The idea of absolute property in domesticated animals was quickly established in the Australian colonies. In part this was due to the fact that "the law as enshrined in Blackstone carried significant weight in the early years of the [NSW] Superior

[27] The transformation of the koala from rural pest and valuable commodity to national icon occurred over many years. In 1896 "one leading fur company held over one million koala skins for export to Russia" (Munro 2001, p. 14). As late as 1927 the Queensland Government, in the face of considerable public outcry, permitted the killing of over one million koalas for the international fur trade (White 2013).

[28] Defence counsel would be appointed for accused animals. On a finding of guilt punishment included excommunication, as well as judicial penalties such as capital punishment.

[29] Francione (1995, p. 39) suggests that "Locke's theory of property had an extraordinary influence on the common law."

Court" (Salter 2009, p. 40).[30] Regardless of developments occurring in animal protection, Salter (2009, p. 40) argues that:

> it was the issue of animal possession, framed by the earlier eighteenth century of the likes of Blackstone, which would pre-occupy the courts of N.S.W. in the late eighteenth and early nineteenth century. Although situations of animal cruelty would continually arise in the facts of civil and criminal superior court trials, the protection of the animal was always legally invisible next to the primary issue of animal possession.

6.5 Animal Protection Law in Post-federation Queensland

The trajectory of the implementation of animal protection law in the various Australian colonies followed a similar structural if not temporal path. This pattern continued post-federation. However, to avoid over-simplification and for practical reasons of space, this Section narrows the focus to a particular jurisdiction in considering post-federation development. Four key stages in the evolution of animal protection law in post-federation Queensland will be addressed. These are the introduction of the first Queensland specific legislation in 1901, a new Act in 1925, a failed reform attempt in the 1990s, and a new Act in 2001. Looking across these legislative developments a consistent approach to the overarching regulatory standards for the protection of animals can be identified, even if the detail of the regulation becomes more detailed over time.

Queensland was established as a colony separate from New South Wales in 1859. However, the animal protection statute passed in New South Wales in 1850, discussed in Sect. 6.4, remained the law of colonial Queensland until 1901, when a new animal protection statute was introduced for the State of Queensland.[31] This legislation included the key characteristics of animal protection legislation developed and applied in most jurisdictions during the nineteenth century, including nominal application to all animals,[32] a prohibition against cruelty,[33] the imposition of duty of care-like obligations,[34] and qualification of the cruelty prohibition effected

[30] Citing Castles (1982).

[31] *The Animals Protection Act of 1901* (Qld). A minor change to the 1850 NSW legislation had earlier expanded the category of "animal" to include camels: *Criminal Law and Evidence Amendment Act 1891* (NSW) s 35.

[32] 'Animal' was defined as "every species of animal whether in a natural or domestic state": s 3.

[33] Section 5(1) provided: "No person shall do any act or observe any forbearance towards any animal which act or forbearance involves cruelty."

[34] Section 7 provided: "No person who has the possession or the custody of any animal which is confined or otherwise unable to provide for itself shall omit to provide such animal during so long as it remains so confined in his possession or custody with proper and sufficient food, drink, and shelter."

through adoption of a "no unnecessary suffering" standard[35] and limited exemptions.[36]

RSPCA Queensland actively lobbied for the passage of the 1901 legislation, concerned to more effectively protect animals and, more prosaically, to burnish its authority to address enforcement issues (Emmerson 2001). However, the 1901 Act left the RSPCA with "no legal standing and its officers no more authority than any other citizen" (RSPCA Qld n.d.). This lack of standing was remedied in a new Act passed in 1925.[37] The 1925 legislation, "modelled on English and Western Australian legislation of the time", specifically conferred powers on officers of the RSPCA to enter premises and check for compliance (Emmerson 2001, p. 13). The 1925 legislation maintained the key characteristics of the 1901 Act.[38] However, the range of exemptions was extended to include religious slaughter[39] and a number of common animal husbandry practices.[40] As well, the legislation was notable for making a further sanction available to a court in dealing with an owner convicted of animal cruelty – depriving the convicted person of future ownership of their animal.[41] In addition to a fine and/or imprisonment, temporary or permanent prohibition against ownership of an animal is a staple of modern sanctions available to sentencing courts.

Apart from some minor amendments, the Queensland legislation remained largely intact into the 1990s. At this time, a major reform effort was initiated by the State Government, at the tail end of animal protection reform projects instigated in a number of other jurisdictions. There had been "a resurgence of legislative activity in the field of animal protection. In the humanitarian spirit of the eighties, the welfare state turned its attention towards the revision of the scope and philosophy of its animal protection legislation" (Emmerson 1993, p. 18). The Animal Protection Act Review Committee was established to review the 1925 legislation, with the Committee drawing on a wide base for input, including the police, animal protection advocacy groups, farm industry sectors, veterinarians and animal welfare sci-

[35] Section 3 defined cruelty as: "The intentional or deliberate infliction upon any animal of pain that in its kind or degree or its object or its circumstances is unreasonable or wanton or malicious."

[36] Exemptions included for "the extermination of rabbits, marsupials, wild dogs or vermin", hunting of wild animals and use of animals in scientific research: s 12(1).

[37] *The Animals Protection Act of 1925* (Qld).

[38] The definition of cruelty was changed in the 1925 Act, to unreasonable, unnecessary or unjustifiable ill-treatment. In a circular approach, ill-treatment was then defined to include a range of cruel practices. As Hill (1985, p. 9) argues, "this definition says little more than that cruelty means unreasonable, unnecessary or unjustifiable cruelty. Surely all ill-treatment is cruel, and certainly in the terms proposed in the Act. The words are synonymous."

[39] *The Animals Protection Act of 1925* (Qld) s 7(1)(a).

[40] Ibid s 7(1)(b) ("dehorning of cattle, or the castration, speying, ear-splitting, ear-marking, or branding of any animal, or the tailing of any lamb, where the operation is performed with a minimum of suffering to the animal operated upon").

[41] Ibid s 19.

entists (Emmerson 2001). Possible reform options were canvassed in a Green Paper in 1990, and draft legislation then prepared. Change was needed to respond to "changing community attitudes towards animal welfare matters, advancements in scientific knowledge and animal behaviour, and to encourage consistency in animal welfare legislation throughout Australia" (Emmerson 1993, p. 35). The ideal of consistency emerged most clearly in October 1991, when the first Animal Welfare Ministers Conference was staged in Adelaide. State and Territory Ministers collectively agreed to strive for uniformity in legislation in each of the Australian jurisdictions (Emmerson 2001). Although slow to take hold, an agenda of consistency was actively pursued by the Commonwealth and the States and Territories, later reflected in the broad policy umbrella of the Australian Animal Welfare Strategy (AAWS) (Commonwealth of Australia 2011). One of the goals of this strategy was to achieve uniform farm animal welfare standards.[42]

Despite the extensive ground work undertaken to prepare an overhaul of the Queensland legislation in the early 1990s, a Bill was not presented to Parliament. In July 1993 drafting proposals for the Bill were leaked to Brisbane's metropolitan newspaper, *The Courier-Mail*, with a range of groups critical of their content, including the police and a number of animal protection advocacy organisations (Emmerson 2001). According to Emmerson (2001, p. 14), the most influential voice was that of the RSPCA:

> The most vocal opponent of the proposed Bill was the RSPCA, which believed it weakened animal protection in the state rather than improving the situation. The RSPCA claimed that the government had ignored the recommendations of the Advisory Committee, releasing drafting guidelines not agreed to by the committee. The RSPCA obtained legal opinion on the drafting guidelines, which suggested the proposals required changing. As a result, the RSPCA began the *Kill the Bill, Not Animals* campaign. This involved mounting public support against the Bill. They sent a public letter of protest to the Minister, and asked that supporters do the same. Acrimonious debate followed. There was a great deal of media attention devoted to the discussion of the proposed Bill. As a result, the animal welfare groups were successful in preventing the Bill becoming law at that time.

Apparently chastened by the strength of the RSPCA response to the proposed Bill, the Government subsequently successfully passed revised legislation in 2001.[43] RSPCA Queensland was supportive of the legislation, especially after the powers conferred on inspectors were extended, an issue that had been a key reason for objection to the 1993 reform attempt (Emmerson 2001).

---

[42] For detailed discussion see Dale and White (2013). The Commonwealth Government has recently retreated from a leadership role in the animal welfare policy field, reflected in the defunding of the AAWS, with the consequences for the national consistency goal uncertain at best (Commonwealth Government 2013).

[43] *Animal Care and Protection Act* 2001 (Qld).

In terms of the key tenets of animal protection legislation discussed already, perhaps the most significant features of the new Act were:

- the imposition of a statutory "duty of care" based on the "Five Freedoms"[44] with breach punishable by a maximum fine of AU$22,500 and/or 1 year's imprisonment[45];
- a significant increase in penalties for a cruelty offence, from a maximum AU$1,500 fine and/or 6 months imprisonment under the 1925 Act to a maximum AU$75,000 fine and/or 2 years imprisonment under the 2001 Act;
- the incorporation of welfare Codes of Practice, particularly for farm animals (under regulations)[46];
- an exemption from prosecution for the cruelty and breach of duty of care offences where there is compliance with a code of practice; and
- the introduction of Animal Welfare Directions, a tool allowing for directions to be given to a person in charge of an animal to comply with their statutory obligations, and "aimed at being proactive, preventing a potential animal cruelty offence from occurring, or resolving an existing problem" (Emmerson 2001).

The same essential elements of this legislation remain in place to the present day.

6.6 Conclusion

This chapter examined changing attitudes to the treatment of animals in nineteenth century Britain, tracing the effects of change through to the first British animal protection legislation in 1822, and beyond to the law of the Australian colonies. By the early part of the twentieth century a number of key facets of animal protection regulation were established: the adoption of the generic "no unnecessary suffering" standard in assessing the extent of cruelty to animals allowed; the use of exemptions from the generic prohibition against cruelty; the imposition of duties to provide for the needs of an animal; and the establishment of one of the key institutional actors

---

[44] These are the freedom from hunger, thirst and malnutrition; freedom from fear and distress; freedom from physical and thermal discomfort; freedom from pain, injury and disease; and freedom to express normal patterns of behavior. The origin of the Five Freedoms can be traced back to the 1960s Brambell Committee inquiry into the treatment of farm animals in Great Britain (Brambell et al. 1965).

[45] In 1991 the Queensland Association for Community and Animal Welfare proposed that a "stewardship" obligation be included in Queensland animal protection law, in terms very similar to the duty of care obligation included in the 2001 Act (Thelander 1991).

[46] Farming industry attitudes to codes of practice seemed to shift dramatically between the mid-1980s, when their adoption under legislation in Queensland was first proposed, and the early 1990s. For example, in 1985 grazing animal farmers were expressing the view that "codes of conduct are not necessary for the grazing animal industry since it operates well, and has done so for many years without any such code in place (Peart 1985, p. 173). By 1991 the United Graziers' Association was arguing that "[i]n our view, the correct strategy is to use the Codes of Practice as a guideline, with the Codes tied to legislation" (Joyce 1991, p. 57).

in the animal protection field, the RSPCA. As well, the Australian colonies faithfully reproduced an understanding of domesticated animals as personal property. This property-based approach was consistent with a Lockean liberal property theory, reflected in William Blackstone's delineation of the principles of the common law. As shown through an exploration of the key animal protection law developments in post-federation Queensland, these facets of animal protection have proven to be enduring, helping to define the nature and limits of contemporary regulation.

The key principles of animal protection regulation established during the nineteenth and early twentieth centuries were a response to the particular socio-political concerns of reformist nineteenth century Britain. They were built on an ethic of humaneness, an ethic which, by and large, continues to animate contemporary orthodox understanding of the significance of animal interests. This "Wilberforce ethic" holds that "while humans had control over animals and could use them for any purpose, such use must be justified and humane" (Wirth 2007). As Wirth (2007, p. 2) points out, "the Wilberforce philosophy of animal welfare remains the keystone of the humane movement today." However, at least since the 1970s, debates in animal ethics have undermined the persuasiveness of the idea that animals are due only humane protection. These debates suggest a greater significance for animal interests than this ethic allows.[47] This has important implications for key principles of contemporary animal protection legislation, including a legal standard such as "no unnecessary suffering", grounded as they are in nineteenth century U.K. responses to changing social conditions, visible animal cruelty and contingent political debates. In particular, does the present day animal protection regulatory framework found in Australian jurisdictions amount, in essence, to a nineteenth century answer to twenty-first century concerns?

References

Anderson, Jerry. 2012. The origins and efficacy of private enforcement of animal cruelty law in Britain. *Drake Journal of Agricultural Law* 17: 263–310.

Armstrong, Susan, and Richard Botzler. 2008. General introduction – Animal ethics: A sketch of how it developed and where it is now. In *The animal ethics reader*, 2nd ed, ed. Susan Armstrong and Richard Botzler, 1–13. London: Routledge.

Bonyhady, Tim. 2000. *The colonial earth*. Carlton: Melbourne University Press.

Brambell, F.W. Rogers, et al. 1965. *Report of the technical committee to enquire into the welfare of animals kept under intensive livestock husbandry systems*, A report to the great Britain minister of agriculture, fisheries and food. London: HMSO.

Castles, Alex. 1982. *An Australian legal history*. Sydney: Law Book Co.

Commonwealth Government. 2013. *Mid-year economic and fiscal outlook 2013–2014*. Appendix A, 97 & 115. http://www.budget.gov.au/2013-14/content/myefo/download/12_appendix_a_expense.pdf. Accessed 30 Oct 2014.

Commonwealth of Australia. 2011. *Australian animal welfare strategy and national implementation plan 2010–2014*. http://www.daff.gov.au/animal-plant-health/welfare/aaws/australian-

[47] For a succinct account of the range of challenges to an ethic of humaneness see Garner (2005).

animal-welfare-strategy-aaws-and-national-implementation-plan-2010-14. Accessed 30 Oct 2014.

Dale, Arnja, and Steven White. 2013. Codifying animal welfare standards: Foundations for better animal protection or merely a façade? In *Animal law in Australasia: Continuing the dialogue*, 2nd ed, ed. Peter Sankoff, Steven White, and Celeste Black, 151–182. Sydney: Federation Press.

Emmerson, Glenda. April, 1993. *Duty and the beast: Animal welfare issues*. Background Information Brief No 25. Brisbane: Queensland Parliamentary Library.

Emmerson, Glenda. 2001. *Duty and the beast: Animal care and protection bill 2001*, Research brief No 2001/23. Brisbane: Queensland Parliamentary Library.

Evans, Edward. 1906, rev ed. 1987. *The criminal prosecution and capital punishment of animals*. London: Faber & Faber.

Favre, David, and Vivian Tsang. 1993. The development of anti-cruelty laws during the 1800s. *Detroit College of Law Review* 1: 1–35.

Francione, Gary. 1995. *Animals, property and the law*. Philadelphia: Temple University Press.

Garner, Robert. 2005. *Animal ethics*. Cambridge: Polity Press.

Griffin, Emma. 2005. *England's revelry: A history of popular sports and pastimes 1660–1830*. Oxford: Oxford University Press.

Harrison, Brian. 1973. Animals and the state in nineteenth-century England. *English Historical Review* 88: 786–820.

Hill, John. 1985. The legality of cruelty. In *Grazing animal welfare symposium*, ed. Brian Moore and Peter Chenoweth. Brisbane: Australian Veterinary Association (Queensland).

Jamieson, Philip. 1991a. Animal welfare: A movement in transition. In *Law and history in Australia: A collection of papers presented at the 1989 law and history conference*, ed. Suzanne Corcoran. Adelaide: Adelaide Law Review Association.

Jamieson, Philip. 1991b. Duty and the beast: The movement of reform in animal welfare law. *University of Queensland Law Journal* 16: 238–255.

Joyce, Jan. 1991. Animal welfare – The Graziers' perspective. In *Proceedings of the animal welfare conference*. Brisbane: Queensland Department of Housing and Local Government.

Kean, Hilda. 1998. *Animal rights: Political and social change in Britain since 1800*. London: Reaktion.

MacCulloch, Jennifer. 1993. *Creatures of culture: The animal protection and preservation movements in Sydney, 1880–1930*. PhD thesis, University of Sydney.

Munro, Lyle. 2001. *Compassionate beasts: The quest for animal rights*. London: Praeger.

O'Sullivan, Siobhan. 2011. *Animals, equality and democracy*. Basingstoke: Palgrave MacMillan.

Peart, W.J. 1985. Codes of conduct for grazing animal welfare: The Graziers' view. In *Grazing animal welfare symposium*, ed. Brian Moore and Peter Chenoweth. Brisbane: Australian Veterinary Association (Queensland).

Petro, Stefan. 2012. Public opinion, private remonstrance, and the law: Protecting animals in Australia, 1803–1914. In *Past law, present histories*, ed. Diane Kirkby, 61–84. Canberra: ANU Press.

Radford, Mike. 2001. *Animal welfare law in Britain: Regulation and responsibility*. Oxford: Oxford University Press.

Ritvo, Harriet. 1987. *The animal estate: The English and other creatures in the Victorian age*. Cambridge, MA: Harvard University Press.

RSPCA Australia. n.d. *Our history*. http://www.rspca.org.au/what-we-do/about-us/our-history. Accessed 30 Oct 2014.

RSPCA Queensland. n.d. *Where we were*. http://www.rspcaqld.org.au/who-we-are/our-history. Accessed 30 Oct 2014.

RSPCA UK. n.d. *Our history*. http://www.rspca.org.uk/in-action/aboutus/heritage. Accessed 30 Oct 2014.

Ryder, Richard. 2000. *Animal revolution: Changing attitudes towards speciesism*, rev ed. Oxford: Berg.

Salter, Brett. 2009. Possess or protect? Exploring the legal status of animals in Australia's first colonial courts: Part I. *Australian Animal Protection Law Journal* 2: 35–48.

Sankoff, Peter. 2013. The protection paradigm: Making the world a better place for animals. In *Animal law in Australasia: Continuing the dialogue*, 2nd ed, ed. Peter Sankoff, Steven White, and Celeste Black, 8–23. Sydney: Federation Press.

Thelander, Andrew. 1991. Positive concepts in legislative reform. In *Proceedings of the animal welfare conference*. Brisbane: Queensland Department of Housing and Local Government.

Turner, James. 1980. *Reckoning with the beast: Animals, pain and humanity in the Victorian mind*. Baltimore: John Hopkins University Press.

White, Steven. 2013. British colonialism, Australian nationalism and the law: Hierarchies of wild animal protection. *Monash University Law Review* 39: 452–472.

Wirth, Hugh. 2007. The animal welfare movement and consumer-driven change. *Farm Policy Journal* 4: 1–9.

Chapter 7
Animal Interests and South African Law: The Elephant in the Room?

David Bilchitz

Abstract Since the inception of constitutional democracy in South Africa in 1994, legislators, policy-makers and courts alike have tended to avoid expressly recognising the interests of animals in law. This chapter will seek to consider this trend in two significant areas – namely, the protection of animals against cruelty and the regulation of wildlife – in which there have been engagements in post-apartheid South Africa between the law and animal interests. Other than complete avoidance, where animal interests are considered, the discourse of courts and legislative bodies avoids the ethical implications of such a recognition and focuses on "objective" scientific matters. I shall contend that this "avoidance" of animal interests and ethics may often be successful in enhancing protections for animals and can be justified, at times, in this light. Yet, if the interests of animals continue to be routinely ignored, legal actors contribute towards the blindness of human beings to their value and thus limit what can be achieved in advocating for better protections. Thus, litigation and advocacy strategies need to develop a manner of ensuring that animal interests are expressly placed on the table and inviting courts (and other actors) to make pronouncements that can alter the status and seriousness with which they are treated. Such an approach, moreover, will be consistent with the ideas that shaped the liberation struggle and new constitutional order in South Africa and recognise that compassion, humanity and a refusal to sanction injustice must not arbitrarily be confined to the human species but extend to other animals too.

7.1 Introduction

> At time, and in fear, I have wondered whether I should concede equal citizenship of our country to the leopard and the lion, the elephant and the springbok, the hyena, the black mamba and the pestilential mosquito. A human presence amongst all these, a feature on the face of our native land thus defined, I know that none dare challenge me when I say – I am an African. (Mbeki 1996)

D. Bilchitz (✉)
Department of Public Law, University of Johannesburg, Johannesburg, South Africa
e-mail: dibilchitz@uj.ac.za

D. Cao, S. White (eds.), *Animal Law and Welfare - International Perspectives*,
Ius Gentium: Comparative Perspectives on Law and Justice 53,
DOI 10.1007/978-3-319-26818-7_7

In his beautiful, moving speech at the adoption of the South African Constitution, Deputy President Thabo Mbeki recognized the inter-relationship between humans and other animals on the African continent, and the consequent far-sighted possibilities of expanding the very notion of citizenship to include them.[1] At the time of the drafting of the Constitution, animal groups have claimed that the second largest number of submissions to the Constitutional Assembly supported express protection for animals in the new Constitution (Van Der Merwe 1996a, b). Unfortunately, the constitution drafters decided not to include any express mention or protection for the interests of animals. Instead, any protections (or entitlements) animals are afforded must be recognized as being implicit in existing provisions or be inferred from the content of specific rights.[2]

This "oblique" approach that was adopted in the Constitution, in many ways, set the trend for the manner in which the law has engaged with animal interests since the advent of constitutional democracy in South Africa. Legislators, policy-makers and courts alike have tended to avoid expressly recognising the interests of animals and the consequent ethical obligations doing so implies. South Africa has, for instance, passed a large amount of environmental legislation, and most of it assiduously avoids the mention of animal welfare. This chapter will seek to evaluate two significant areas – namely, the protection of animals from cruel treatment and the regulation of wildlife – in which there have been engagements in post-apartheid South Africa between the law and animal interests. The examples discussed will also highlight the possibilities and pitfalls of the new constitutional framework. In the second section, I shall consider and evaluate two case studies that arose in connection with animal protection legislation: the one deals with the constitutionality of the legislative framework governing performing animals; and the other deals with a Zulu cultural practice that was challenged on grounds of animal cruelty. In the third section, I focus on two case studies that have arisen in the context of wild animals: the one deals with the regulations adopted by the government to limit the practice of "canned" lion hunting; and the other deals with the regulations relating to elephant culling. Throughout I shall seek to analyse the reasoning of the courts and government bodies in order to determine the extent to which they were willing to engage expressly with the interests of animals and the justifiability or otherwise of not doing so. In the conclusion, I shall contend that these examples demonstrate that avoidance strategies may often be successful and can be justified in this light. Yet, if the interests of animals continue to be routinely ignored, legal actors contribute towards the blindness of many human beings to their value and importance and thus limit what can be achieved in advocating for better protections. Thus, litigation and advocacy strategies need to develop a manner of ensuring that animal interests

---

[1] Recently, Kymlicka and Donaldson (2011) have indeed provided a strong argument for recognising animals as citizens (of differing types).

[2] Bilchitz (2012) makes the case for why the existing Constitution should be interpreted to include protection for animals. The piece acknowledges that these protections are not explicit and must be drawn from an exercise of purposive constitutional interpretation. For a contrary view, see Metz (2012).

are expressly placed on the table and inviting courts (and other actors) to make pronouncements that can alter the status and seriousness with which they are treated. Such an approach, moreover, will be consistent with the ideas that shaped the liberation struggle and new constitutional order in South Africa and recognise that compassion, humanity and a refusal to sanction injustice must not arbitrarily be confined to the human species but extend to other animals too.

7.2 The Constitution and Animal Protection Legislation

Shortly after South Africa became a single state with four provinces in 1910, it passed the first animal welfare legislation titled the Prevention of Cruelty to Animals Act 8 of 1914. This Act was replaced by the Animal Protection Act ("APA") 71 of 1962 which has been amended on several occasions subsequently. The APA essentially criminalizes a range of offences that cause "unnecessary suffering" to animals and provides for law enforcement powers to address animal cruelty. A further Act – the Performing Animals Protection Act 24 of 1935 ("PAPA") –was passed in 1935 which dealt specifically with animals in the entertainment and security industries. Whilst there are other pieces of legislation that engage animal welfare and regulate specific practices, these two acts form the main explicit protections afforded to animals in South African statutory law. Whilst sporadically there have been indications that these Acts would be replaced, the hand of the legislature has now been forced in respect of the PAPA by a ruling of the Constitutional Court in 2013.

7.2.1 The Separation of Powers and Animal Protection Legislation

7.2.1.1 NSPCA v Minister of Agriculture, Forestry and Fisheries

The PAPA regulates a particular sub-set of the human engagement with animals, namely, the "exhibition or training of performing animals and the use of dogs for safeguarding."[3] It essentially creates a system whereby those who wish to use animals in the entertainment and security industries must apply for a license from a magistrate.[4] The application for a license allows the magistrate to investigate whether the person who applies for it is fit and proper: in turn, this allows for an examination as to whether the person has any record of cruelty to animals and the nature and type of animals being used.[5] In 2012, the National Society for the Prevention of Cruelty to Animals (NSPCA) applied to the High Court to declare key

[3] PAPA Long Title.

[4] Sections 2 and 3 of the Act.

[5] Regulations GNR 1672 (1 September 1993).

segments of the PAPA (sections 2 and 3) unconstitutional. Their legal representatives argued that these sections were inconsistent with the doctrine of separation of powers as it had been developed by the South African Constitutional Court. The legal problem they identified was that magistrates were empowered by the PAPA to grant these licenses; yet, that function, they argued, was an administrative one rather than a judicial one.

The High Court found that the "functions of issuing of licenses and certificates as envisaged in sections 2 and 3 are executive or administrative functions which have nothing to do with the core judicial functions of magistrates" (*NSPCA High Court* 2012, para. 27). These provisions therefore violated the separation of powers enshrined in the Constitution and were, therefore, unconstitutional. The judge (Legodi J) ordered that they be revised by parliament within 6 months. He also made an interim order in which he created a committee to exercise the licensing function temporarily that would include an expert team comprising animal welfare experts from the NSPCA and Veterinary Council as well as members of the Ministry of Agriculture (*NSPCA High Court*, para. 46).

The judgment then went to the Constitutional Court for confirmation. The Court engaged in some detail with the separation of powers doctrine and summarized key elements that have emerged from its jurisprudence thus far. The key question for the court was whether there is any compelling reason why a non-judicial function of this kind should be performed by a member of the judiciary and not a member of the executive. If there is no good reason, then the separation of powers is offended (*NSPCA CC*, para.38).[6] A unanimous court could not find any good reason why a magistrate should perform the licensing functions granted to them in the PAPA. The central reasoning that led to this conclusion is included in the following paragraph:

> I do not see why, if, for example a non-judicial body or officer can be given the power to issue casino or liquor licenses, a judicial officer such as a Magistrate should be assigned the function of issuing animal training and exhibition licenses. If we were to hold that it accords with this country's model of separation of powers for a statutory provision to require a member of the Judiciary to issue animal training and exhibition licenses and that does not offend the separation of powers, where will the requirement for the performance of administrative functions by Magistrate's stop? (*NSPCA CC*, para. 39)

This reasoning led the court to find that the separation of powers doctrine was violated by the PAPA and the provisions in question were declared unconstitutional. The court gave parliament 18 months to cure the defect but suspended the declaration of validity to allow the existing provisions to continue to operate pending the required revisions.[7]

[6] This is a summary of the requirements outlined by the court.

[7] I have drawn elements of the description of this case and the analysis of its defects from Bilchitz (2014).

7.2.1.2 Animal Interests and Institutional Reasoning

This case, significantly, was the first in which animal welfare legislation was addressed by the Constitutional Court in the new democratic era. Moreover, the court placed parliament and the executive on terms to amend the PAPA within the next 18 months. That is in itself an important development given that, for several years now, the Ministry of Agriculture has been suggesting that animal welfare legislation would be reviewed. For the first time, the executive and parliament now have a time period within which they have to amend at least the PAPA legislation.[8]

Unfortunately, the judgment, however, provides no guidance whatsoever as to the direction in which such reform should head. The case is notable for the fact that the interests of animals – despite being clearly implicated – were not mentioned at all. The case was argued and decided simply on the basis of the case law that has developed surrounding the separation of powers in South Africa since the advent of the Constitution. The Constitutional Court conceived of the matter simply as one of the institutional propriety of magistrate's granting PAPA licences without mentioning the substantive interests of animals that are at stake. The focus on institutions is important and often not adequately addressed by organizations and academics seeking to protect animals; the court, however, made a determination on an institutional question in a very abstract manner without considering the impact of different institutional arrangements on the interests of animals. This is worrying for a several reasons.

There was clearly a substantive dispute between the parties underlying the case concerning the protections to be afforded to animals. The NSPCA is itself a body specifically set up by statute to protect animals and, it can thus reasonably be inferred, that its decision to bring this case to court related to the fact that the provisions of the Act were offering little protection for performing animals. Magistrates anywhere in the country were empowered to grant such licenses and very little guidance is provided in the law as to the factors that must be taken into account in doing so. Support for this inference can be gleaned from the request in the papers that the NSPCA be granted the power to decide about license applications in relation to performing animals (*NSPCA High Court*, para. 28). Presumably, this was because of a concern that had arisen in practice concerning the issuing of licenses by magistrates in this regard. The Minister of Agriculture's submission also recognized the need for "expertise" in deciding on matters relating to license permits regarding performing animals (*NSPCA High Court*: para. 39). Those opposing the order included a collective association of those who train or exhibit animals or use dogs for safeguarding as well as those in the advertising and television industry. Clearly, performing animals are used by all these bodies: it is not entirely clear why they opposed the order, but, a reasonable inference would be that the current regime worked reasonably well for them in acquiring licenses and they could have been concerned about a stricter regime replacing it. By focusing solely on the abstract question of separation of powers, it could be argued that the courts failed to address the substantive underlying questions.

---

[8] Parliament failed to amend the bill within the initial time period stipulated by the Constitutional Court and asked for an extension of time to complete the amendment. The Performing Animals Protection Amendment Bill was passed by the Portfolio Committee on Agriculture, Forestry and Fisheries on 20 November 2015 and was due to be introduced to the National Assembly shortly thereafter.

Yet, it could well be objected that courts must only address the arguments presented to them. Given that the case was argued on the basis of the separation of powers there was no need for the constitutional court to address any issues relating to animal welfare. This would be true – as a matter of adjudicative practice – if the court could coherently avoid questions of animal welfare in reaching its conclusions concerning the separation of powers. However, I shall attempt briefly to show why it could not do so.

The Court ultimately ruled that there was no good reason why magistrates (as part of the judiciary) should exercise the licensing functions granted to them in the PAPA where non-judicial bodies are tasked with deciding upon casino and liquor licenses. The court essentially holds that there is no relevant difference between such licenses and licenses relating to animal training and exhibition. However, the court provides no reason to support this proposition and simply assumes it to be the case. Yet, there are some obvious points of dissimilarity between these various licenses: licenses under the PAPA relate to an extremely vulnerable group of creatures (non-human animals) who are often the subject of terrible abuse when required to perform in these industries. The license system under the PAPA was developed to protect vulnerable sentient creatures from harm by very powerful owners. Moreover, the needs and capacities of animals require detailed understanding and knowledge.

These elements are distinguishable from licenses relating to casinos and liquor. These licenses also regulate industries that involve social ills: however, regulation in this area importantly governs practices that are engaged in by human beings who have the choice whether to do so or not. Regulation in this area often rather involves protecting human beings from harming themselves (having a paternalistic justification). Gambling and alcohol abuse can, of course, also create harms to others but it is unlikely that a licensing system in and of itself can address these harms. Given these dissimilarities in the licensing powers in question, it could well be argued that a different regime would be justifiable for licenses granted in relation to performing animals and those relating to casinos and liquor.

The court also assumes in its reasoning that the license function in relation to performing animals is a very basic administrative power and that, in fact, it cannot be distinguished in any meaningful way from the wide range of administrative functions that can be conceived. Yet, again, however, it may be argued in response that in fact the licensing power in this case concerns a very important and grave matter: it relates to the protection of creatures that have deep needs and capacities and may be the subject of strong abuse in the entertainment world. As such, there could be a justifiable distinction drawn between other administrative powers and the granting of licenses in these cases.[9] Cameron J, in a minority judgment in 2008, held that the APA recognises that animals,

> are sentient beings that are capable of suffering and of experiencing pain. And they recognise that, regrettably, humans are capable of inflicting suffering on animals and causing

[9] This does not lead me to conclude that the court was wrong to find a violation of separation of powers in this case; rather, its reasoning was flawed and should have been based on the better protection an executive committee would offer for animals: see Bilchitz (2014, pp. 190–192).

them pain. The statutes thus acknowledge the need for animals to be protected from human ill-treatment. (*Openshaw*, para. 38)

The fact that the Constitutional Court reasoned so thinly and failed to address these (rather obvious) arguments is worrying for the development of more expansive protections for animals in the new constitutional order. The Court saw the matter as purely one of licensing: the protection of performing animals was no different from the regulation of casinos and liquor. Underlying this view, I would contend, is an assumption deeply rooted in the South African common law tradition that animals are "things" or "legal objects" and not "persons" or "legal subjects" and so can be treated in a similar way to other "things" or "legal objects" (Sinclair 1999, p. 3).[10] Yet, it is deeply disappointing to see the Constitutional Court uncritically adopting the assumptions underlying this common law tradition. Indeed, the post-apartheid constitutional order precisely requires a reflection on common law categories which may no longer be justifiable. Section 39(2) of the Constitution expressly requires the development of the common law in light of the values underlying the new Constitutional order.[11] Its ethos has been understood to push South African law in the direction of a more caring society that takes account of the needs of the most vulnerable.[12] It is high time that these values are applied to animals. Doing so, would require the recognition of their particularities and vulnerabilities which would involve particular institutional and substantive responses in the law. By failing to engage with the sensitive and difficult subject matter with which this case was concerned, the Constitutional Court not only reasoned poorly but did a disservice to the new constitutional order it is developing. It also failed to provide any guidance about the status of animal interests in the new dispensation, a matter that would become of importance in the *Smit* case to which I now turn.

7.2.2 Cultural Rights and Animal Protection

7.2.2.1 Smit v His Majesty King Goodwill Zwelithini

This case concerned an application for an urgent interim interdict to prevent the slaughtering of a bull or any animal at a traditional festival – known as Ukweshwama or the First Fruits festival – celebrated by the Zulu king and members of his community. The festival is one in which a bull is killed by young warriors with their bare

[10] I have criticized the coherence and justifiability of this common law approach Bilchitz (2009, pp. 41–50).

[11] It also prohibits the arbitrary exercise of state power (see *Merafong*: para 62): treating animals simply as 'things' could be regarded as arbitrary – see Bilchitz (2014).

[12] Famously, in *Makwanyane*: para 88, Justice Chaskalson said that 'it is only if there is a willingness to protect the worst and the weakest amongst us, that all of us can be secure that our own rights will be protected'. Mahomed CJ also stated in that judgment: para. 262 that the Constitution commits us to a 'democratic, universalistic, caring and aspirationally egalitarian ethos'.

hands. The application was brought by Animal Rights Africa (ARA) which was a non-profit organisation seeking to act for the benefit of animals and the environment. ARA contended on the basis of reports it had received that the animal is subjected to terrible cruelty: "the bulls eyes, genitals and tongue are ripped out whilst it is still alive, and sand or mud is thereafter forced down its throat in an apparent attempt to suffocate it while it is trampled, kicked and beaten to death. The bull dies after being subjected to such treatment for approximately forty minutes" (*Smit*, 5). This practice had raised condemnation from a range of animal welfare organizations around the world and ARA sought an urgent interim interdict to prevent the killing and the cruelty on the basis of the Animal Protection Act. It argued that the APA "represents, in this context, a codification of basic norms and standards which are binding upon all members of society" (*Smit,* 8).

The respondents contended, however, that the important constitutional rights of the Zulu nation to practice its religion and culture were at stake in this case. The ceremony itself had a primary religious and cultural significance and was rooted in a practice whereby the corn or fruits of the new harvest could only be consumed after the King had given permission to do so in a public feast. The occasion was also one where older soldiers in the military were allowed to marry and younger men were recruited. As part of the ceremony, young warriors were required to kill a bull with their bare hands. "To achieve this aim the animal is overpowered to disable it by closing its airways and thereafter its neck is broken in a specific manoeuvre that causes a quick and painless death. No bloodletting of any kind is allowed nor is dismemberment of any kind whatsoever part of the ritual slaying" (*Smit,* 13). The killing of the bull symbolizes the killing of the old king, "so that his power may be regained and he would be revitalized. It follows that the bull could not be mutilated as in form he represents the King, hence the killing by bare hands. The colour of the bull is of particular importance, it must be black. No blood must fall on it" (*Smit,* 13) The respondents thus claimed that the description by ARA of the manner in which the bull was killed was inaccurate and based on hearsay. They contended that no barbaric or inhumane cruelty in fact existed in the ceremony and that to suggest this was inflammatory and hurtful to those who are part of Zulu culture.

Van der Reyden J dismissed the application. He pointed out that clearly there was a significant discrepancy between the parties in the factual account of how the bull was in fact slaughtered. Given that the case was based on an urgent application, the judge – according to the rules of South African procedure and evidence – followed a rule set out in the *Plascon Evans* case which allowed him to accept the evidence of the respondent. He therefore found that the application by ARA was based on nothing more than a "conjecture by animal rights activists from various parts of the world about the manner in which the bull is slaughtered during the ceremony" (*Smit*, 10). The judge thus dismissed the case for the lack of a sufficient factual foundation. He went on to express deep concern about the applicants having accepted the reports of severe cruelty in the ceremony and the "consequences of vilifying the cultural and religious beliefs of communities and the polarization it causes" (*Smit*, 14). ARA had also, in his view, not shown sufficient respect for the King by attempting to enter into direct correspondence with him and not respecting the appropriate protocol. The judge also made comments which cast the application as evincing intolerance and a lack of respect for African culture and religion, particularly in light of past discrimination in South Africa. If the application was upheld, the judge was also concerned that this could spark "unrest and civil disobedience in the province with grave harm to the public" (*Smit*, 16) and be the "proverbial match under the powder keg" of South African society (*Smit*, 18).

7.2.2.2 Evidence and the Balancing of Interests

The judge in this case made his decision without directly addressing the legal questions that arose. The failure to disclose a sufficient factual foundation for the application was determinative in his view and allowed him to avoid having to weigh up the infringement of the laws protecting an animal against cruelty against the right of the community to practice their religion and culture. This may have been a deft mechanism of avoiding deciding the case on legal grounds; yet, upon closer analysis, the judge's failure to provide any relief is not justifiable even on the basis of the respondent's evidence.

The respondent attested to the fact that a bull is killed with the bare hands of young warriors and without any weapon of any kind. Killing a large animal with one's bare hands is not an easy task and this should have raised questions for the judge as to how it would take place. The respondent's version of the method of killing is also vague in crucial respects. First, it mentions that the animal is overpowered and disabled by closing its airways: how exactly does this take place? Wrestling an animal such as a bull to the ground and closing its airway would be very difficult and the animal is likely to struggle. This process itself could be regarded as cruel and in contravention of the APA which proscribes ill-treating, infuriating or terrifying an animal (Rutherford-Smith 2012, p. 75).[13]

It is thus not clear that the applicant's evidence is in direct contradiction with the respondent's own version here. The respondent then claims that its neck is broken in a specific manoeuvre: this is very vague and should have raised further questions for the judge about the method of doing so. Breaking the neck of a bull is not an easy task to accomplish and, without further information, it is difficult to see how cruelty is not involved. The reports provided by the applicant – whilst hearsay – could well have raised concerns that the vague elements of the respondents' account needed to be more fully addressed.[14] Whilst accepting the respondent's own version thus could have been sufficient to disclose a contravention of the APA, the judge could also have found that he needed to have further evidence in order to make a decision in

[13] See section 2(1)(a) of the Animal Protection Act.

[14] Rutherford-Smith (2012, fn. 34) gives a range of anthropological studies of the ceremony but states that '[p]hotographic evidence of the ceremony suggests that the slaughter is far more cruel, abusive and unorganized than is proposed by the respondents'. In *Ramakatsa:* para. 131, the Constitutional Court also refuses to accept the version of the respondent where it failed specifically to refute the irregularities complained of by the applicant in that case.

the matter. South African civil procedure is sufficiently flexible to have allowed him to refer the matter for oral evidence.[15]

The other comments made by the judge indicate that his primary worry lay in protecting the rights to religion and culture of the Zulu people and avoiding major social disruption that could have been caused by the application. These were clearly important concerns, particularly in light of South Africa's divisive past: yet, in dealing with the disadvantage and harms in this scenario, the judge placed very little emphasis on the suffering of the bull. In order to succeed in an application for an interim interdict, the balance of convenience must favour the granting of the interdict (*OUTA,* para. 41). In this context, the judge claims that the harm to the "life of the bull" is "far outweighed by the prejudice that can arise to the general public" (*Smit*, 16). Here, the judge misrepresents what was at stake: the concerns raised by the applicants were not about whether the life of the bull would be lost but rather about the cruelty that would be imposed in the course of causing its death. If the possibility of severe cruelty was seriously to have been considered, the judge should not have been so quick to dismiss the application. Indeed, underlying the case is once again a sense in which the significant interests of the animal are not given the weight they deserve. If the judge had seen the statutory framework as imposing an obligation to protect animals against serious abuse,[16] he would need to have taken stronger measures to ensure that the bull's interests were protected. Consider, for instance, if a complaint of serious child abuse had reached the court and that the relevant practices were performed in the name of a culture. The court, in such a case, would be unlikely to hide behind rules of procedure to avoid finding out the nature and extent of the abuse. Animals – whilst not the same as children – are also extremely vulnerable and courts should therefore require strong proof that they are not being abused where allegations surface that this is the case.[17]

The under-valuing of animal interests is not simply a matter of judicial ideology. Sadly, it is enshrined in the South African legal framework through the common law categorization of animals as "things" and the failure expressly to recognise protection for animal interests in the Constitution. This case provides an important instantiation of this problem. The South African Constitution expressly protects the rights of individuals to freedom of conscience, religion, thought belief and opinion.

[15] Rule 5(g) of the Uniform Rules of Court allow a judge to 'make such order as it seems meet with a view to ensuring a just and expeditious decision…[i]n particular, but without affecting the generality of the aforegoing, it may direct that oral evidence be heard on specified issues with a view to resolving any dispute of fact….'.

[16] The exact reasons for the APA are a matter of dispute in South African law. Early decisions such as *Moato* focused on human sensibilities as the reasons for the protection. However, this rationale has been challenged in a minority decision by Justice Edwin Cameron in *Openshaw* and in Bilchitz (2009).

[17] A similar problem can be detected in *Openshaw* where the majority of the court is willing too readily to accept that animals would not be abused on the basis of a mere allegation in court papers. See Loggerenberg and Farlam (2014, Rule-B1-p50A) where it is recognized that courts must take a stronger line on the credibility of evidence on affidavit to avoid injustice and that 'testimony which is contrary to all reasonable probabilities or conceded facts…goes for nothing'.

Section 31 protects the rights of persons belonging to linguistic, cultural and religious communities "to enjoy their culture, practice their religion and use their language." Where individuals or a community engages in a cultural or religious practice which exhibits cruelty towards animals, they immediately can reference these rights in order to be allowed to continue to do so. Importantly, there is no countervailing express constitutional rights on the part of animals.[18] As such, if the human rights are not always to trump animal interests, it would be necessary to argue that they meet the criteria of the limitation's clause in section 36(1) of the South Africa constitution. This would require showing that a restriction on cultural rights in the name of avoiding animal cruelty is reasonable and justifiable in an open and democratic society based on human dignity, equality and freedom (see Rutherford-Smith 2012, p. 77). Most open and democratic societies in the world do provide protection for the interests of animals and thus section 36(1) would require the application of the proportionality enquiry in this context. This mechanism could potentially offer a structured means to balance the respective human and animal interests if the latter are given the recognition and weight they deserve.[19]

Without an understanding and deep engagement with animal interests, however, cases such as the present one become a clash between the rights of the *human* animal defenders against those members of a community who wish to practice their culture and religion. It is not then difficult to sec why thc matter can then be constructed as one about cultural intolerance. The judge in the present case stated that "[t]he Applicants proceed from the premise that they have a right to interfere with the religious and cultural practices of others that they find intolerable to their own beliefs. The Applicants are completely misguided in their contention that they have such a right" (*Smit*, 15). The judge here constructs the question as about the applicant's own beliefs and interests rather than that of the animal in question. If the matter becomes a matter of one set of beliefs seeking to dominate over another set of beliefs, and we take account of the history of subjugation of African cultures, it is not difficult to sympathise with the judge's decision.[20] We could go further and see the protection of animals as simply a cultural belief of some which must not seek to prevail over the cultural beliefs of those who have differing attitudes towards animals. Values such as tolerance and understanding can be marshalled to require respect for the practices of different cultures, no matter how cruel.[21]

Yet, this perspective is fatally flawed: it essentially adopts the position that animal interests matter only insofar as humans of different cultures and beliefs think that they matter. If this position were to be adopted, it would be unjustifiable to have laws prohibiting animal cruelty at all as such ill-treatment could always be justified

[18] For the case for reading the existing bill of rights to apply to animals, see Bilchitz (2012).

[19] This is a problem evidenced by examples in Germany which suggest that where a constitutionally-protected human interest clashes with statutorily-protected animal interests, human interests tend to triumph (Nattrass 2004, pp. 283ff). This led to a constitutional amendment in Germany and the express recognition of animal interests in article 20a of the Constitution.

[20] See Rutherford-Smith (2012, pp. 81–83) who essentially makes this argument.

[21] An argument to this effect is made in Mnyongani (2012, p. 102).

according to individual or culturally different beliefs. The only justification for proscribing cruelty to animals in a diverse constitutional democracy must be that animals have interests that matter in and of themselves and that the invasion of their most fundamental interests is objectively a serious moral wrong. If this is the case, animal interests do not automatically trump human interests but they must be given the weight they deserve in any balancing exercise. Moreover, a concern for those who are weakest and disadvantaged must include animals: "[t]he very concern for the worst off which justifies our support for disadvantaged groups also prohibits sacrificing their weakest members" (Casals 2003, p. 22).

Cases where there is a clash between cultural rights and animal interests are often not productive: positions become entrenched and often actively harm the desire to reach a win-win solution. The problem raised by this case is also not unique to Zulu culture. It occurs across South African society and the world in that cultures often exhibit blind-spots in relation to marginalised groups such as animals.[22] The law in a constitutional democracy can help uncover these blind-spots: animal interests should not simply be ignored or under-valued. At the same time, there is an important question of how legal remedies can help to diffuse the social conflict that may arise in these cases. "Weaker" remedies which draw on the internal resources of a community may be more effective in changing attitudes and practices towards animals (and show appropriate respect to the culture) than strong, coercive interventions. An example of such a remedy would have been for the judge in this case to order that an observation team be composed of relevant experts who would attend the *ukweshwama* ritual and provide a report to the court. This would have allowed for a deeper investigation of the factual manner in which the animal is killed. It would also have avoided prohibiting the killing but have focused the minds of the community leaders on ensuring that their practices were seen in the best possible light and did in fact conform to the very standards they pleaded to in court. Clearly, an ethos of respecting different cultures and practices should be inculcated without accepting the notion that the branding of a practice as "cultural" is sufficient to render it acceptable. Indeed, "what we should celebrate, and struggle for, is the existence of a multitude of practices that are both diverse *and* good, not a varied collection of cruelties and crimes" (Casals 2003: 22).

7.3 Environmental Law and the Interests of Wild Animals

South Africa is fortunate to have a diverse array of wildlife which draws in tourists from around the world. Section 24 (2)(b) of the Constitution is often seen to have relevance to wild animals through requiring protection for the environment:

---

[22] See the examples in Casals (2003, pp. 2–7). There is much discussion about discrimination that often occurs within religion and cultures against women and LGBT people (for instance). For an interesting case in the SA context relating to African culture and gender equality, see *Bhe* (2005).

> for the benefit of present and future generations through reasonable legislative and other measures that – (i) prevent pollution and ecological degradation; (ii) promote conservation; and (iii) secure ecologically sustainable development and use of natural resources while promoting justifiable economic and social development.

In giving effect to this section of the Constitution, the legislature has since the advent of constitutional democracy, passed a large amount of environmental legislation. The main piece of framework legislation is known as the National Environmental Management Act 107 of 1998 ("NEMA"). Further pieces of legislation which have an impact on wildlife include the Protected Areas Act 57 of 2003 and the Biodiversity Act 10 of 2004. Whilst these pieces of legislation are, in some respects, progressive developments that link with advances in international environmental law, it is notable that they almost completely omit any direct mention of the interests of individual animals including providing protection for their welfare or their lives. This has led to animals being considered in very abstract terms as part of the "environment" and so classified along with plants, rivers and mountains. As such, the particular interests of individual animals as sentient creatures are often not captured by this legislative framework. In this context, an abstracted, scientific discourse around animals which focuses on ideas such as "biodiversity", and the "management of populations" becomes of central significance and determines policy surrounding them. The following two examples highlight the trends, challenges and possibilities of this framework for animals.

7.3.1 *Canned Lion Hunting*

7.3.1.1 Predator Breeders Association v Minister of Environmental Affairs and Tourism

This case involved a challenge to the validity of regulations that sought severely to restrict the practice of "canned lion" hunting. Canned hunting in South Africa generally involves allowing hunters to shoot an animal who has been reared by humans (and thus domesticated) often in a wild setting. Given that the animal is habituated to humans, it usually does not run away and the kill is virtually guaranteed. Canned hunting facilities thus "share the same basic concept: hunters pay for a guaranteed kill – they never have to walk away empty-handed" (Ireland 2002, p. 227). South Africa had seen the extensive growth of canned lion hunting which led to the introduction of these regulations.[23]

The Minister of Environmental Affairs and Tourism passed the Threatened or Protected Species Regulations in 2007 ("ToPS Regulations") in terms of powers granted to him under the Biodiversity Act. Section 57 of that Act authorizes the Minister to "prohibit the carrying out of any activity – (a) which is of a nature that may negatively impact on the survival of a listed threatened or protected species;

[23] See Lindsey et al. (2012) for discussion on the increase of canned lion hunting in South Africa.

and (b) which is specified in the notice…". The ToPS Regulations prohibited the hunting of a "put and take animal" which is defined in section 1 as a "live specimen of a captive bred listed large predator…that is released on a property irrespective of the size of the property for the purpose of hunting the animal within a period of 24 months." The regulations, however, contained an exception: a listed large predator which had been bred in captivity could be hunted provided it had been rehabilitated to live naturally in an environment that was subject to minimal human intervention and had been fending for itself for at least 24 months.[24] This effectively prevented the hunting of animals that had not been living freely for 2 years and vastly increased the costs of raising animals for hunting purposes. Whilst lions, at the time case was lodged, had temporarily been removed from the published category of "listed large predators", the Minister had clearly expressed the intention that they were to be included in the near future, with the consequent implications that these regulations would apply to them.

The Predator Breeders Association (PBA) – a group of breeders of predators and of hunters of those animals that were bred in captivity – challenged the regulations and, particularly, the exception, on the grounds that they were irrational. In terms of South African law, all exercises of public power must be rational which requires that there be a rational relationship between the scheme adopted and the achievement of a legitimate government purpose; that a decision is rationally related to the purpose for which the power was given; and that there is a rational connection between the information available to a government functionary and the actions taken by him on the basis thereof (*PBA case*, para. 28). The PBA argued that the period of 24 months bore no rational connection to any legislative purpose of the Act; no rational basis or evidence existed for the underlying assumption that a captive-bred lion could be rehabilitated at all; and the period of 24 months was not justified by information in the possession of the Minister (*PBA case*, para. 29).

The Supreme Court of Appeal first considered the legislative basis for the prohibition on canned lion hunting. In terms of the Biodiversity Act, the focus of any regulations had to be on the purpose of ensuring the survival of a threatened or protected species. This meant that the prohibition on canned lion hunting or any exception thereto had to be justified in terms of its impact on the survival of lions as a species. The court then went on to analyse the reasons provided by the Minister for the prohibition and the exception.

The Minister, the court found, was motivated by the recommendations of an expert panel convened to discuss the regulation of the hunting industry; public opinion which was opposed to captive-bred lions being hunted; and concepts of ethics surrounding hunting which included the notion of a "fair chase". The Minister argued that he was initially in favour of an outright ban on canned lion hunting. However, as a compromise position, he decided to enact the exception which the court construed as an attempt to allow a sufficient period in which the animal could become self-reliant. The Minister, however, admitted that there was no scientific basis for the

[24] Regulation 24(2).

assumption that captive-bred lions could be rehabilitated in the wild or for the specific time period in the regulations.

In analyzing these justifications, Heher J began by stating that "[i]t is by no means clear to me how either ethical hunting (whatever its limits may be) and fair chase fit into a legislative structure which is designed to promote and conserve biodiversity in the wild and, more especially in relation to captive-bred predators that are not bred or intended for release into the wild" (*PBA case,* para. 37). Apart from the lack of a clear legislative basis for the regulations in the statutory powers granted to him, the court found that the 24 month period was "an arbitrary attempt to cut the Gordian knot which linked the two irreconcilable protagonists, without a justifiable basis in fact or expert opinion for choosing that cut-off point" (*PBA case,* para. 38).

The Minister had also misconstrued the findings of an expert panel which stated that hunting should only be permitted once an animal was self-sustaining. It had not suggested any specific period of delay before an animal could be regarded as reaching such a point. There was also a complete lack of expert evidence that captive-bred lions could in fact be rehabilitated into the wild. The court found that "if there was no real prospect that such lions will be able to look after themselves then there will be as little prospect of hunting being permitted" (*PBA* case, para 40). Regulation 24(2) thus purported to create an exception which would never be capable of being invoked and thus was meaningless and irrational. On these grounds, the court ruled that the regulations in their present form were irrational and would be invalid should lions be included in the definition of a "listed large predator".

7.3.1.2 The Inadequacy of the Environmental Law Framework

This case has had a severely detrimental impact upon the regulation of canned hunting in South Africa, which currently is wide-spread. It appears that the Department of Environmental Affairs and Tourism lost the nerve to address the matter further after these regulations were declared invalid by the court. The case highlights a number of key deficiencies in the legal framework governing wildlife in South Africa and also, once again, demonstrates an unacceptable disregard on the part of the judges for the interests of the animals concerned.

As has been mentioned, the environmental laws which were passed in post-apartheid South Africa largely avoid directly addressing the interests of individual animals and focus on large collective goals relating to the environment. The objectives of the Biodiversity Act illustrate this well: section 2 (a) states that the purpose of the act is:

…to provide for –

(i) the management and conservation of biological diversity within the Republic and of the components of such biological diversity;

(iA) the need to protect the ecosystem as a whole, including species which are not targeted for exploitation;

(ii) the use of indigenous biological resources in a sustainable manner;
(iii) and the fair and equitable sharing among stakeholders of benefits arising from bioprospecting involving indigenous biological resources.

The key concepts clearly are "biological diversity" and the protection of "ecosystems" and "species". Animals are to be regarded simply as part of the "biological resources" of South Africa which may be utilized in a sustainable manner. The notion of the sentience of individual animals and their welfare is not explicitly mentioned.[25] This framework is itself deeply flawed and ill-suited to addressing important questions relating to the regulation of hunting. The PBA case illustrates the difficulty of capturing the Minister's interest in the matter through the current legislative framework. In reaching its conclusion, the court found that the Minister was entitled to take account of the strong opposition expressed by the public to canned lion hunting (*PBA case*, para 44). Yet, the empowering provision upon which the Minister relied, only allowed regulation for the purpose of ensuring the survival of a threatened or protected species. Canned hunting by its nature involves the rearing of animals specifically for the purpose of being killed. It could of course have severe genetic effects through in-breeding but this would not necessarily affect wild populations of lions unless there was some mixing with them. The ethical problems of rearing a mammal such as a lion in a domesticated environment and then placing it outdoors to be shot could not, as the court notes, be adequately expressed through the current legislative framework.

Interestingly, in this context, a more complex understanding of animal interests is required to capture the ethical issues. On the one hand, there is the serious question of the animal's welfare in this process: is the animal fed before it is released? Is it killed quickly or in an agonizing manner? The APA could be used to prosecute those operators who can be shown to treat an animal in a cruel manner.[26] On the other hand, ethical objections and revulsion in relation to canned hunting are often not simply about the possible cruelty involved in the killing of the animals; they are rather about the very idea of raising domesticated animals and then allowing them to be shot. Underlying this concern appears to be a worry that the lives of magnificent creatures such as lions are not treated with respect and that they are simply reduced to objects to be reared and shot for profit.[27] Such a lack of respect for animals conflicts with the purpose of ensuring the survival of a threatened or protected species in the Biodiversity Act: the court, unfortunately, did not explore these issues.

The court judgment, however, focuses very little on the question of canned hunting and how it affects the interests of animals. Instead, large segments of the

---

[25] This fact does not mean that Minister and Department concerned cannot address animal welfare in regulation. Many of the concerns of the Biodiversity Act are necessarily tied to questions of animal welfare which would found the basis for legislative intervention in this regard: see Bilchitz and Finn (2014).

[26] Ireland (2002) argues that existing animal cruelty statutes can help address some of the abuses in the canned lion hunting industry in the United States context; similar points would apply in the SA context.

[27] See Francione (2012, pp. 29–35) which draws on previous extensive work on this subject.

reasoning are focused upon the absence of a scientific basis upon which it can be proved that lions can be rehabilitated into the wild. The current regulations were clearly a compromise but as the court states "in providing an alternative he [the Minister] was bound to rely on a rational basis. The evidence proves that he did not do so" (*PBA* case: para 44). The court thus places at the centre of its reasoning a question of evidence and science that ostensibly is amenable to clear, objective proof. Given the state of scientific enquiry in this area and the fact that there have been few attempts to rehabilitate lions in the wild, there was little evidence upon which to base the minister's finding. The court's finding that the exception in the regulations was not well-conceived and that the 2-year period was unfounded appears thus to be justifiable on scientific grounds.

The court's engagement with the evidence helps to expose the fact that the regulations displayed a lack of conviction on the part of the Minister who wished to ban canned hunting. Yet, whilst the finding may be reasonable, we need to be cautious about the level of evidence required of the government in developing policy relating to animals. There have indeed been significant advances in the study of animal welfare yet there is still great uncertainty around large areas of animal behavior. In making policy, the government should not be held to an unrealistic standard. As Choudhry (2006, p. 524) states in the context of justifying a limitation on fundamental rights by the government, "[p]ublic policy is often based on approximations and extrapolations from the available evidence, inferences and comparative data, and, on occasion, even educated guesses. Absent a large-scale policy experiment, this is all the evidence that is likely to be available." Similarly, the Canadian Supreme Court has recognised that government policy decisions that may limit fundamental rights "must inevitably be the product of a mix of conjecture, fragmentary knowledge, general experience and knowledge of the needs, aspirations and resources of society" (*Mckinney* 1990, 304). Adopting a realistic standard to assess government policy relating to the protection of animal interests is also necessary given the state of science in this area.

Moreover, there should be great caution exercised in accepting the evidence of those involved in exploiting animals. In this case, for instance, an expert for the PBA sought to argue on welfare grounds that it was better for the animals to be shot shortly upon being released as they would otherwise suffer from loneliness and being withdrawn from having social bonds with the other lions in captivity. Whilst the social needs and behaviours of animals of course must be taken into account, this was an entirely self-serving argument, seeking to provide a reason for why the owner of the animals should have them shot as quickly as possible and so avoid the costs of maintaining them for a long period. Caution should thus be exercised in accepting arguments which are purportedly made for the animal's sake but are themselves self-interested. In this vein, one silver lining in the judgment was the court's finding that the PBA need not have direct representation on the scientific authority which had been formed to assist the Minister in dealing with threatened and protected species (*PBA case,* paras 46–49). The court found that the industry would have to be consulted on matters that related to its interests: to have insisted that they had to have a seat on the scientific authority, would have allowed self-interested businesses to affect the integrity of a body set up to advise on scientific matters relating to threatened and protected species.

Whilst the court's conclusion relating to the irrationality of the exception in the regulations was justifiable, its final remedy – simply to declare the regulations invalid should they be applied to lions – was not. Merely declaring the regulations invalid allowed canned lion hunting to continue unabated and left the room open for the government to fail to regulate at all. The court should have adopted an approach that integrated the legislative concerns which prompted the regulations as well as a recognition of the animal interests involved to ensure that lions received protection from these unethical practices. The court could well have suspended its declaration of invalidity and required the Minister to produce revised regulations within a set time frame that met the concerns of the court. It could also have ordered that the PBA would have to maintain the lions in a decent condition of welfare until then.

7.3.2 Norms and Standards on Elephant Management

7.3.2.1 The Process and the Regulations

A similar approach towards wildlife was evident in the debate surrounding the resumption of elephant culling in 2004. During colonial times, there was a wholesale destruction of elephants in South Africa through hunting and the desire for ivory. To prevent the wholesale destruction of South Africa's elephants, the government introduced protection measures with the result that elephant populations in South Africa increased, particularly in its flagship Kruger National Park (KNP). By 1967, scientists in the KNP, given the scientific knowledge of the time, worried that elephants would become too numerous and destroy the biodiversity in the park. They therefore motivated for a policy whereby elephant herds would be "culled" to maintain a population of around 7000 elephants in the park which they claimed was its "carrying capacity". This practice became highly controversial and with the advent of the new democracy in South Africa in 1994, lobbying intensified. In 1995, the government issued a moratorium on culling pending further scientific investigation and debate. In 2004, South African National Parks (SANParks) organised a meeting to discuss the resumption of culling. The meeting was deeply divided between a range of "conservation" organizations and the traditional establishment which favoured the resumption of culling and a group of scientists, ecologists, lawyers and animal rights activists who were firmly opposed to its resumption.[28]

In March 2005, SANParks went on to convene a group of scientists for a meeting at Luiperdskloof to advise on the matter. Despite a large amount of uncertainty and disputation among scientists about culling and its impact on elephant populations, SANparks still advocated for culling. Later that year, the Minister of Environmental

[28] I was amongst the lawyers who opposed the resumption and presented a paper arguing why this was unjustifiable on a legal basis: Bilchitz (2004).

Affairs convened a meeting with relevant stakeholders which included a range of scientists, lawyers and activists (from both sides of the debate).

What became clear was that scientists could not show that elephants in fact necessarily harmed biodiversity. Traditional concepts such as "carrying capacity" were questioned and shown to depend on an outdated notion that eco-systems were static rather than dynamic (Gillson and Lindsay 2003, p. 414). Elephant impacts also had to be considered over the longer term (Van Aarde and Jackson 2007, p. 293). Scientific studies were being done to consider a range of possibilities for the management of elephants which included range expansion, and immuno-contraception (see Fayrer-Hosken et al. 1999; Delsink et al. 2006). There was also evidence that culling as an intervention was in fact harmful to elephant populations, causing trauma to these animals and nearby populations (Bradshaw and Schore 2007, p. 37). It also kept breeding rates at the highest possible level (see Whyte et al. 2003) and did not allow natural restraints on population numbers to develop (Ben-Shahar 1997).[29]

In the face of the inability of science to support the resumption of culling, the Minister lacked a rational basis upon which to do so without qualification. I presented an argument to the Minister as to why the precautionary principle (that had been relied on by proponents of culling) in fact required a cautious approach to be adopted before allowing culling to be re-introduced (Bilchitz 2006). The result of all the consultations and engagements were the National Norms and Standards for the Management of Elephants in South Africa. These were issued in terms of the Biodiversity Act which in terms of section 9 allows the Minister to set norms and standards for the "management and conservation of South Africa's biological diversity and its components".

The Norms and Standards sets out a range of objectives which include ensuring the long-term survival of elephants populations in the wild as well as "sustainable use". Significantly, though, they include the recognition that elephants must be managed in a manner that is "ethical and humane…and recognizes their sentient nature, highly organised social structure and ability to communicate".[30] Persons who carry out any function in relation to an elephant are required to have regard to a range of guiding principles which include:

(a) elephants are intelligent, have strong family bonds and operate within highly socialised groups and unnecessary disruption of these groups by human intervention should be minimized; …
(h) management interventions must, wherever practicable, be based on scientific knowledge or management experience regarding elephant populations and must-
  (i) take into account the social structure of elephants;
  (ii) be based on measures to avoid stress and disturbance to elephants;

[29] See further Whyte et al. (1999) for the historical development on elephant culling in the KNP.
[30] Regulation 2(2)(vii).

(i) where lethal measures are necessary to manage an elephant or group of elephants or to manage the size of elephant populations, these should be undertaken with caution and after all other alternatives have been considered;
(j) while efforts should be made to ensure that elephants continue to play an important role in an already well established nature- based tourism sector this should not occur in an inappropriate, inhumane or unethical form or manner;...
(l) every effort must be made to safeguard elephants from abuse and neglect.[31]

The Norms and Standards essentially require a set of management plans to be adopted for the keeping of elephants which take into account a range of factors relating to their habitat and needs. They place a duty of care on those who are responsible for elephants not to subject them to any form of abuse or neglect. The norms and standards also recognise that there a range of interventions that can be used to address the concern that elephants are becoming too numerous including contraception and range expansion. Culling is once again allowed but under strict conditions: importantly, a key restriction is that it can only take place in terms of a culling plan which is drawn up with the assistance of an ecologist who is an elephant management specialist. Such a culling plan requires the applicant to provide a justification as to why culling is necessary; and also, importantly, requires evidence to be provided why all other population management option have been rejected. Essentially, the norms and standards thus render culling as an option of last resort.

7.3.2.2 The Subordination of Ethics to Science

All indications during 2004 were that the government agencies were strongly committed to the resumption of wholesale culling in the KNP and other parts of South Africa. SANparks – which was a key stakeholder and effectively controlled the national parks – came out in favour thereof. Yet, the eventual Norms and Standards, whilst allowing culling, place stringent conditions on when it can be utilized. They also depart from the trend in environmental legislation to include specific mention of an elephant's sentience and intelligence. How did this happen?

The campaign surrounding elephant culling was perhaps the most successful and sophisticated that the movement concerned with animal rights has conducted in post-apartheid South Africa.[32] Whilst a perspective supportive of animal rights was articulated, a wide-ranging, multi-disciplinary group of people was engaged which culminated in the "Elephant's Alive" workshop run by Xwe African Wildlife in July 2005. What became increasingly clear was that the science was strongly on the side of those against the resumption of culling. Scientists from overseas – the ecologist

[31] Section 8(f).

[32] The two key individuals – Michele Pickover and Steve Smit – deserve specific mention for their dogged determination and excellent organisation on this matter through Xwe African Wildlife and then Animal Rights Africa. Sadly, these organizations have now collapsed.

Keith Lindsay and the welfarist Gay Bradshaw – as well as those conducting research locally – including Rudi Van Aarde and Bruce Page – were able to show why, for a variety of strong scientific reasons, the resumption of wholesale culling could not adequately be justified. Large animal welfare organizations internationally also presented and made representations to the minister, and indicated the dire effect that a resumption of culling could have on South African tourism. The ethical side was not forgotten: with lawyers, philosophers and activists drawing attention to the crucial interests of the elephants that were at stake.

The campaign also showed the power of the democratic process. Whilst the other branches of government were generally, at the beginning of the consultation process, favourable to the lifting of the moratorium on culling unconditionally, the discussions and evidence presented indicated that this could not be justified. The open, consultative processes required by South African constitutional democracy thus facilitated the engagement of diverse viewpoints but also meant that the science could not be ignored. The government minister was willing to listen and learn and eventually produced a policy which took into account the results of the scientific engagement that took place. The ethical dimension which had been stressed throughout by animal rights groups was not fully embraced; yet, significant strides were made by including in the policy express recognition of the complexity of the awareness and societies of elephants and the obligation to treat them humanely.

The Norms and Standards perhaps also once again reflect a negative side of the openness of the government– namely, its failure to commit to a particular point of view.[33] The norms and standards display – as in the case of the canned lion example – an attempt to make all parties happy and thus to allow culling but only under severely restricted circumstances. Were the government properly to have taken on board the ethical implications of the scientific understanding of elephants and their societies that was presented to them, it would only have allowed culling truly as a matter of emergency. As Moss (1988) states, "[elephants are not so many rodents to be exterminated: they deserve something better than that and I am not afraid to say that ethics and morality should be essential considerations in our decisions for their future".

---

[33] Unfortunately, the strong protections for elephants in Norms and Standards have led to those affected by this stringent standard to lobby government for a weakening of these protections. The Department of Environmental Affairs announced in 2014 that it would be seeking to re-draft the Norms and Standards and try and remove provisions relating to the welfare of elephants (as it claimed it lacked a mandate in this regard). Once again, this demonstrates the artificiality of attempting to separate environmental protection and welfare concerns in relation to animals: for a critique, see Bilchitz and Finn (2014).

7.4 Conclusion: Trends and Possibilities in South African Animal Law

The four examples that I have discussed in this paper are illustrative of several trends and possibilities in the engagement between the law and animals in post-apartheid South Africa. First, the new constitutional order – despite lacking express protection for animals – opens up a wide range of possible avenues for the defenders of animal interests to challenge their abuse. We have seen how often gains can be achieved for animals without directly addressing their interests through arguments based on, for example, the separation of powers doctrine. I have sought to show the insufficiency of the reasoning in these instances and the fact that these arguments often do not lead to optimal results for animals. Yet, given the trend, is it justifiable to argue a case that implicates animals without engaging directly with their interests?

In a deeply imperfect legal system, lawyers must use the arguments that work. If some gains can be achieved for animals without directly referencing animal interests, it does not appear justifiable to insist on doing so. At the same time, it is important for lawyers and activists to be aware of what they are doing and also to be concerned about the potential ideology underlying such a move. Ignoring animal interests directly in some sense suggests that animals themselves do not count (Perlo 2007, p. 12). If this trend is maintained in the future, South African law will further entrench a view that animals are simply expendable resources, who only matter to the extent that they engage human interests. Such a perspective is – in the longer-term – likely to be detrimental to animals in that cases will arise where it is not possible to subsume their concerns under other legal doctrines. Lawyers thus need to be careful not to win the battle but lose the war. Legal strategies should thus – if possible – at least include some argument engaging animal interests directly and invite the courts to make concrete statements that can be of important precedential value in this area.

The second trend that has been noticed is the increasing attempt to subordinate ethical and normative decision-making in favour of reasoning that is "scientific" and based on firm "objective evidence". There is of course nothing wrong with having reference to the best science available and, increasingly, it is bearing out the complexity of animal life and behavior. Animal activists and lawyers thus need not be afraid of science and should marshal it to support the protection of animal interests. Yet, importantly, it is also necessary to recognise the limitations of science: ultimately, it will describe the world but not prescribe how we should engage with it. The language of ethics and justice is necessary to transform scientific understanding into normative and legal principles and rules. Legal cases and advocacy campaigns should engage both elements in seeking to advance the interests of animals.

There remains much in South Africa that has not been explored in the potential that the new constitutional framework offers for the protection of animals. Provisions relating to access to information and administrative justice, for instance, open the door to many possible challenges that still need to be utilised more fully. The Constitutional revolution in South Africa sought to shake up the legal system and require a fundamental revaluation of its basic assumptions. The Constitutional court has clearly emphasized the need to challenge arbitrary categorizations and to protect those who are most vulnerable in society. As of yet, the implications of these important holdings have not been properly applied to animals and the protection of their interests. It is time for that to change.

References

Ben-Shahar, R. 1997. Elephants and woodlands in Northern Botswana: How many elephants should be there? *Pachyderm* 23: 41–43.

Bilchitz, David. 2004. Making democracy work: The impact of the constitution upon the 'Management' of elephants in Post-apartheid South Africa. http://www.sanparks.org/docs/events/elephants/2004-10-17_DB_Paper_Elephant_Indaba.pdf. Accessed 28 Apr 2014.

Bilchitz, David. 2006. SANParks's report on elephant management strategy and the precautionary principle (unpublished).

Bilchitz, David. 2009. Moving beyond arbitrariness: The legal personhood and dignity of non-human animals. *South African Journal on Human Rights* 25: 38–72.

Bilchitz, David. 2012. Does transformative constitutionalism require the recognition of animals rights? In *Is this seat taken? Conversations at the bar, the bench and the academy about the South African Constitution*, ed. Stu Woolman and David Bilchitz, 173–208. Pretoria: Pretoria University Law Press.

Bilchitz, David. 2014. What was left unsaid: The unconstitutionality of the Performing Animals Protection Act in NSPCA v Minister of Agriculture, Forestry and Fisheries [2013] ZACC 26. *South African Journal on Human Rights* 30: 183–195.

Bilchitz, David, and Megan Finn. 2014. Animal welfare and environmental protection go together daily Maverick (14 October 2014). Available at http://www.dailymaverick.co.za/article/2014-10-14-analysis-animal-welfare-and-environmental-protection-go-together/.

Bradshaw, G., and A. Schore. 2007. How elephants are opening doors: Developmental neuroethology, attachment and social context. *Ecology* 113: 426–436.

Casals, P. 2003. Is multiculturalism bad for animals? *The Journal of Political Philosophy* 11: 1–22.

Choudhry, S. 2006. So what is the real legacy of Oakes? Two decades of proportionality analysis under the Canadian Charter's Section 1. *Supreme Court Law Review* 34: 501–525.

Delsink, A.K., J.J. van Alterna, D. Grobler, H. Bertschinger, J. Kirkpatrick, and R. Slotow. 2006. Regulation of a small, discrete African elephant population through immunocontraception in the Makalali Conservancy, Limpopo, South Africa. *South African Journal of Science* 102: 403–405.

Fayrer-Hosken, R.A., H.J. Bertschinger, J.F. Kirkpatrick, D. Grobler, N. Lamberski, G. Honneyman, and T. Ulrich. 1999. Contraceptive potential of the Porcine Zona Pellucida Vaccine in the African Elephant (Loxodonta africana). *Theriogenology* 52: 835–846.

Francione, G.L. 2012. The problems of animal welfare and the importance of vegan education. *Southern African Public Law* 27: 28–39.

Gillson, L., and K. Lindsay. 2003. Ivory and ecology: Changing perspectives on elephant management and the international trade in Ivory. *Environmental Science and Policy* 6: 411–419.

International Fund for Animal Welfare (IFAW). 2005. The debate on elephant culling in South Africa: An overview. http://www.ifaw.org/united-states/node/6283. Accessed 28 Apr 2014.

Ireland, L. 2002. Canning canned hunts: Using state and federal legislation to eliminate the unethical practice of canned "hunting". *Animal Law* 8: 223–241.

Kymlicka, Will, and Sue Donaldson. 2011. *Zoopolis*. Oxford: Oxford University Press.

Lindsey, P., R. Alexander, G. Balme, N. Midlane, and J. Craig. 2012. Possible relationships between the South African captive-bred Lion Hunting Industry and the hunting and conservation of lions elsewhere in Africa. *South African Journal on Wildlife Research* 42(1): 11–22.
Mbeki, Thabo. 1996. *I am an African*. http://www.anc.org.za/show.php?id=4322. Accessed 23 Apr 2014.
Metz, Thaddeus. 2012. Animal rights and the interpretation of the South African Constitution. In *Is this seat taken? Conversations at the bar, the bench and the academy about the South African Constitution*, ed. Stu Woolman and David Bilchitz, 209–220. Pretoria: Pretoria University Law Press.
Moss, Cynthia. 1988. *Elephant memories*. New York: William Morrow and Company.
Myongani, Freddy. 2012. The status of animals in African cosmology: A non-legal perspective. *Southern African Public Law* 27: 88–102.
Nattrass, K. 2004. Und Die Tiere: Constitutional protection for Germany's animals. *Animal Law* 10: 283.
Perlo, K. 2007. Extrinsic and intrinsic arguments: Strategies for promoting animal rights. *Journal for Critical Animal Studies* 5: 1–14.
Rutherford-Smith, Sarah. 2012. Balancing the bull: Smit NO v His Majesty King Goodwill Zwelithini Kabhekuzulu [2010] JOL 25699 (KZP). *Southern African Public Law* 27: 75.
Sinclair, June. 1999. Introduction. In *Boberg's law of person and the family*, 2nd ed, ed. Belinda Van Heerden, Alfred Cockrell, and Raylene Keightley. Western Cape: Juta and Company.
Van Aarde, R.J., and T.P. Jackson. 2007. Megaparks for metapopulations: Addressing the causes of locally high elephant numbers in Southern Africa. *Biological Conservation* 134(3): 289–297.
Van der Merwe, L. 1996a. Animal rights and the constitutional assembly. *Animal Voice*, April–June 1996.
Van der Merwe, L. 1996b. The constitution leaves animals out. *Animal Voice*, August–November 1996.
Van Loggerenberg, D.E., and P.B.J. Farlam. 2014. (Revision Service 45) *Erasmus: Superior court practice*. Cape Town: Juta.
Whyte, I.J., H.C. Biggs, A. Gaylard, and L.E.O. Braack. 1999. A new policy for the management of the Kruger National Park's elephant population. *Koedoe* 42: 111–132.
Whyte, I.J., R.J. van Aarde, and S.L. Pimm. 2003. Kruger's elephant population: Its size and consequences for ecosystem heterogeneity. In *The Kruger experience: Ecology and management of savanna heterogeneity*, ed. J.T. Du Toit, K.H. Rogers, and H.C. Biggs, 332–348. Washington, DC: Island Press.

Cases Cited

Bhe v Magistrate Khayelitsha 2005 (1) SA 580 (CC).
Mckinney v University of Guelph [1990] 3 SCR 229 (Canada) ('Mckinney').
Merafong Demarcation Forum v President of the Republic of South Africa 2008 (5) SA 171 (CC) ('*Merafong*').
National Society for the Prevention of Cruelty to Animals v Minister of Agriculture, Forestry and Fisheries [2012] ZAGPPHC 329 ('*NSPCA High Court*').
National Society for the Prevention of Cruelty to Animals v Minister of Agriculture, Forestry and Fisheries [2013] ZACC 26 ('*NSPCA CC*').
National Treasury v Opposition to Urban Tolling Alliance ('OUTA') [2012] ZACC 18.
NCSPCA v Openshaw (462/07) [2008] ZASCA 78 (RSA) ('Openshaw').
Plascon-Evans Paints Ltd v Van Riebeeck Paints (Pty) Ltd 1984 (3) SA 623 (A).
Predator Breeders Association v Minister of Environmental Affairs and Tourism [2010] ZASCA 151 ('*PBA case*').

Ramakatsa v Magashule [2012] ZACC 31.
Rex v Moato 1947 (1) SA 490 (O).
S v Makwanyane 1995 (3) SA 391 (CC) ('*Makwanyane*').
Stephanus Smit v King Goodwill Zwelithini Kabhekuzulu [2009] ZAKZPHC 79 available at http://www.saflii.org/za/cases/ZAKZPHC/2009/75.html.

Legislation

Animal Protection Act 71 of 1962 ('APA').
National Environmental Management Act 107 of 1998 ('NEMA').
National Environmental Management Act: Biodiversity Act 10 of 2004 ('Biodiversity Act').
National Environmental Management Act: Protected Areas Act 57 of 2003 ('Protected Areas Act').
National Norms and Standards for the Management of Elephants in South Africa GN 251 (29 February 2008) ('Norms and Standards').
Performing Animals Protection Act 24 of 1935 ('PAPA').
Prevention of Cruelty to Animals Act 8 of 1914.
Threatened or Protected Species Regulations GN R.152 (23 February 2007) ('ToPS Regulations').

Chapter 8
Animal Protection Under Israeli Law

Yossi Wolfson

Abstract Israeli law, in some aspects, is rather progressive as to the protection of non-human animals. While "animal abuse" is interpreted as causing animals "unnecessary suffering", the content given to this phrase defeats some characteristics attributed by Gary Francione to "legal welfarism". In some instances the interests of nonhuman animals override substantive human interests, anchored in institutionalized forms of exploitation. An example is the ban on force-feeding of geese and ducks. Animal protection organizations were granted a broad mandate by the Israeli legislature and courts to represent the interests of animals in civil, administrative and criminal procedures. This solves the problem of standing which undermines animal protection in other jurisdictions. While Israeli law protects the interests of animals for the sake of the animals and gives them some weight, social power relations are still based on the cruel exploitation of nonhumans by humans. This chapter describes the principles of animal protection in Israeli law, discusses aspects that may inspire animal protection in other countries, and goes into some detail regarding the specific provisions of Israeli laws that are difficult to access without the knowledge of Hebrew.

8.1 Background: Israeli Society and Animal Exploitation

The land between the Mediterranean Sea and the Jordan River saw some of the earliest centres of human civilization, including some of the first stages of the subordination of non-human animals. This was later also the site of development of some of the oldest laws protecting non-human animals from their human exploiters – the laws contained in the Bible. These laws took for granted hierarchy among humans and between humans and non-humans. Despite some dissenting voices in the Bible that view meat-eating negatively (*Genesis* 1, 29; *Numbers* 11, 4–35; *Deuteronomy* 12, 20–28; *Amos* 6, 4 and more), it generally approved of the exploitation of animals for labor and of their use and killing for food and for raw materials.

Y. Wolfson (✉)
Attorney-at-law, Jerusalem, Israel
e-mail: reyo@netvision.net.il

D. Cao, S. White (eds.), *Animal Law and Welfare - International Perspectives*,
Ius Gentium: Comparative Perspectives on Law and Justice 53,
DOI 10.1007/978-3-319-26818-7_8

Nevertheless, it perceived them as living beings, sharing the world (on a non-equal basis) with humans, according to an order dictated by God (*Genesis* 9, 8–17). True to this ideology, the Bible allows animal exploitation, but puts limits on it: Animals must be allowed to rest on the Sabbath (*Exodus* 20, 10; *Deuteronomy* 5, 14); an ox may not be muzzled while treading out cereals (*Deuteronomy* 25, 4); one must unload a donkey that fell under the weight of his burden (*Exodus* 23, 5); one may take a bird's eggs or fledglings – but not together with the mother and not while she is watching (*Deuteronomy* 22, 6–7). These laws and other similar ones reflect real situations in everyday life of a society that thrives on traditional agriculture and some hunting and gathering.

Well into the twentieth century, the diet of the land was virtually vegan, based on wheat and barley, olive oil, vegetables and fruits, with little dairy products. Meat was consumed on special occasions, and was not a daily routine even among the urban rich. Animal husbandry was small, traditional and extensive (Vickers 1944; Kligler et al. 1931). Jewish and German immigrants from Europe were the main agents of change in this respect. They brought with them appetite for animal-based products, imported European chickens and cows and implemented more intensive production systems (though from today's perspective they may look extensive and nostalgic). Since the creation of the State of Israel, animal agriculture has been virtually in the hands of Jewish farmers in villages established by the Zionist Movement.[1] In terms of technology, today's animal industries in Israel have adopted all the worst systems of Western factory farming. In some areas, such as the milk industry, Israel takes pride as being a leader in intensive production.

The identification of animal agriculture with the Zionist project still has influence today: Most animal-industries are well-connected in the political elite; and there is high governmental involvement in these industries and intimate relations between the government and production councils. Producers often use national ethos to promote sales and avoid criticism and reforms.

On the other hand, Israel experiences a strong social tide of animal rights consciousness. This manifests, *inter alia*, in the prevalence of veganism and vegetarianism. 13 % of the Israelis define themselves as vegetarians or vegans according to a survey done in late 2013. 25 % said they reduced meat consumption in 2013, 13 % considered changing to veg diet, 10 % had a relative that became vegan in 2013 (12 % had a relative that became vegetarian in 2013) and 14 % had a close friend who made the move (16 % had a close friend that became vegetarian in the same period) (Aharoni 2014). Other surveys also indicate major support to animal causes. The sociological reasons for this are still to be examined, but it does have an impact on legal norms.

---

[1]There are some exceptions regarding pigs, sheep and goats, industries that also have smaller political influence.

8.2 What Is "Animal Abuse" Under Israeli Law?

8.2.1 The Prohibition of Animal Abuse and the Hamat Gader *Case*

Since the time of the English mandate on Palestine, a prohibition of animal abuse was contained in the Criminal Code. In 1994, this clause was replaced by the Animal Welfare (Protection of Animals) Act (henceforth: the Animal Protection Act).[2] The main provision in the Act, article 2(a), states: "A person shall not torture an animal, act cruelly towards one or abuse one in any other way." These strong, yet unspecific terms, needed interpretation, which soon came in the framework of a ruling of the Supreme Court in the case of *Let the Animals Live v. Hamat Gader Recreation Enterprises* [1997] (henceforth: the *Hamat Gader* case). The case deals with a short portion of a show using alligators (47 s out of a 30 min show). This portion, titled "a man-alligator battle" consisted of a number of stressful manipulations on the alligator's body. Let the Animals Live, an animal protection organization, sought an injunction against this portion of the show, using its authorization under the Animal Protection Act (this authorization will be discussed later in this chapter). It won in the magistrate court, lost an appeal to the district court, and asked for a leave to appeal to the Supreme Court. The leave was granted and the appeal accepted.

The court interprets the charged words of the Act to mean prohibiting the causing of any unjustified suffering to animals. The court rejects an interpretation that limits the scope of the ban to severe suffering only: "We will not set out a measure for suffering, nor will we determine the threshold of this 'suffering-metre', which magically indicates whether an individual has tortured an animal, treated one cruelly or abused one. Pain or suffering – though not severe – is sufficient to satisfy the second element establishing torture, cruelty, or abuse." The court continues to determine that mental suffering counts as well as physical suffering. The court also rejects an interpretation that takes into account the mental motivation of the transgressor: "[f] or the purpose of punishment, as per section 20(a) of the Penal Law, it suffices that one has 'awareness of the nature of the deed, of the circumstances and the possibility of the ensuing consequences.' As for obtaining the civil remedy of an injunction, it is sufficient that the actions can objectively be deemed to constitute torture, cruelty or abuse. The law's purpose is evident from its wording, and it is to protect animals from harm. From the animal's point of view, does it make any difference whether the person harming him is a sadist, deriving pleasure from the suffering of another, an emotionless person who doesn't recognize another's pain, or a person that might anguish with the pain of another, but is ready to be cruel to an animal as long as his will is done?"

The element that differentiates abuse from a lawful infliction of pain and suffering upon an animal is the legal element of justification. Here the court adopts the

---

[2] The title of the law uses the Jewish Law phrase '*tzaar baalei hayim*', literally 'animal suffering', and hence is sometimes translated as 'Prevention of Animal Suffering' or in similar ways.

same test that is used in checking the constitutionality of laws that restrict human rights: Does the act have a worthy objective? Is there a rational connection between the act and the objective? Can this objective be pursued in less injurious ways? Is there due proportionality between the suffering of the animals and the objective?

In *Hamat Gader*, the case was clearly decided upon the first portion of the test. The Court felt that the show was serving no legitimate objective. It describes it as anti-educational, and argues that it can harden people's hearts and cause indifference to the suffering of humans too. "[T]he performance", the Court adds, "is essentially a violent one. Violence is supplied to us in abundance and no more of it is needed."

This reasoning and rhetoric seem to fit into what Garry Francione (1995) termed "legal welfarism". According to Francione, the prohibition of causing "unnecessary suffering" to animals covers only those cases where animal suffering fails to facilitate the human objectives of animal exploitation. The prohibition maximizes the value of animals as property, and sanctions inefficient use of their suffering. According to Francione (1995), one of the characteristics of legal welfarism is that it will deem as "necessary" any animal use that is part of a generally accepted social institution. In other words, what the court did in the *Hamat Gader* case was to ban wasteful use of animal suffering, and defend human society from possible negative effects to humans. The judges would have wanted to censor this uncivilized show anyways, and animal protection was a mere pretext to do so.

However, some characteristics of the case counter this kind of criticism. The abuse in this case was institutional one, done for profit by a commercial company held by four *Kibutzim* – an elite group in Israeli society. The respondents did claim that banning this portion of the show will cause them significant financial damage. The acts of abuse in this case are not a manifestation of criminal anti-social subculture (like in a case of a teenager setting a cat on fire) nor can the court decision be interpreted as a disguised attack on a marginalized group based on class or ethnicity bias. The identity of the animal – an alligator and not an animal socially favored as "pet" – is also worth noting: this shifts the focus of the case from protecting the injured feelings of "animal lovers" to protecting the feelings of the animals themselves.

Indeed, the court's rhetoric, while mentioning human utility, also puts emphasis on the protection of animals for their own sake: "An animal, like a child, is a defenseless creature. Neither are able to defend themselves, nor can either stand up for their rights, honor and dignity. Would we stand idly by upon witnessing a person harming a helpless child? If we would come to the child's rescue – and surely we would – an animal being made to endure suffering deserves no less."

8.2.2 The Ban on the Force-Feeding of Geese and Ducks

The interpretation of the prohibition against animal abuse came to a more difficult test when animal protection organizations sought a ban on the force feeding of geese and ducks for the production of fatty liver (*foie gras*). Here the social benefits of the targeted practice were clear. According to the data given to the Supreme Court, about 100 family-farms around the country were engaged in the production of fatty liver. The livelihood of hundreds of families was dependent on this industry. It was a well-established industry which had existed for about 40 years, and had received financial aid and guidance from the government. At that time, Israel was one of the leading producers of fatty liver with an annual production of 500 tons. The annual income of the industry was tens of millions of Israeli Shekels.

The campaign to ban the force-feeding of geese and ducks was long and complex. In different phases the matter shifted between the Ministry of Agriculture, the full Cabinet, the Parliament, the courts and other governmental agencies. For the sake of the current discussion, though, I will focus on the Supreme Court verdict in *"Noah" – Israeli Federation of Animal Protection Organizations v. The Attorney-General* [2003] (henceforth the Force-Feeding case). At the time that the verdict was delivered, there were regulations in force, issued under the Animal Protection Act, which regulated – rather than banned – the force-feeding of geese. These regulations incorporated existing practices, prohibited the opening of new force-feeding facilities, and included some other provisions too vague to have any effect. The Ministry of Agriculture had drafted similar regulations pertaining to mallards, but these had been blocked already in the Parliamentary committee that needs to approve of any regulations under the Animal Protection Act.

The question facing the court was whether the regulations on the force feeding of geese were a reasonable implementation of the Animal Protection Act, and thus legal; or whether they were in contradiction to the Act, and thus exceeding the authority granted to the secondary legislature and void. To answer this question, the court asked whether force-feeding geese to produce fatty liver was in violation of the prohibition against animal abuse, as it had been interpreted in the *Hamat Gader* case. The answer was that it did violate this prohibition.

Most legal reforms in animal-agriculture are based on replacing the cruelest practices by other practices that cause less suffering. Hens may be held in industrial non-cage systems rather than in battery cages. Sows may be imprisoned in enriched group enclosures or on deep litter, rather than restrained in individual stalls. Disabled chicks, as well as surplus ones and male chicks of laying strains, may be gassed or crushed alive rather than thrown to the garbage to die slowly. The "alternative" systems usually incur some extra costs on the industry, but allow for inexpensive mass production of the same products. This is not possible in the case of the fatty liver industry. Fatty liver, the production of which is the *raison d'etre* of the industry, is by definition a liver in pathological state of hypertrophy and steatosis – enlarged and fat. This pathological condition can only be reached by coercive process of forcibly introducing large quantities of food, high in carbohydrates, to the birds' intestines. This is generally done by shoving a pipe into the bird's esophagus, and compressing the food through it using mechanical, hydraulic or pneumatic pressure. The enlarged liver presses other organs, which causes difficulties in breathing and in walking – another problem which is inherent and unavoidable. While there may be some variations of technique, animal suffering cannot be substantially reduced while still producing the desired product. This has dramatic impact on the application on force-feeding of the tests set out in the *Hamat Gader* case.

As to the suffering caused to the birds, the court followed the conclusion of the E.U. Scientific Committee on Animal Health and Welfare (1998) that force feeding is "detrimental to the welfare of the birds." But is this suffering justified according to the *Hamat Gader* tests? This time the court identified the objective of the practice as legitimate. As to defining the objective, the court put emphasis on the livelihood of the people working in the industry, rather than any consumers' interest in fatty liver. To the contrary, the court took into consideration that fatty liver is not a basic food but a luxury. The second question in the justification test is whether there are less injurious ways to pursue the legitimate objective. The answer, as we saw, is negative: it is an all-or-nothing situation. The court had to confront the last element of the test, where human and nonhuman interests directly clash and are weighed against each other.[3] In this clash, the court favored the nonhuman interest: "Indeed, one should consider the legitimate interest of farmers in preserving their livelihood within an industry promoted by the authorities. But this interest cannot automatically override the conflicting interest of protecting animal welfare. The legislator considered both interests, but it appears that he didn't give each of them the right weight .... One received too much weight, the other too little." The regulations were found to be void, the practice illegal. A phase out time of almost 2 years was granted. The Ministry of Agriculture and the industry did not give up, but after some more rounds of campaign the force-feeding industry in Israel came to its complete end.

In describing the approach of Israeli law to the welfare of animals used in agriculture, the court explicitly rejects the doctrine expressed in the "common farming practices" exemption which appears in many U.S. state animal protection statutes (Wolfson 1996). While "common farming practices" exemptions give agricultural practices immunization from the law just on the basis of being widespread, the Israeli court emphasized that "[l]ong established agricultural practices are not immune from the applicability of clause 2(a) of the Law" (i.e. the prohibition of animal abuse).

The court further says that this approach allows flexibility. The legality of a certain practice may change based upon "cumulative scientific knowledge and changing social perceptions". While the scientific knowledge parameter is prominent and explicit in the E.U. legal system, it is revealing that the Israeli court explicitly mentions developing social values as a parameter in applying the law.

---

[3] Justice Strasberg-Cohen does comment that the regulations do not even implement all the measures that may reduce the birds' suffering. However, this is not the ratio of the decision. In a later verdict (HCJ 7713/05 *'Noah' – Israeli Federation of Animal Protection Organizations v. The Attorney-General* [2006], no English translation available) the court rejected a claim by the industry that there are available techniques that allow for the production of fatty liver without violating the Animal Protection Act.

8.3 Animals in Agriculture

8.3.1 The Numbers of Animals Used in Israeli Agriculture

With but marginal exceptions, animals exploited for food in Israel are held in industrialized facilities. In these facilities, the animal's life and death are under total control, conditions are artificial, genetics is distorted, and everything is designed to intensify the economically-desirable functions of the animal's body (reducing and even nulling other functions) – to maximize profits rather than facilitate animal needs and allow for their personal happiness.

How many animals are used in the Israeli food industry? According to the Israeli Egg and Poultry Board (2014), 216,311,000 chicks were marketed to the chicken-meat industry in 2013, and 7,101,000 to the turkey-meat industry. 4,488,000 female chicks were marketed to the egg industry, which (according to Israeli Central Bauru of Statistics (ICBS) 2014) imprisons at any given time about 9,000,000 laying hens. Virtually all hens are in battery cages. These numbers do not include chicks killed in the hatcheries. According to the Israeli Veterinary Services (IVS) (2013), the reported numbers of mammals killed for food in Israel in 2012 were 197,390 pigs, 129,534 sheep and goats, 105,789 individuals of cattle and 1590 rabbits. These numbers do not include unreported unauthorized killings, on which estimations vary. The Israeli Milk Council (2013) estimates that 40 % of the cattle meat from local slaughterhouses comes from calves and cows from the milk industry, 13 % from cattle raised on pasture and 47 % from imported cattle. The number of cows used for milking at any given time in Israel is estimated to be 119,000. The number of adult pigs (virtually all sows used to bear piglets for slaughter) is estimated to be 24,000 (ICBS, 2014). There is no statistics available on the number of individual fishes used in the fish industry, only of the total weight of those who were marketed (excluding fishing bycatch and those that did not survive the conditions in industrialized pools and sea-cages). The numbers for 2012 were 22,000 tons of marketed fish meat – of which 17,100 tons were from pools, 2200 tons from cages in the sea and 2700 tons from catching both in sea and lakes (ICBS 2014).

According to the Force-Feeding case, each and every practice in these industries is subject to judicial review. However, the main instrument of regulating these industries is secondary legislation issued by the Minister of Agriculture, subject to the approval by a parliamentary committee. Existing regulations are far from being comprehensive or systematic. Issues covered reflect the agenda of animal protection organizations that had to push hard for each set of regulations: at least six petitions to the Supreme Court were filed throughout the years asking for regulations on different issues. The regulations are detailed and complex, and most of them are not available in English. I will therefore go into some details but focus on those provisions that may be of international interest.

8.3.2 *Calves in the Veal Industry*

The Israeli regulations on calves in the veal industry were enacted in 2006. They generally follow the E.U. model[4] with some differences.

Veal crates are banned. It is allowed to hold young calves (under the age of 2 months and under the weight of 90 kg) in isolation, as long as they can see each other. Isolated calves should be held in enclosures whose length are at least double the calf's length and whose width are at least the calf's height at the withers. That is twice the minimum space according to the E.U. regime. The permission to isolate these young calves does not pertain to imported calves (the argument at the time of enacting the regulations was that these calves are imported in group-boxes, so that the health pretext to isolating them cannot hold). As all calves in the current Israeli veal industry are imported, isolation is de-facto banned (except for cases of disease).

Calves older than 2 months should be held in groups of at least four calves. The minimum measures of group enclosures are slightly higher than in the E.U. regime: 1.5 m^2 for calves under 150 kg, 1.8 m^2 for calves of 150–220 kg and 2 m^2 for calves over 220 kg.

Water should be supplied ad-libitum at all times, with the exception of water withdrawal 1 h before feeding and 1 h after feeding, with a maximum of 4 h of water withdrawal per day.

Tiny quantities of solid food are mandatory from the age of 4 weeks: at least 100 g per day up to the age of 15 weeks and 250 g afterwards. Hemoglobin blood concentration should be above 7.3 g per deciliter (4.5 mmol/l) – a degree of anemia that is supposed not to have clinical expression under the holding conditions of the veal industry. This minimum value, however, is calculated as an average of the individual values of the calves in the facility who are of the age of 12–24 weeks.

Other provisions require that light during the daylight hours of the day is no lower than 40 lux (in practice, to the best of the author's knowledge, light is natural), that sick isolated calves are provided with litter, and that ventilation is at least in accord with certain minimal technical standards.

8.3.3 *Pigs in the Pig-Meat Industry*

The regulations on keeping pigs for the meat industry were enacted in 2015. They follow the E.U. regime[5] with some significant improvements.

---

[4] Council Directive 91/629/EEC of 19 November 1991 laying down minimum standards for the protection of calves as amended.

[5] Council Directive 2008/120/EC of 18 December 2008 laying down minimum standards for the protection of pigs (Codified version).

Sows in the pig-meat industry go through a repetitive cycle of enforced pregnancy, giving birth, losing their offspring at the early age of 3 or 4 weeks and enforced pregnancy again. One of the worst torturous practices is that all this time they are isolated and restricted. They spend the pregnancy periods in pregnancy stalls where they cannot even turn around. Before farrowing and until their piglets are taken away they are held in farrowing stalls, with metal bars restricting almost any movement and separating them from their young.

The regulations ban pregnancy stalls altogether. Gilts and sows may be isolated for a maximum of 1 week for insemination – but in enclosures that are small but allow turning around (the minimum size of these, 2.2×1.4 m is twice the size of conventional pregnancy stall). Afterwards, the pregnant sows should be held in groups. Farrowing stalls are allowed, but only until the 14th day after farrowing. After this, the restricting bars should be lifted. The net size of the farrowing enclosure should be at least 4 m^2 (which is clearly too small), out of which 1 m^2 should be devoted to the piglets, and protected by a partition.

Space is another critical issue. The regulations specify the net minimum space for gilts sows and boars depending on group size. As to piglets from weaning to slaughter, the specification given for each range of average weight applies the formula $0.047X(W)^{0.67}$ when W is the pigs' average weight. This formula should allow all pigs lie simultaneously on their sides.[6] It is not enough to allow separate area for dunging, separate space for moving and standing by the food and water troughs and appropriate space for social interaction.[7] Unsatisfactory as they are, these minimum specifications are higher than those in the E.U. regime. Separate (and higher) specifications are given regarding piglets grown on deep litter.

Pigs are very intelligent, curious and playful animals. One of the cruellest features of industrial pig farming is that they are generally held in barren environments. Therefore, the regulations require some amount of environmental enrichment to all pigs. Environmental enrichments should include both objects such as chains and pieces of wood and materials that allow investigation such as straw and compost. This, of course, is only a pity bit relative to what pigs enjoy in deep litter system let alone natural conditions.

Surgical castration, tail-docking and tooth clipping are scandalously allowed. Up to 7 days of age, analgesics should be use. After that age, local anesthesia should be used before the operation, analgesics given after it, and the operation should be done by a veterinarian.

Maximum gas concentration in a pig facility is 10 ppm NH_3; 3000 ppm CO_2 and 2.5 ppm H_2S. Continuous noise may not be over 85 dB. Other provisions give specifications on slated floors; require inspection of automatic equipment, give instructions regarding food and water (including a minimum of 10 % fiber for pregnant gilts and sows – more if "according to her behavior it seems that she needs more");

[6] The Welfare of Intensively Kept Pigs, Report of the Scientific Veterinary Committee, Adopted 30 September 1997, Section 4.4.6.

[7] EFSA (2005) Opinion and Report on the Welfare of Weaners and Rearing Pigs: Effects of Different Space Allowances and Floor Types pp 56–59.

ban the tethering of pigs, beating them and using electric prods on them; deal with veterinary care and more.

8.3.4 Transport of Cattle, Sheep, Goats and Camels

The regulations on the transport of cattle, sheep, goats and camels were enacted in 2006.

The regulations limit the transport time of animals to 6 h (4 h in pigs). This may only be extended by the Director of the Veterinary Services for special reasons to be registered, if circumstances justify this. Transport time is measured from the time the first animal is caught in order to be transported to the time the last animal is unloaded from the vehicle. Transport should be continuous and without stops – under the circumstances.

The transport of unfit animals is prohibited. These include any sick or injured animal unless the transport is for medical treatment or the sickness and injury are slight and do not influence the animal's welfare during transport. The transport of animals who gave birth in the 48 h before transport is prohibited except for medical treatment only. There is a conditioned exception for transporting animals for slaughter, subjected to veterinary approval – however, unambulatory animals cannot be transported under this provision. The Director of the Veterinary Services may allow a transport of an unfit animal for special reasons to be registered, if circumstances justify this.

Food and water should be available to the animals up to the time of loading (except for food for pigs destined to slaughter) and water should be available immediately after unloading.

A series of provisions prohibit certain violent acts towards animals and determine characteristics of vehicles, ramps and other loading and unloading facilities, as to reduce stress, slipping and injury. There are provisions on stocking densities, ventilation, separation of animals that might be dangerous to each other, environmental conditions, training of workers, inspection, documentation and more.

An issue that is not properly covered is the international transport of animals by sea and air. Though the regulations do not explicitly limit their application to land-transport, they lack provisions relevant to sea- and air-transport. In 2013, Israel imported from Australia and from European countries a total of 206,771 calves and sheep (IVS 2014) in horrendous journeys that contradict common sense and basic principle of animal protection.

8.3.5 Transport of Poultry

The regulations on the transport of poultry[8] were enacted in 2011, and cover the transport of chickens, turkeys, ducks, geese and quail used for commercial agricultural purposes.

These transports, especially the transports of birds to the slaughterhouse, are of special concern. Over 1,000,000 individual birds are found dead each year in Israel when slaughterhouse workers open the cages in order to hand them to the slaughterer (IVS 2013).[9] This is a mere indication of the unimaginable suffering of the birds who die on the way to slaughter and of those that survive just to be killed. Professional literature describes the many causes of stress in such transports, including handling by humans during loading and unloading (physically painful and frightening), injuries caused during loading, hunger and thirst, crowded and restrictive conditions in the transport cages, noise, motion and more. Heat stress is a major problem. Even in a cold winter day, when chickens at the side-cages suffer from cold winds, rain or snow, those in hot-spots in the middle of the cargo may suffer from extreme heat, created by the crowded mass of birds.[10]

Journey time is crucial. Maximum journey time to slaughter, according to the regulations, is 6 h (4 h between May and October if the transport is between 10 am and 5 pm), measured from the time the first bird is put in a transport cage to the time the last one is taken out. For poultry not sent to slaughter the maximum journey time is 8 h. This maximum time can be extended: 1 h can be added at the loading location, if the birds are kept in conditions of shading, protection from rain, cooling, noise-prevention, and air movement through the cages (ventilators placed adjacent to the cages, which suck the air and remove the humidity and heat created by the birds, reduce heat stress). The birds can also be held in the destination for 8 h (six between May and October) under the same conditions. Note that these 8 h are measured from the time the birds reach the destination, and not just added to the basic allowed journey time. The objective of this set of rules is to allow catching and transporting the birds in the late night, bringing them to the slaughterhouse in the very early morning and slaughtering them during regular work hours. This keeps transport hours to the minimum while not changing the work hours of slaughterhouse employees and preserving the advantages of night transports.

An important provision prohibits the transport of birds for killing other than for human consumption. This is actually – and was meant to be – a ban on the transport of "spent hens" from the egg industry. At the young age of 2 years (or a few months more) it becomes cheaper for the industry to replace these hens with new ones, and they are killed. At this point the hens suffer from significant osteoporosis due to the

[8] For a more comprehensive overview of these, see Wolfson (2012).

[9] For religious reasons, slaughter of poultry in Israel is done by hand. A worker takes the birds one by one out of the transport cage, and hands them to a religiously authorized slaughterer, who cuts their throat. Only then the bird is put upside down in a cone-shaped container or hanged on a conveyor to bleed and then be processed.

[10] For more details and scientific references see Wolfson (2012).

high laying rate and lack of movement, and are in special risk of bone fractures. The Israeli practice is to kill them with electric shock in special mobile units that move from farm to farm.

Other provisions require a trained attendant to accompany each transport, ban the transport of unfit birds (a strange provision given the poor state of "broiler" chickens at slaughter age), give instructions regarding the catching of birds (e.g. no more than three at each hand), and deal with cage construction, environmental conditions, ventilation, documentation and reports.

8.3.6 Hens in the Egg Industry

As this chapter is written, Israel still has no detailed regulations on minimum conditions for hens in the egg industry. The debate has been going for several years, and the issue has been the subject of cabinet decisions, implemented in draft-regulations. The regime proposed by the government is gradual adoption of the European regime,[11] including a prohibition on forced molting. Animal protection organizations oppose this regime which allows keeping hens in "furnished" battery cages. These cages were proven to have failed in accommodating the few behavioral needs they were designed to satisfy.[12]

8.3.7 Killing Chicks in Hatcheries

About 5.5 million male chicks are killed annually in Israeli hatcheries producing female chicks for the egg industry (Egg and Poultry Board, 2014).[13] Millions more (the number is not known) are killed in hatcheries working for both the egg and meat industries because they are disabled, weak or late-hatchers. Hatcheries used to dispose of these unwanted chicks by throwing them away in garbage bins or plastic sacks. Public outcry after the exposure of these practices, led the Ministry of Agriculture issue instructions that prohibit this practice and requires the usage of killing machines approved of by the Officer in Charge of the Act. Hatcheries that were caught breaching these instructions were charged with animal abuse. As this

[11] Council Directive 1999/74/EC of 19 July 1999 laying down minimum standards for the protection of laying hens.

[12] For analysis of 'furnished' cages see for example: Compassion in World Farming (2002, 2007). The failure of 'furnished' cages was shown also by scientific research in Germany, which later banned them (private communication from Provieh).

[13] The number is higher than the number of female chicks marketed to Israeli breeders, mentioned above, because some female chicks are marketed to the Palestinian egg industry.

article is written, draft regulations on the killing of chicks are pending approval. While comprehensive regulations on the killing of all animals[14] were drafted and debated, their legislation process was never completed.

8.3.8 *Dehorning and Tail Docking in Cows*

Dehorning cows is cruel, unnecessary and therefore criminal – so ruled an English court back in the nineteenth century (*Ford v. Wiley* 23 QB 203 [1889]). Yet, it is a common practice in the milk industry. While the industry sometimes justifies the practice on the pretext of preventing danger to handlers and to other cows, it seems that the 1889 decision, rejecting such claims, still stands. Dehorning is, however, conducive in altering the cow's body structure to fit farm structures – especially stanchions – though one might think that farm structures should be fitted to the animals and not vice versa. In January 2002, a prime time TV program exposed graphic footage of dehorning – the workers restrained the cows using nose rings, cut the horns with a plier, and stopped the blood with open fire. The next day, the Director of the Veterinary Services issued instructions banning the dehorning of cows without anesthesia after the age of 10 days. The alternative used is caustic paste spread on the horn buds of the young calves – a painful method in itself. At the same time, the Director of the Veterinary Services also banned tail-docking of cattle.

8.4 Vivisection

About 300,000 animals are used in Israel annually for vivisection.[15] The largest groups are mice and rats (83 % in average in 2004–2013) and animals used in agriculture (chickens, turkeys and cattle – 9.29 % in 2013). Smaller groups of animals come from varied species from monkeys to fish. Virtually all animals are killed at the end of experiment. In 2013, 48 % of the research projects using animals were categorized as "basic science" and 43 % as "medical research". 70 % of research was categorized in the three highest degrees of suffering to the animals (out of five). This data does not include vivisection in the army and other security agencies. According to anecdotal data, experiments in these agencies include high pressure and explosion experiments in pigs, and research in infectious agents that can be used in biological weapons.

[14] That is killing not for human consumption. The act of killing of an animal for human consumption is not covered by the Animal Protection Act to avoid debate on ritual slaughter.

[15] The average number of the years 2004–2013 was 309,621 animals per year. The number for 2013 was 299,144 (The Israeli Board for Experiments on Animals 2014).

Vivisection in Israel is regulated through the Animal Welfare (Experiments on Animals) Act of 1994 (henceforth the Animals Experiments Act).

The agency overseeing the matter is the Board for Experiments on Animals. The board consists of 23 members, representing different stakeholders. The composition of the board has been criticized for creating automatic majority to vivisectors and representatives of groups with vested interest in vivisection. Nine members come from the academia and three represent unions of doctors, veterinarians and industrialists. Seven members represent governmental ministries (the Ministries of Health, Science, Education, Religions, Environment and Education) and the Veterinary Services. One is a veterinarian who heads the committee on animal experiments in the security system. Three represent animal protection organizations.

Unbalanced as it is, the Board's functions are limited. Its main functions are to create rules on the performance of vivisection and issue permits for institutions that perform vivisection. The authorization of specific experiments is handed to smaller committees. While there is a national committee for this purpose, most institutions are authorized to have internal committees which discuss experiments to be performed in the same institutions. Institutions with internal committees include universities, hospitals, research institutes and many biotechnological companies.

The material provisions of the Animals Experiments Act prohibit the conduct of an experiment if its objective may be achieved through a reasonable alternative way. The number of animals in any experiment should be kept to the minimum. The suffering of the animals should be minimized. The testing of cosmetics and household products on animals is prohibited. The Pharmacists Order generally prohibits the marketing of cosmetic products which were tested on animals after January 1, 2014. Experiments in schools (including dissections) are prohibited following a decision of the Minister of Education in 1999.

The rules made by the Board generally incorporate the "Guide for the Care and Use of Laboratory Animals" of the National Research Council (NRC) in the United Sates. The NRC guide is mostly worded as recommendations, allowing vivisectors broad discretion. The rules implement the prohibition on experiments if their objective may be achieved through a reasonable alternative way, by merely requiring the vivisector to declare that the experiment has no alternative.

It is not surprising, thus, that the legal regulation of vivisection in Israel has been the subject of harsh criticism – regarding both the legislation and the way it is applied. A number of reports by the State Comptroller harshly criticized the work of the Board for Animal Experiments and the application of the law in institutes that perform vivisection. The issue has been dealt with also through numerous cases in the courts and bills to change the law. However, major advancement in this field has not yet been achieved.

8.5 Particular Provisions in the Animal Protection Act and Regulations

Particular vs Framework Provisions The prohibition of animal abuse, which was interpreted by the court in the *Hamat Gader* case, is a "framework" prohibition, a principle rule that can be applied in any concrete set of circumstances. Besides this framework-prohibition, the Animal Protection Act includes a number of particular prohibitions.

Animal Fights All animal fights are banned, as well as coursing, making an animal attack another animal.

Mutilations Any cutting in the living tissue of an animal for cosmetic reasons is banned. This includes tail-docking and ear-clipping of dogs to make them confirm to "race" rules. It also includes the tattooing of fish and changing their colors in other ways before they are sold as ornaments. These bans are complimented by regulations banning the holding of animals that had been thus operated or painted in pet-shops and showing them in exhibitions. Israel also prohibits the import of tattooed and colored fish. Cat declawing is prohibited, with a narrow exception in case of a medical need of the cat or a documented severe health risk to the holder of the cat. Each declawing operation should be reported.

Abandoning Abandoning an animal is prohibited. This prohibition is complemented by an assumption that if an animal was found stray, she had been abandoned by her last holder – unless the person proves otherwise. An animal protection organization, a municipality and a vet who cared for an abandoned animal may collect their expenses for up to 60 days from the former owner or holder of the animal.

Working Animals The Animal Protection Act prohibits the working of an animal who cannot work because of her physical condition and working an animal to exhaustion. While these provisions are vague and deal with extreme situations, regulations give more specific rules on exploiting animals to draw carts, carry burden or be ridden. Sick animals, young ones (up to 3 years) and mothers whose young ones are younger than 30 days may not be used. Afterwards, they may be used for a limited time each day. A mother whose sibling accompanies her may only be used for uncommercial riding. To commercially use an animal for drawing carts, carry burden or be ridden, the animal should be implanted with a microchip and the user should have an up-to-date veterinary certificate declaring the animal fit for the work demanded from her. The weight that the animal is made to carry should be within her physical capability; the total weight when drawing a cart should not be more than 1000 kg per horse or 500 kg per donkey. Appropriate breaks should be given – no less than half an hour every 3 h. Specifications on cart structure are given. The Traffic Regulations ban animal drawn carts from urban areas. An exception may be made by the municipality, under some conditions, for carts used for touristic and leisure purposes.

Regulations on Holding Conditions A major set of regulations deals with the holding conditions of animals in any framework except for agriculture and vivisection. These regulations apply in varied circumstances – from animals who live in people's homes, to municipal pounds, shelters, pet-shops, breeding facilities, zoos etc. An important statement is included in the regulations' objective clause:

> The objective of these regulations is to advance the welfare of animals, since they possess physical, health, behavioral, mental and social needs; for the satisfaction of such needs, animals held by humans need appropriate conditions and environmental enrichment; Since the needs of animals, as well as the possibility to satisfy them, are dependent on the circumstances, these regulation stipulate minimal conditions for holding them.

This declaration of principle compensates for the impossibility to define in advance what may be considered as "appropriate conditions and treatment" in so many varied and complex situations. However, the regulations do go some way applying their declared objective.

Physical Needs The regulations include provisions on food, water, ventilation, climatic conditions, cleanliness and prevention of harm. For animals living in water, water quality should be appropriate to the animal's needs, including its temperature and oxygen and mineral concentrations. The provision on light includes requirements to a number of darkness hours appropriate to the animal's nature, and in relevant cases sunlight or ultra-violet light – for some animals a physical need, for other a sensory need.

Health Health needs are served by preclusion of health hazards and mandatory treatment of any sick or suffering animal.

Behavioral Needs A pitiful number of behavioral needs are protected to some extent. Animals that by nature perform one of a number of listed behaviors, should have means that facilitate this behavior. These behaviors are hiding in a shelter, digging, climbing, perching and staying in water, on land or both. Animals may not be constantly tethered. Dogs in shelters, pet-shops and breeding facilities should also be released to a yard or taken for walks on a regular basis. Cats are entitled to scratchers and a litter box.

Enclosure/Cage Sizes The regulations set minimum sizes for enclosures and cages. While designed to be the minimum that one cannot go beneath even in temporary or difficult circumstances, practice proves that they are insufficient even in such cases. The absolute minimum water-volume in an aquarium is 1 l, and the minimum net area of a terrestrial enclosure is 600 cm^2. While the size of an enclosure may not be less than these values, it should exceed them according to calculations based on the sizes of the animals held in it. If there are only one or two animals in the enclosure, it should be at least 2X long and 1.5X wide, if X is the length of the longest animal (all animal lengths are measured without the tail). For any additional animal $1.5X^2$ of area should be added, based on that animal's length. The height of the enclosure should be twice the height of the tallest animal in it, or one time the length of the longest one – according to the larger value. All these values refer to net area, not used for equipment. Smaller values apply to snakes, and smaller height values to horses. Specific sets of measures relate to dogs and cats.

The calculation for the minimum size of bird cages is a little different. For cages of up to three birds the minimum length and width is 3X, if X is a bird's length without the tail, calculated based on the longest bird. For any additional bird $3X^2$ of net area should be added (X being the length of the additional bird). The height of the cage should be at least 5X, calculated based on the longest bird. The cage should allow the largest bird stretch her wings.

Freedom from Fear In terms of mental needs, the regulations principally deal with the issue of fear. A series of provisions attempt to mitigate fear: a prohibition to hold an animal in cage exposed to the public from all angels; mandatory provision of resources in number and manner that prevent competition and a number of rules that aim to prevent harassment.

Early Weaning and Imprinting Social needs are almost totally ignored. Only two provisions protect such needs. One prohibits the separation of a mammal from her mother before she reaches weaning age (except for reasons of the protection or welfare of the mother or offspring). The second prohibits growing a bird, in the first 4 days after hatching, not in the company of adult birds of the same species (the same exception applies). The first provision aims to prevent the suffering and long-term harms of early weaning. The second provision aims to prevent the social disability and other harms that come with imprinting on humans.

Animal Holding Facilities Additional rules apply to pet-shops, breeding facilities, shelters, municipal pounds, petting zoos, zoos, pensions and commercial riding sites. These rules include provisions on capability, sufficiency and training of staff; stringent rules on veterinary care and disease prevention; control of visitors and prevention of harassment; daily checks; documentation; registration with the authorities and periodic reports to them. Specific provisions limit the exposure of animals to humans in petting zoos. Provisions on the selling or giving away of animals seek that the buyer or adopter will receive the needed information on the animal and basic written guidance regarding her treatment.

8.6 Wild Animals

While the Animal Protection Act protects every vertebrate animal, wild animals are also the subject of the protection of the Wild Animal Protection Act of 1955. One of the Act's objectives is to protect nature, which may imply treating individual animals as mere specimens of their species. However, it has been decided by the Israeli Supreme Court that the Act also aims to protect individual animals as such.[16]

---

[16] *Rishon LeZion Municipal Company for Culture and Sport vs. The Nature and Parks Authority* (2009).

The Wild Animal Protection Act defines "wild animal" as "mammal, bird, reptile, amphibian or any part of the above, or any derivative of the above, originating in the State's territory or outside the State's territory, whose nature is not to live in the company of a human." Accordingly, the act applies to wild animals both in nature and in captivity, including those bred in captivity and those who, by their individual history, are accustomed to humans. It also applies to objects such as eggs and fur of wild animals, ivory, etc.

Hunting Hunting is subjected to a licensing regime (except for the hunting of animals from a limited number of species which were declared pest). Some species of birds may be hunted as "sport" by people granted a hunting license. The hunting of animals from other species is permitted on the pretext of prevention of damage to agriculture, maintaining ecological equilibrium, prevention of danger or disease to humans and to animals and for breeding or scientific objectives. Prominent victims of legal hunting include boars, jackals, porcupines, some species of crows, rose-ringed parakeets and more. Prominent victims of illegal hunting include again boars, porcupines, and also gazelles and passerines (for trading).

Hunting methods are limited. While some limitations (such as the prohibition of explosives and poison and rules regarding the usage of guns) are aimed to ensure the safety of people or to prevent collateral damage to the environment, some limitations protect animals from cruel or "unfair" practices. These limitations include prohibitions on the use of traps, nets and glue; on dazzling animals; on chasing animals and hunting them with a motorized vehicle; and on setting dogs on animals (using dogs for retrieving animals that have been hunted is allowed, and using them to locate boars may be permitted by the Nature and Parks Authority on a case-to-case basis).

Trade and Possession Trading in wild animals and holding them is subjected to a permit regime. The Park and Nature Authority uses its powers regarding captive animals to implement the Convention on International Trade in Endangered Species of Wild Fauna and Flora (CITES) and its nature protection policies. Its powers regarding pet-shops and zoos are used (sometimes) to enforce improvements in holding conditions. However, the dramatic applications of the Wild Animal Protection Act on captive animals are in the fields of circuses and vivisection.

Animal Shows The policy of the Park and Nature Authority is not to allow the use of wild animals in circuses. This policy is anchored in the Israeli stricter domestic measures concerning import and export of wild fauna and flora, as reported to the parties to the CITES. While in the first half of the 1990s circuses with wild animals were regular visitors in Israel, now they are forgotten. Animals who are not defined as wild are also virtually non-existent in circuses: horses have not been seen in a circus in Israel for years, and when a Russian cat "theater" made it to the country in 2011, the protests and legal proceedings it faced were such that it did not come back. The disappearance of non-wild animals from circuses is legally based on regulations under the Animal Protection Act that stipulate that any use of animals in a show, exhibition or competition (including a TV show, a movie or an advertisement) requires a permit from the Officer in Charge of the Act. The Officer in Charge may grant a permit only if persuaded that the animals will not suffer during the show, in the preparation to it or in the way they are held. This is also an implication of the *Hamat Gader* case.

Exporting Monkeys for Vivisection Another application of the Wild Animal Protection Act is the ban on exporting commercially bred primates for vivisection. Starting from January 2015, such monkeys may be exported only for purposes of nature preservation, rehabilitation or education and exhibition. This lead to the closing of a breeding facility for macaque monkeys in the village Mazor which had been in the focus of protest by animal rights organizations since its establishment in the early 1990s.

8.7 Feral Cats

The climate in Israel allows for a large population of feral cats. They flourish on the plenty of food thrown or fed to them by humans. Many feeders also care for veterinary treatment when needed. Cat life in the yards and streets of Israeli cities can be satisfactory – but can also be dangerous and miserable. Overpopulation is a serious problem of feral cats in Israel.

In the past, municipal authorities used to deal with the "problem" of feral cats by mass killings – that just invited new populations of cats, new mass killings and so on. Two important decisions of the Supreme Court[17] put an end to this. The decisions distinguish between different risks raised by cat populations. In cases of a real and concrete risk that the cat population will transmit dangerous disease (such as rabies) to humans – killing them, following an orderly procedure, may be allowed as a last step. The procedure should ensure that killing will only be done to a small extent, under clear limitations of time, place and circumstances, and only when the safety of humans cannot be protected by other reasonably-costing means. In cases where the harm is not to human health but to human convenience and well-being, killing of cats or causing them significant suffering is forbidden.

A related issue is attempts by municipalities to restrict the feeding of cats in public areas. In Jerusalem it was banned by law. The Jerusalem Court for Municipal Issues decided that such a ban could be deemed legal only if interpreted as applying only to feeding that causes nuisance or sanitary risk.[18] This interpretation is now accepted nationwide.

---

[17] *The Society for the Cat vs Arad Municipality* (1998) and *Let the Animals Live vs The Director of the Vetertinary Services in the Field* (2004).

[18] *The State of Israel vs Yurovsky* (2001).

8.8 Legal Personhood and Standing

The above description of different provisions of Israeli law is enough to conclude that interests of non-human animals are recognized under Israeli law, that non-human animals are protected for their own sake, and that their interests may sometimes override significant human interests. While they can be held as property, the interests of non-humans put restrictions on the property rights in them. Nevertheless, it is also clear that while they carry some weight, the most basic interests of animals are regularly and legally sacrificed for minor human interests.

This means that animals are not merely property. As an Israeli court put it in *Doe vs Doe* (2004):

> In correlation with the moral obligation of us as humans not to abuse animals, not to hunt them in prohibited and painful ways, not to unnecessarily hurt them etc, there exists the right of animals not to be treated these ways. True, this right of the animals is relative and not absolute, as they may be deprived of the most basic right granted to a living creature – the right to life – provided that it is done for an appropriate objective and in a way permitted by law while minimizing the pain and suffering caused. However, these relative rights, granted by the above mentioned legislation, clarify that the Israeli positive law does not treat animals *in all situations* as objects, but treats them in certain issues as living creatures, with a soul, and granted them protection for their own sake.

The case dealt with a dispute between two ex-spouses on the possession of a dog and a cat that they had adopted when they had been together. While the parties' lawyers based their arguments on property law, the court preferred to decide the case based on the best interests of the animals. The Court nominated a specialist that checked not only the objective conditions in the two places, but (like with older children) also on the preference of the animals, or, to be exact – of the dog. The dog was more enthusiastic when meeting the woman.

An important provision in the Animal Protection Act allows animal protection organizations, approved for this purpose by the Minister of the Environment, seek court injunctions against acts violating the Animal Protection Act. The *Hamat Gader* case was brought under this clause. Another clause allows also for private criminal indictment, but has never been used. Supreme Court Justice Grunis remarks in the force feeding case that "those two alternative ways, i.e. private criminal indictment and a request for an injunction, that allow pro-animal organizations to represent their interests, are a solution to a problem sometimes encountered in other countries when it is argued that animals have no standing in court (see Sunstein 2000)."

When it comes to administrative law, Israel has a liberal doctrine of standing that opens a wide door to public interest petitioners, including animal protection organizations. However, when a group of petitioners put "the Mountain Gazelle" as a sixth petitioner (they actually referred to thirty individual gazelles), the court was reluctant to accept it. In a preliminary decision in *Reise vs The National Board for Planning and Zoning* (2005), the court remarks that petitions for protection of animals can be filed by animal protection organizations, but not by the animals themselves: "If you wish, the animals are provided with guardians that deal with their interests and seek their good, but they are not themselves a 'party' in the legal meaning."

The standing of animals got a precedential upgrade in *Let the Animals Live vs Israel Institute of Technology* (2012), when a District Court decided to add animal protection organizations to a pending civil procedure between private parties. The case was brought by Israeli universities against El-Al Airlines, seeking an injunction that will obligate the company to ship animals destined for vivisection for these universities. El-Al had adopted a policy of not shipping animals for vivisection. Animal protection organizations asked to join the procedure. It is irregular that the court intervenes in a plaintiff's decision as to whom to choose as defendant, but an exception is when the ruling will affect a legal right of a third party. In adding the organizations as defendants, the court reasoned that animals hold protected rights, and that the ruling might affect the legal rights of animals that the universities wished to ship. Under an alternative test, the court stipulated that the ruling might affect legally protected interests of the animals that were not appropriately represented by any of the existing parties. Whatever the test, the voice of the animals should be heard in the court-room for the court to make a full and efficient ruling. This voice is brought by animal protection organizations.[19]

8.9 The Israeli Experience

Much of the legal discourse about animal law sees the key to animal exploitation in their status as property and the key to liberation in granting them legal personhood. Thus, many legal efforts are aimed at presenting courts with legal test-cases that challenge the courts to directly recognize the personhood of animals.

Israeli law may serve as "laboratory experiment" to question this approach. As I have shown, when presented with a direct challenge to recognize mountain gazelles as a party to a procedure, the court felt reluctant. It was not only conservativist reaction. The approach of the petitioners in that case raised questions such as: What makes a human or a corporation a legitimate representative of an animal? How would a non-human party technically act within a procedure? However, when the legislator and the courts were faced with the pragmatic need to account for recognized interests of animals, they found tools that de-facto granted animals personhood and standing.

The Israeli experience also reminds us that like virtually all human rights, nonhuman animal rights, even when recognized, are relative and do not act like trump cards.[20] The extent of these rights in the legal sphere derives from the general

---

[19] The case itself was closed later without a ruling, after the plaintiffs withdrew their claim. In another case, brought by the Mazor breeding facility for monkeys destined to vivisection, the Supreme Court upheld El-Al's policy not to ship animals destined for vivisection.

[20] E.g. even the human right to life is not absolute. Life may be legally taken in war, and its protection by traffic laws, torts and the like is not absolute.

attitude and the materialistic relations within society. While standing-rights are important tools, the challenge is to expand these rights and make them weigh more in social and legal deliberations. Animal protection organizations that work both in the public sphere and in the legal sphere are thus the best agents in the legal evolution towards animal liberation.

References

Aharoni, Efrat. 2014. Survey: A quarter of Israelis reduced meat consumption. 10% became vegetarian. *Globes*, 2 January 2014 (In Hebrew).

Compassion in World Farming. 2007. Alternatives to the barren battery cage for the housing of laying hens in the European Union. https://www.ciwf.org.uk/media/3818829/alternatives-to-the-barren-battery-cage-in-the-eu.pdf. Accessed 11.12.2015

Compassion in World Farming Trust. 2002. Laid Bare – The case against enriched cages in Europe. https://www.ciwf.org.uk/includes/documents/cm_docs/2008/l/laid_bare_2002.pdf. Accessed 11.12.2015.

E.U. Scientific Committee on Animal Health and Welfare. 1998. Welfare aspects of the production of Foie Gras in Ducks and Geese. http://ec.europa.eu/food/fs/sc/scah/out17_en.pdf. Accessed 11.12.2015.

Francione, Garry L. 1995. *Animals, property and the law*. Philadelphia: Temple University Press.

Israeli Central Bauru of Statistics. 2014. http://www.cbs.gov.il. Accessed 13 Sept 2014.

Israeli Milk Council. 2013. *Annual report for 2012*. (In Hebrew).

Israeli Veterinary Services. 2013. *Annual report for 2012*. (In Hebrew).

Israeli Veterinary Services. 2014. http://www.vetserv.moag.gov.il/NR/rdonlyres/F3F80D77-8B72-4737-AB6B-D547EC72EB2B/0/import_12_2013.pdf. Accessed 13 Sept 2014.

Kligler, I.J., A. Geiger, S. Brombeg, and D. Gurevitch. 1931. An inquiry into the diets of various sections of the urban and rural populations of Palestine. *Bulletin of the Palestinian Economic Society* 3: 5–58.

Sunstein, Cass R. 2000. Standing for animals. *UCLA L.Rev* 47:1333.

The Israeli Board for Experiments on Animals. 2014. http://www.health.gov.il/Services/Committee/animax/Pages/default.aspx. Accessed 19 Sept 2014.

The Israeli Egg and Poultry Board. 2014. http://www.ofotm.org.il. Accessed 13 Sept 2014. (In Hebrew).

Vickers, W.J. 1944. *A nutritional economic survey of wartime Palestine 1942-1943*. Palestine: The Government Printer.

Wolfson, David J. 1996. Beyond the law: Agribusiness and the systemic abuse of animals. *Animal Law* 2: 123–154.

Wolfson, Yossi. 2012. Reporting from the field: The Israel adoption of transportation regulations for poultry. *Journal of Animal & Natural Resource Law* 8: 171–177.

Israeli Cases Cited

CA 4398/06 *Rishon LeZion Municipal Company for Culture and Sport vs. The Nature and Parks Authority* [2009] (Available only in Hebrew).

Civil Case Central District 54789-12-11 *Let the Animals Live vs Israel Institute of Technology* [2012]. Translation available at https://www.animallaw.info/

Criminal Case Jerusalem 897/01 *The State of Israel vs Yurovsky* [2001] (Available only in Hebrew).

Family Case Tel-Aviv 32405/01 *Doe vs Doe* [2004]. Available only in Hebrew.

HCJ 466/05 *Reise vs The National Board for Planning and Zoning* [2005]. Available only in Hebrew.

HCJ 4884/00 *Let the Animals Live vs The Director of the Veterinary Services in the Field* [2004] (Available only in Hebrew).

HCJ 6446/96 *The Society for the Cat vs Arad Municipality* [1998] (Available only in Hebrew).

HCJ 9232/01 *'Noah' – Israeli Federation of Animal Protection Organizations v. The Attorney-General* [2003]. Two different English translations exist: one in the Israeli Supreme Court website court.gov.il and one on CHAI's website http://www.chai.org.il/en/compassion/foie-gras/foiegras.pdf (Both accessed 11 Aug 2014).

LCA 1684/96 *Let the Animals Live v. Hamat Gader Recreation Enterprises*. 1997. An English translation of the decision is available in the Israeli Supreme Court's website court.gov.il. accessed 11 August 2014. While citations in this article are based on this translation, I did edit it when found to divert from the original text of the decision.

Chapter 9
The Constitutional Defense of Animals in Brazil

Tagore Trajano de Almeida Silva

Abstract This chapter offers a framework of animal law and its development in Brazil. The purpose is to explain the trajectory of animals from objects to subjects in the Brazilian Constitution and the response of the Brazilian judiciary, legislative, and executive branches to the matter of animal law. The chapter considers the contributions of animal and human rights activists as well as the evolution of the legal protection of animals in Brazil. Furthermore, the chapter explores the effects of the article of the Brazilian Constitution that both prohibits cruelty to animals and requires governmental protection for fauna and animal species. Thus, this chapter shows that the discussion on animal welfare law offers insight into how the Brazilian legal system functions and how it affords an opportunity to build in the interest of animals.

9.1 Introduction

This chapter discusses the debates and development of animal legal protection in Brazil and the legislation that evolved over the years to recognize animals as subjects of law. Some questions will be important to consider in order to understand the Brazilian system: How does the legal system deal with issues related to the recognition of rights for animals? Does the Brazilian legal system recognize the rights of individuals to animals? What does the Brazilian Constitution say about this matter? And what direction have the new laws taken in the animal law debate?

The initial answer is provided throughout this work through the elucidation of the Brazilian legal system and the laws that involve the interests of non-human animals. Over the centuries, humans built up a wall to artificially divide humans from other species. This myth, constructed and reinforced by legal doctrine, created impediments for legal systems to recognize the constitutional rights for the non-human (Wise 1999, p. 61). However, subjective and constitutional rights give individuals a defense against violations perpetrated by other individuals or by the

T.T. de Almeida Silva (✉)
Tiradentes University, Aracaju, Sergipe, Brazil
e-mail: tagoretrajano@gmail.com

D. Cao, S. White (eds.), *Animal Law and Welfare - International Perspectives*,
Ius Gentium: Comparative Perspectives on Law and Justice 53,
DOI 10.1007/978-3-319-26818-7_9

state itself, and these constitutional rights can be extended to the protection of animals (Tribe 2001, p. 2).

The Brazilian legal system has shown it is possible to protect animals as individuals and as a collective, bringing to legislation the non-human interests. This paper examines the trajectory of legislation geared toward recognizing animals as subjects of law. The conclusion is that Brazil has already developed a Constitution and some legislation that identify animal interests. However, the economic, social, and judiciary policies have not, as yet, been able to effect these legislative provisions(Silva 2014c, p. 512).

Consequently, this chapter first discusses the progress of the Brazilian legal system as it evolved from equating animals as things (*res nullius*) to recognizing fauna as deserving protection as sentient beings. It then summarily explores the role of animal rights associations to support and expand animal protection in Brazil, followed by an examination of anticruelty laws in Brazil and their influence on the way animals are regarded in Brazilian society at large as well as the way in which sensitivity to animal cruelty has led to animal rights. Finally, this article proposes a framework for legal professionals who work in this field and are interested in understanding the Brazilian animal legal system.

9.2 *Res nullius:* Animals as State Property

The 1988 Brazilian Constitution carried out a gradual change in animal law concerns, advancing individual rights to these beings by creating legal, constitutional protection. The Constitution states:

> Article 225 – All people have the right to an ecologically balanced environment which is an asset of common use and essential to a healthy quality of life, and both the Government and the community shall have the duty to defend and preserve it for present and future generations.
>
> Paragraph 1 – In order to ensure the effectiveness of this right, the Government shall:
>
> VII – protect the fauna and the flora, with prohibition, in the manner prescribed by law, of all practices which represent a risk to their ecological function, cause the extinction of a species, or subject an animal to cruelty.

Environmental legislation protecting wildlife and endangered species represents the beginning of the establishment of a legal framework for the treatment of nonhumans. These statutes on animal interests recognize that animals are part of a group on a continuing biological (and ecological) existence and form a collective defense under the name group species (Favre 2000, p. 485). Brazil has signed environmental treaties that reflect a concern for animals, but the ecological resource of animals can now be protected under the label of fauna.

This purpose was promoted by the signing of three conventions by Brazil:

1. the Convention on Nature Protection and Wild Life Preservation in the Western Hemisphere, ratified by Legislative Decree No. 3 of 1948, in force in Brazil since

26 November 1965. It was promulgated by Decree No. 58,054, of 23 March 1966, and establishes, through its Article VII, that countries adopt appropriate measures "to prevent the extinction that threatens a species", thereby buttressing Article IX, which provides that each country shall take the necessary measures for the oversight and regulation of imports, exports, and transit of protected species of flora and fauna (Cassuto & Reed 2011. p. 3);
2. the Convention on International Trade in Endangered Species of Wild Fauna and Flora (CITES), of which Brazil is a signatory, ratified through Decree Law No 54/75 and promulgated by Decree No. 76,623, November 1975, provides protection for a range of plants and animals by regulating, monitoring, and preventing their international trade, particularly those threatened with extinction, as well as preventing them from reaching unsustainable levels; and finally,
3. the Convention on Biological Diversity (CBD), ratified by Brazil through Legislative Decree No. 2 on 8 February 1994. In its Article 8 (f), the CBD states that countries should "recover and restore degraded ecosystems and promote the recovery of threatened species through the development and implementation of plans or other management strategies."

The Brazilian Law on Protection of Fauna, Law N° 5197, was enacted on 3 January 1967, and established in Article 1 that "animals of any species at any stage of their development and naturally living out of captivity, constituting the wildlife and their nests, shelters and natural breeding are the property of the State, its use, persecution, destruction, hunting or harvesting is prohibited." This law changed the legal status of wild animals, refusing to consider them as *res nullius*, and therefore allowing wild animals, from that moment, to become state property (Benjamin 2001, p. 150).

These statutes use the system to discourage criminal behavior and to adopt recovery plans as an incentive for the protection of animals. The laws aim to protect species and guarantee biological diversity while preserving the habitat, fostering a supportive attitude of protection by public authorities responsible for the recovery of endangered species (Favre 1997–1998, p. 365).

Two critical aspects that need to be considered are the economic and social (Duckler 2002, p. 200). Both influence the legal protection of fauna in Brazil, while scientific and biological discoveries complement the legal provisions in the regulatory text.

The concept of fauna under Brazilian law has been influenced by private law (Civil Code of 1916), distinguishing between domesticated animals and wild animals. Domesticated animals had economic value as personal property whereas wild animals had no such consideration (Huss 2002–2003. p. 69). This scenario, however, could change automatically when a person appropriated or possessed an animal and made use of it. According to Brazilian legislation, a human obtained ownership of a wild animal at the moment that the animal was tamed or simply confined. The animal was still regarded as something to be used and possibly even destroyed at the pleasure of any individual possessing ownership.

Indeed, animals in Brazil that have social and economic valuation were under private ownership because of their benefit as goods (Mellon 2012). In contrast, wild animals receiving protection of the State would be ensured this protection by being considered living beings that were essential environmental resources of wildlife (Dias 2000, p. 93).

In this context, the protection of endangered wild species represented direct, indirect, aesthetic, and moral benefits to the state itself, which could, therefore, direct its resources toward a particular group within a variety of beings (Favre 1997–1998. p. 344). The state could focus its budget on preservation of genetic and biological diversity by allocating incentives to industry and to scientific research, thereby contributing to the conservation of wild species (Giménez-Candela 2008, p. 26). This positive picture generated an ethical valuation directed to citizens and the state itself, in that both had responsibilities to other species of life on Earth. These obligations were distributed through three points: (1) environmental resources must be preserved for present and future generations, (2) animal cruelty is defined and prohibited by law, and (3) all species have a right to exist thrive (Rosiers 1990–1991, pp. 828–834).

Environmental laws protecting wildlife warned of the importance of balance between humans and other species of interest, establishing the need for limits to human action. A different picture began to be drawn in Brazil. The legal construct had to consider a balance of political, regulatory, and social interests while identifying groups of non-humans deserving of protection (Silva 2010, p. 85).

The challenge now is to develop a legal status and protection of animals (Wise 1996, pp. 525–530). This would necessitate certain considerations, such as: (1) the political perspective, which would define the protection directed toward non-human animals, protecting them in the legal field and serving as a guide for decision making by their operators; (2) the normative aspect, which would balance various interest groups, functioning as a frame of conduct to human actions; and (3) the social facet, which would link the political to the rule of law and potential regulation (Favre 2005, p. 342).

A relevant body of law was built around the protection of the non-human as environmental goods without any individual consideration of these beings. Although an analysis persists in the doctrine of non-humans through the ownership status, the process of post-humanization has helped advance the consideration of subjective rights to animals (Silva 2013a, p. 162). The recognition of the individuality of each non-human animal has recently occurred with the enactment of anti-cruelty laws and the decisive roles of animal protection societies in this debate.

9.3 Contributions of Animal and Human Rights Activists

The debate on the recognition of animal interests found fertile ground in England in the eighteenth and nineteenth centuries, influenced by Eastern culture's respect for animals during the period of British colonization in India. Because of the rapid development of agriculture, industry, and science, Britain was at a propitious stage for the construction of law on animal protection (Kean 1998, pp. 33–34). The subject was taken up by Richard Martin in 1822 with the passing of the anti-cruelty law in England: The Cruel Treatment of Cattle Act, known as the Martin's Act. Richard Martin was a representative in the House of Commons from 1801 to 1826 and actively supported the emancipation of Catholics and the defense of humanitarian causes such as the abolition of the death penalty as well as the establishment of legal assistance to the needy and the recognition of non-human animal interests (Favre and Tsang 1993, p. 4). In 1824 England, the Society for the Prevention of Cruelty to Animals (SPCA) was founded, contributing to a worldwide improvement in debates on animal protection. The work of the SPCA inspired the creation and expansion of animal protection societies in several countries such as Australia, New Zealand, the U.S.A., Ireland, Germany, Belgium, the Netherlands, Argentina, and Brazil (Ryder 1989, p. 79. See also White, Chap. 6 in this volume).

In North America, Henry Bergh was responsible for the creation of new laws to protect animals. After a visit to England, where he was able to see the work of the Royal Society for the Prevention of Cruelty to Animals (RSPCA), Bergh returned to the United States and founded, on April 10 1866, the American Society for the Prevention of Cruelty to Animals (ASPCA). (Ryder 1989, p. 80). The ASPCA collaborated on the development of legislation to prevent the suffering and abuse of animals. Before the existence of the ASPCA, the cause and its defenders endured much opposition. The establishment of the organization advanced more discussion of the issue and influenced the creation of anti-cruelty law in New York, which then served as the model for several U.S. states (Silva 2013c, p. 61).

The development of Latin American legislation on protection of animals was due primarily to a lawyer Lucas Ignacio Albarracín. He founded the Argentine Society for the Protection of Animals (*La Sociedad Argentina Protectora de los Animales*) in 1881, served as its president, and championed campaigns against cockfights, sport hunting, and bullfighting. That association supported the enacting of the National Law for the Protection of Animals Argentina No 2786 on July 25 1891, with the goal of prohibiting cruelty to animals in Argentina (Santana and Oliveira 2006, p. 13). The debate in Brazil started soon after, on May 30, 1895. A Swiss citizen called Henri Ruegger, with the collaboration of Furtado Filho, a reporter for People's Daily (*Diário Popular*), defended the need for the creation of a Brazilian association against animal cruelty: The International Union for Protection of Animals – UIPA (*União Internacional de Proteção Animal*) was founded in 1895 by Senator Ignácio Wallace Cochrane, a Brazilian of English descent, in order to protect non-humans against abuse and mistreatment (Silva 2012a, pp. 43–45). The creation of UIPA helped form a national animal protection movement that would aid in the preparation and editing of protective laws such as: (1) Decree 16.560/1924, which prohibited the use of bullfights, roosters, and canaries in public places of entertainment; and (2) Decree 24.645/1934, which is considered the Brazilian anti-cruelty legislation that established the measures for the protection of animals (Cochrane 1895).

Only in 1934 could Brazilian law absorb elements of foreign laws, advancing the criminal, administrative, technical, and procedural areas of the legal consideration of animals. Enacted after the establishment of UIPA, two decrees provide that animals will be protected by the State, with 31 designations of mistreatment and abuse of animals (Levai 1998, pp. 27–28). Decree 24.645/1934 was the beginning of consideration of individual interests of non-human animals, that they deserved legal valuation, welfare, and quality of life, and that their legal interests could be represented in court by a public prosecutor as well as animal associations (Benjamin 2001, p. 150).

A parallel can be established in the evolution of the human rights process, and a look back into history indicates that the status of lands, chattels, ideas, slaves - indeed, even the question of who could own or inherit that property (individuals or married women, for example) - was undergoing a change of legal status, with each change of the legal concept stimulating a change in the consciousness of society (Silverstein 1996, pp. 162–164).

According to Peter Singer (2004, p. 252), the argument that there is a boundary between human and non-human was used for many years as an excuse not to do anything with respect to both human and non-human animals. However, there is no incompatibility between human and non-human interests (Wise 1999, p. 61); both try to make life on Earth better for everyone.

In fact, activists such as Henry Bergh were responsible for the advancement of public policies in favor of children. Sometimes in his defense of the child, Bergh used the standards of animal protection established in New York to support his pro-children argument. In the same way, Marie Françoise Martin and Mary Wollstonecraft, both well-known feminists, advocated for animal welfare. In 1883 Marie Martin created a society for the protection of abandoned animals called the French Anti-Vivisection Society, and Mary Wollstonecraft engaged in debate with Thomas Taylor on the occasion of the publication of her book: *A Vindication of the Rights of Woman* (Singer 2004, pp. 252–254). The movement created by these animal protection societies was responsible for the emergence of law in defense of animals as well as humans. These societies have brought pressure to bear on the law to consider individually all animals, both human and non-human.

9.4 Evolution of Legal Protection of Animals in Brazil

Most importantly, animal protection societies contributed to the development of animal law in Brazil by (1) individualizing protection for animals, meaning that animals need to be considered without any human constraints because they exist as individual species, and by (2) helping to develop a range of Brazilian legislation in the early 1900s in support of non-human animals (Silva 2014a, p. 154).

Before the work of the associations, a few laws did exist that offered some protection of animals in certain cases. For example, at that time some regional laws existed that criminalized intentional mistreatment of animals that had financial value, a vestige of the intrinsic consideration of the interests of non-humans (Lourenço 2008, p. 491).

In Brazil, one of the first regulations to protect animals was the Code of Postures in São Paulo, promulgated on October 6, 1886. That law said in Article 220 that: *"coachmen are forbidden to mistreat animals and barbaric excessive punishments shall be fined."* The law makes clear the commercial importance of animals to the economy of larger cities such as São Paulo.

In the early 1900s, popular support started pushing the legislative to make new laws to (1) expand the number of protected animals, (2) propose a differential treatment for domestic animals, and (3) guarantee to animal cruelty societies and public prosecutors the right to a means of protection against any violation of those laws (Silva 2012a, p. 85). Some of these elements can be found in the 1867 New York Law, in which Decree No. 24.645/1934 enabled these interests to defend animals against cruelty in Brazil. For instance, "excessive work" and "unnecessary killing" as well as the "abandonment of sick animals" and "constraint to rest and movement of non-human" were considered cruelty to animals, and violators would be punished with severe penalties. Prosecutors and members of animal protection societies were given the responsibility to represent these animal cases in court.

In 1941 another law was enacted to defend animals. In Article 64, the Misdemeanors Law (Decree Law 3688) moved to ban cruelty to animals and the overworking of animals:

> Article 64 – Every person who shall exercise cruelty to animals or submit them to excessive work shall be punished by imprisonment of 10 (ten) days, but not to exceed one (1) month, or fined.
>
> Paragraph 1 – Penalties will be exacted for administering pain and suffering, even in the name of science and research, where there are alternative resources available.
>
> Paragraph 2 – Penalties shall be increased by 50 %, if the animal is subjected to excessive labor or mistreated public entertainment.

Notably, the effects of the law were far reaching, covering animal treatment not only in entertainment but also in educational and scientific research. At the same time penalties were greatly increased in cases in which violations of the law were committed to provide public spectacles, or entertainment. Moreover, such anti-cruelty laws were not limited to the criminal law sphere. Identical regulations to protect animals were enacted in administrative, civil and procedural laws. This kind of hybrid model of legislation to traverse on a specific theme with various forms of liability was again used by the Environmental Crimes Law in 1998 (Rodrigues 2008, p. 70).

Indeed, among these perspective new acts about to be passed by the legislative branch, the most important were Environmental Criminal Law (Act 9605/1998), and Laboratory Animals Act (Act 11,794/2008), both of which penalize animal cruelty. The Environmental Criminal Act (1998) is a federal law that defines environmental felonies and misdemeanors (Sirvinskas 2005, p. 60). Although not a broad anticruelty regulation, the law focuses on various environmental debates as well as cruelty law (Silva 2010, p. 85). Ten years after the new Constitution, the

Environmental Criminal Law Act was enacted, the major article of which addressed the animal law debate:

> Article 32 – Every person who shall abuse, mistreat, maim or injure wild animals, domestic or domesticated, native or exotic… shall be punished by imprisonment of 3 (three) months not to exceed one (1) year and fined.
>
> Paragraph 1 – Shall be punished with the same penalties as those who exact pain or cruelty to living animals, even for educational or scientific purposes, where there are alternative resources.
>
> Paragraph 2 – Penalties shall be increased by 1/6 (one sixth) to 1/3 (one third), if death occurs in animals.

The Environmental Crimes Law sought to unify the environmental offenses and administrative violations in a single document, unifying the crimes against animals. In turn, Article 32 of the Environmental Law alternatively criminalized the conduct of performing the act of abuse, mistreatment, injuring or mutilation of wild, domestic, or domesticated animals. This clause forbidding conduct detrimental to animals had as its purpose the regulation of the Constitutional clause of cruelty that prohibits cruelty to animals (Custódio 1997, p. 59). "Cruelty" - *crudelitas* in Latin, *crudus* originating in *cruor* (blood live) - is used with the meaning of practicing intentional, ruthless, and insensitive conduct against animals (Silva and Baretta 2007, p. 320).

In this way the Supreme Court has understood that spectacles that cause suffering, injury and death of the animal shall be prohibited in the Brazilian legal system by characterizing these behaviors as acts of cruelty, such as certain practices in the festival of the oxen. Similarly, cockfights have been, since 1957, consistently configured as practices of animal cruelty by submitting the animals as objects, instead of as aspects of a sport (Cadavez 2008, pp. 88–120).

The 1988 Constitution imposes a ban on direct practices to prevent inappropriate behavior against all kind of animals. Fábio Feldmann, Brazilian constituent legislator responsible for writing the chapter on the environment in the 1988 Constitution, asserts that the will of the new Constitution was to deter the manifestations of animal cruelty, such as in the Festival of the Oxen, and to prohibit such unworthy and cruel practices against animals. According to Feldman, the 1988 Constitution provided a means for the interpreters to disallow any kind of cruelty to animals in Brazil (Feldmann 2013).

In this way, it is possible to say that the development of animal law in Brazil evolved from that of: (1) protecting property, by giving animals a legal *status* as objects, in which it would be the duty of the owner or keeper to care for them; to (2 implementing anti-cruelty laws and the introducing a new perspective in the Brazilian society by encouraging an obligation of compassion from people toward animals, making it a duty to protect animals from any cruel practices; (3). advancing the standards of animal law to understand that those who abuse animals are more susceptible to doing harm to humans than those who do not, therefore, a practice that need to be discouraged; and 4. providing the Constitutional interpretation of animal dignity that prohibits the practice of cruelty to all animals: humans and non humans (Chiesa 2008–2009, p. 15).

It goes without saying that non-human animals have a history of being victims harmed by unlawful human conduct. The 1988 Constitution gave a new status to animals, as subjects protected by law.

9.5 Animal as Subject in 1988 Constitution of Brazil

The 1988 Brazilian Constitution was responsible for allowing a paradigm shift on the legal *status* of the animals. Prior to this Constitution, nonhuman animals were classified as public or private property, with particular animal or public interest dependent on the interests of the humans involved. However, even with the new Constitution, most law professors understand that animals should be part of the "*common property of the people use*", i.e. a natural resource of hybrid property, owned by all members of the community on an equal basis (Antunes 2000, p. 68). Nevertheless, it is necessary to make a systematic interpretation of the constitutional text and to realize that it became possible in the 1988 Constitution to redefine the legal *status* of the animal, and move the Brazilian legal system forward in the direction to recognize animals as subjects of law, that is, there is a legal relationship between human and non-humans (Oliveira 2013, p. 11346).

Before the 1988 Constitution, the law adhered to a recognition of the existence of entitled human subjects (or tangible), and non-human (or intangible), demonstrating the failure to understand that legal relationships happen only among humans. Human subjects are men and women, coming from the time of implantation, in which some rights are already guaranteed to the unborn child (*quasi-property*) with a special property interest. The subjects of non-human law are others, possibly including this categorization of animals (Silva 2010, pp. 83–86). Scientific advances in the fields of language, neuroscience, genetics and even philosophy have developed the possibility of expanding the circle of legal consideration through the evolutionary interpretation of the legal provisions (Cassuto 2004, p. 81).

In fact, a post-humanist interpretation of the 1988 Brazilian Constitution leads one to understand that it is a fundamental right to comprehend that non-humans have fundamental rights that need to be protected by the judiciary. Art. 5°, XXXV combined with art. 225, § 1°, VII, guarantees access to justice for animals individually. The animal is perceived as a singular being, and the Judiciary must analyze the breadth of animal law and determine whether a case falls under that law and has broader implications or import. The analysis will depend on the appropriateness, necessity, and proportionality of each case. There should be no preponderance of human or non-human interests, but rather the attempt to provide equal protection of the fundamental rights of humans and non-humans (Silva 2013b).

The 1988 Constitution, by proposing principles around the consideration of animals, has to be understood as an "umbrella" of protection of human and non-human interests. A post humanist interpretation of article 225, §1°, VII, enables legal consideration of the interests of animals in their individual and collective basis (Ayala 2012, p. 380). Thus, an interpretation is possible that includes all animals as beings deserving of dignity (Bolliger 2007, pp. 1–2), an interpretation that expands the legal categorizations to include humans and non-humans with the goal of guaranteeing fundamental rights for present and future generations, without species distinction. An interpretation of the text of the 1988 Constitution must apprehend this context: The Brazilian Constitution allows a rupture with the environmental perspective of animals as natural resource and provides the opportunity to develop and create new paradigms to protect all animals.

The individual protection that the Constitution gave animals in stating that they cannot be subjected to cruelty establishes a major constitutional law, that is, all Brazilians and the government, need to guarantee the protection of animals. This declaration changed the legal status of animals from trans-individual goods to individual consideration in the legal system (Rodrigues 2008, p. 70), thereby establishing a new legal relationship between humans and non-humans in the Brazilian legal system. Even in the moments of the discussion and writing of the 1988 Constitution, the representatives said that the time had come to recognize the right of any human or animal swimming, flying, hunting, running, or being brought into the state to have the right to enjoy their own existence (Gordilho et al. 2006, p. 263). Such was the evolution in constitutional value of the new status of animals. Furthermore, the recognition of the new status of animals in the Brazilian legal system rejected the narcissistic human perception of the legal system as only a human structure or as a manifestation of human culture. The former interpretation created an abyss between the world of man and the world of non-human animals, being a true rhetorical exaggeration, unaware that the human soul is, likewise, animal (Dawkins 1993, p. 81).

Indeed, the Brazilian judiciary branch has taken a leading position in the legal consideration of animals with an effective system and ground-breaking decisions establishing international precedents such as: (1) *Salvador/Bahia Zoo v. Swiss Chimp*, which recognized the possibility of the great apes to be as considered persons by the legal system (Cruz 2006, p. 281); and (2) *Animal Protection Associations v. Circus Portugal*, which recognized that animals are subjects of fundamental rights such as life, liberty, and physical and psychological integrity, all guaranteed by the Constitution of Brazil (Ferreira 2013, p. 187). In the last case, the Judge Ana Barbuda Ferreira (2013, p. 187) held that "the Brazilian legal system should ensure an effective and individualized protection of the non-human animal, since the Constitution itself precludes any form of animal cruelty, the judge must protect all forms of dignity, respect and care for life." Definitely, the constitutional and infra-constitutional legislation assigns moral and legal valuation to these beings, not forgetting to address fundamental rights as perceived in the above sections. The legal relation between State /guardian/owner and citizen/owner/keeper of the animal has been redesigned by the legal system, enabling individual and collective recognition (Silva 2014b, p. 162). The responsibility does not rest solely with the State, but all - State and citizens - have the duty to protect and refrain from committing cruelty to animals (Medeiros 2013, p. 70). The Constitution of Brazil provides a basis to enable the system to dissolve the limits of applicability of fundamental rights only to "State-humans" and to widen the circle of legal consideration to encompass non-humans. Luis Roberto Barroso (1999, p. 277) and Laurence Tribe (2001, p. 2) teach that in every case the legal system chooses values to be protected and the ends to be pursued. The indication of these values shows what beliefs have social transit in the legal relations.

9.6 Conclusion

As shown in this chapter, the Brazilian Constitution has allowed a change of perspective toward animals. Any practice that subjects animals to cruelty will not be allowed inside the Brazilian legal system in this new post-humanism age. Now is the time to develop and debate new alternatives to push forward this system to protect all animals and resolve animal concerns. The way will not be easy. Economic and social problems remain that take away the public attention and energy from the matter of animal rights. Nevertheless, the Brazilian society has lobbied the legislative, judiciary, and executive branches to take on these issues pertaining to animals and cruelty to animals. The legal system of Brazil has given constitutional interpreters the possibility to go forward in the direction of regarding animals as subjects of law. The 1988 Constitution is the foundation to understanding the Brazilian system about animal law. All the amendments to the Constitution must adhere to it and, therefore, must protect the present and future generations of animals. This is the desire of the representatives to the Constitutional Convention, to create a Brazil where humans can learn to walk together with non-humans.

References

Antunes, Paulo de Bessa. 2000. *Direito Ambiental*, 4th ed. Rio de Janeiro: Lumen Juris.

Ayala, Patryck de Araújo. 2012. Deveres ecológicos e a regulamentação da atividade econômica na Constituição brasileira. In *Direito constitucional ambiental brasileiro*, 5th ed. José Joaquim Gomes Canotilho and José Rubens Morato Leite (Org.), 294–351. São Paulo: Saraiva.

Barroso, Luís Roberto. 1999. *Interpretação e aplicação da Constituição*: *fundamentos de uma dogmática constitucional transformadora*. São Paulo: Saraiva.

Benjamin, Antônio Herman de Vasconcellos e. 2001. *A natureza no direito brasileiro: coisa, sujeito ou nada disso*. *Caderno jurídico*. Escola Superior do Ministério Público, n°. 2, julho de 2001.

Bolliger, Gieri. 2007. Constitutional and legislative aspects of animal welfare in Europe. *Stiftung Für das Tier im Recht* 1 February 2007: 1–2.

Cadavez, Lília Maria Vidal de Abreu Pinheiro. 2008. Crueldade contra os animais: uma leitura transdisciplinar à luz do sistema jurídico brasileiro. *Direito & Justiça* 34(01): 88–120. Porto Alegre: Jan/Jun, 2008.

Cassuto, David N. 2004. The law of words: Standing, environment, and other contested terms. *Harvard Environmental Law Review* 28: 79–128.

Cassuto, David N., and Steven Matthew Reed. 2011. Water law and the endangered species act. In *Whose drop is it, anyway? Legal issues surrounding our nation's water resources*, ed. Megan Baroni, 51–80. Chicago: ABA Section of State and Local Government Law, ABA Book Publishing.

Chiesa, Luis E. 2008–2009. Why is it a crime to stomp on a goldfish? Harm, victimhood and the structure of anti-cruelty offenses. *Mississippi Law Journal* 78:1–67.

Cruz, Edmundo. 2006. Sentença do *Habeas Corpus* impetrado em favor da chimpanzé Suíça. *Revista Brasileira de Direito Animal*, vol. 01. Ano. 01. Jan/Dez. Salvador: Instituto de Abolicionismo Animal.

Custódio, Helita Barreira. 1997. Crueldade contra animais e a proteção destes como relevante questão jurídico-ambiental e Constitucional. *Revista de Direito Ambiental*, vol. 10. abril/junho. São Paulo: Revista dos Tribunais.

Dawkins, Richard. 1993. Gaps in the mind. In *The great ape project: Equality beyond humanity*, ed. Paola Cavalieri and Peter Singer, 80–87. New York: St. Martin's Press.

Dias, Edna Cardozo. 2000. *A tutela jurídica dos animais*. Belo Horizonte: Mandamentos.

Duckler, Geordie. 2002. The economic value of companion animals: A legal and anthropological argument for special valuation. *Animal Law* 8: 199–221.

Favre, David. 1997–1998. The risk of extinction: A risk analysis of the endangered species act as compared to CITES. *N.Y.U. Environmental Law Journal* 6: 341–366.

Favre, David. 2000. Equitable self-ownership for animals. *Duke Law Journal* 50: 473–502.

Favre, David. 2005. Judicial recognition of the interests of animals: A new tort. *Michigan State University College of Law Michigan State Law Review* 2: 333–367.

Favre, David, and Vivien Tsang. 1993. The development of the anti-cruelty laws during the 1800's. *Detroit College of Law* 1: 1–36.

Feldmann, Fábio. 2013. Os tribunais e o direito à vida do pato mergulhão. *Brasil Econômico*. Disponível em: http://brasileconomico.ig.com.br/noticias/os-tribunais-e-o-direito-a-vida-do-pato-mergulhao_134442.html. Acesso em: 01 de Agosto de 2013.

Ferreira, Ana Conceição Barbuda Sanches Guimarães. 2013. Ministério Público do Estado da Bahia, Associação Brasileira terra verde viva e associação célula mãe v. Portugal produções artísticas ltda. "Circo Portugal". *Revista Brasileira de Direito Animal* 8(12). Jan/abr. 2013.

Giménez-Candela, María Teresa. 2008. New rules to ensure the protection of animals in Spain: Spanish Animal Welfare Act. 32/2007. *Revista de Bioética y Derecho*. n° 14: 25–28. Septiembre, 2008.

Gordilho, Heron José de Santana, Luciano Rocha Santana and Tagore Trajano de A Silva et. al. 2006. Habeas Corpus impetrado em favor da chimpanzé Suíça na 9ª Vara Criminal de Salvador (BA). *Revista Brasileira de Direito Animal*. Vol. 01. Ano. 01. Jan/Dez. Salvador: Instituto de Abolicionismo Animal.

Huss, Rebecca J. 2002–2003. Valuing man's and woman's best friend: The moral and legal status of companion animals. *Marquette Law Review* 86: 47–105.

Ignacio Wallace da Gama Cochrane, 1895. *Presentation to the First General Assembly of União Internacional Protectora Dos Animales* (30 May 1895).

Kean, Hilda. 1998. *Animal rights: Political and social change in Britain since 1800*. London: Reaktion Books.

Levai, Laerte Fernando. 1998. *Direito dos animais*. O direito deles e o nosso direito sobre eles. Campos do Jordão: Editora Mantiqueira.

Lourenço, Daniel. 2008. *Direito dos Animais: fundamentação e novas perspectivas*. Porto Alegre: Sergio Antonio Fabris.

Medeiros, Fernanda Luiza Fontoura de. 2013. *Direito dos animais*. Porto Alegre: Livraria do Advogado.

Mello, Marcos Bernardes de. 2012. *Teoria do fato jurídico – plano da existência*, 18ª ed. São Paulo: Saraiva.

Oliveira, Fábio Corrêa Souza de. 2013. Direitos da Natureza e Direito dos Animais. Um Enquadramento. *Revista do Instituto do Direito Brasileiro da Faculdade de Direito da Universidade de Lisboa – RIDB* 10: 11325–11370..

Rodrigues, Danielle Tetü. 2008. *O Direito & os Animais: uma abordagem ética, filosófica e normativa*, 2ªed. Curitiba: Juruá.

Rosiers, Jared des. 1990–1991. The exemption process under the endangered species act: How the "God Squad" works and why. *Notre Dame Law Review* 66: 825–862.

Ryder, Richard. 1989. *Animal revolution: Changing attitudes towards speciesism*. Cambridge, MA: Basil Blackwell.

Santana, Luciano Rocha, and Oliveira, Thiago Pires. 2006. Guarda Responsável e Dignidade dos Animais. In: *Revista Brasileira de Direito Animal*. Vol. 01. N°. 01. jan/dez. Salvador: Instituto de Abolicionismo Animal.

Silva, Tagore Trajano de Almeida. 2010. Brazilian Animal Law Overview: Balancing human and non-human interests. *Journal of Animal Law* 6: 81–104.

Silva, Tagore Trajano de Almeida. 2012a. *Animais em juízo*. Salvador: Evolução.

Silva, Tagore Trajano de Almeida. 2013a. Teoria da Constituição: Direito Animal e Pós-Humanismo. *Revista do Instituto do Direito Brasileiro da Faculdade de Direito da Universidade de Lisboa* – RIDB 10: 11683–11732.

Silva, Tagore Trajano de Almeida. 2013b. Direito Animal e Pós-humanismo: formação e autonomia de um saber pós-humanista. *Revista Brasileira de Direito Animal* 14: 161–262.

Silva, Tagore Trajano de Almeida. 2013c. O ensino do Direito Animal: um panorama global. *Revista de Direito Brasileira* 6: 232–272.

Silva, Tagore Trajano de Almeida. 2014a. Brazilian animal law overview: Balancing human and non-human interests. *Journal of Animal Law* 6: 81–104.

Silva, Tagore Trajano de Almeida. 2014b *Direito Animal & Ensino Jurídico: formação e autonomia de um saber pós-humanista*. Salvador: Evolução.

Silva, Tagore Trajano de Almeida. 2014c. Origins and development of teaching animal law in Brazil. *Pace Environmental Law Review* 31:501–530.

Silva, Luciana Caetano da., and Gilciane Allen Baretta . 2007. Algumas considerações sobre a crueldade contra os animais na Lei 9.605/98. In *Direito Penal Contemporâneo:* Estudos em homenagem ao Professor José Cerezo Mir. 1: 320–327, Luiz Regis Prado (Org.). São Paulo: Editora Revista dos Tribunais.

Silverstein, Helena. 1996. *Unleashing rights. Law, meaning, and the animal rights movement*. Ann Arbor: University of Michigan Press.

Singer, Peter. 2004. *Libertação Animal*. Trad. Marly Winckler. Porto alegre: Lugano.

Sirvinskas, Luís Paulo. 2005. *Manual de direito ambiental*, 3rd ed. São Paulo: Saraiva.

Tribe, Laurence H. 2001. Ten lessons our constitutional experience can teach us about the puzzle of animal rights: The work of Steven M. Wise. *Animal Law* 7: 1–8.

Wise, Steven M. 1996. Legal thinghood of nonhuman animals. *Boston College Environmental Affairs Law Review* 471(23): 525–530.

Wise, Steven M. 1999. Animal thing to animal person – Thoughts on time, place, and theories. *Animal Law* 5: 61–68.

Chapter 10
Regulatory Capture and the Welfare of Farm Animals in Australia

Jed Goodfellow

Abstract Recent controversies over the treatment of animals within Australia's agricultural sector have raised questions over the adequacy of current governance and regulatory arrangements for farm animal welfare. Concerns have been expressed over perceived conflicts of interest on behalf of State and Federal Departments of Agriculture in administering animal welfare law. The theory of regulatory capture is applied to assess the veracity of such claims. Capture theory concerns the private distortion of public purposes. It occurs when a regulatory agency acts in the interests of the industry it is charged with regulating in a way that is inconsistent with the public interest the regulation is designed to serve. Key to the analysis is a consideration of the relationship between farm animal welfare and on-farm productivity. It is argued that due to the structure of the Departments' reward system, measurable economic goals associated with productive and profitable primary industries are prioritised over the more elusive, less determinate public interest in animal welfare. This has led the Departments to a community of interests with the regulated parties where they have adopted the same instrumental approach to animal welfare as that advanced by the livestock industries. Consequently, the Departments have deviated from serving the public interest in farm animal welfare as demonstrated by various regulatory failures and process deficiencies presenting a strong case for the existence of regulatory capture. The chapter concludes with exploring options for regulatory reform to circumvent the capturing influences and to improve the accountability and democratic legitimacy of the framework.

The views expressed do not represent the views of RSPCA Australia. Thanks are due to Steven White and Deborah Cao for constructive comments and Mariko Kawase for assistance with diagrams.

J. Goodfellow (✉)
Department of Law, Macquarie University, Sydney, NSW 2109, Australia
e-mail: jed.gooodfellow@mq.edu.au

D. Cao, S. White (eds.), *Animal Law and Welfare - International Perspectives*,
Ius Gentium: Comparative Perspectives on Law and Justice 53,
DOI 10.1007/978-3-319-26818-7_10

10.1 Introduction

The welfare of farm animals has been attracting an increasing degree of public attention in Australia in recent years. A continuous stream of prominent controversies concerning the mistreatment of farm animals within livestock production and processing facilities,[1] and particularly within Australia's live animal export trade,[2] has attracted national public outrage and sparked heated political debates.[3] A key focus of the public dialogue has related to the adequacy of current governance and regulatory arrangements for farm animal welfare. In particular, a number of politicians have raised concerns over the role of Federal and State departments of agriculture or primary industries (Departments of Agriculture) in administering animal welfare legislation on the basis that such arrangements give rise to an inherent conflict of interest.

This concern formed part of the basis for the Australian Labor Party's decision to adopt a policy for the establishment of an "independent Office of Animal Welfare" at its national conference in 2011 (Australian Labor Party 2011, p. 179). When explaining the rationale for the proposed Office in Federal Parliament, Melissa Parke MP (2013) stated that "public faith in Australia's animal welfare system had been undermined in recent years by revelation after revelation of cruelty to livestock", and that "[d]epartments of agriculture…are not widely regarded by the community or animal welfare groups as impartial when it comes to animal welfare." Such perceptions exist, she claimed, because these departments are "inherently conflicted…improvements in animal welfare are often not consonant with increased productivity and profitability, and vice versa." Similar sentiments have been expressed by the Australian Greens (2013) in giving their support for the establishment of such an institution.

These issues have also been identified within Australia's animal law scholarship (see, Caulfield 2008, p. 11; Ellis 2010, pp. 14–21; Sharman 2010, pp. 220–221; Bruce 2011, pp. 83–84; McEwen 2011a, b, p. 5; Dale and White 2013, pp. 166–168). Researchers and practitioners in the field have long criticised the "self-evident" conflict of interest that arises when "[t]hose who are responsible for the maintenance of [the animal welfare] legal regime in its enforcement and formulation view themselves as the 'friend of industry'" (McEwen 2011a, b, p. 5). This perceived partiality is said to create a regulatory environment in which it "is almost exclusively those who have a stake in profiting from animals who continue to draw the

---

[1] See list of incidents in Sect. 10.3.2.

[2] In 2011, national demonstrations were held to protest against the export of animals to Indonesia following an investigative news program depicting horrific slaughter practices. The Commonwealth Department of Agriculture reported receiving 284,415 items of ministerial correspondence during the 2011–2012 Financial Year. This represented a 556 % increase from the previous year with over 97 % of the correspondence relating to the live export trade (Commonwealth Department of Agriculture 2013a). See also, further discussion in Sect. 10.3.2.

[3] Since May 2011, six separate Bills have been introduced to Federal Parliament to end the live export trade.

line on what is necessary or unnecessary in the treatment of animals" (Dale and White 2013, pp. 166–168). While there has been quite widespread acceptance of this view within the literature, it would appear that the "self-evident" nature of the conflict has removed any perceived need for further inquiry and scrutiny. This is a problem because other voices within the debate do not see the same self-affirming evidence. Indeed, it is common for leaders of livestock industries and their political representatives to assert precisely the opposite claim – that there is no such conflict of interest because animal welfare and productivity actually go "hand in hand" (see for example, AgForce Queensland 2013).

This chapter seeks to go behind the claims and provide a considered account of the issue, to assess whether Departments of Agriculture are appropriate custodians of Australia's farm animal welfare regime, or whether they suffer from the irreconcilable conflicts of interest as suggested. It is a question that is of critical importance to the effectiveness and legitimacy of Australia's current governance and regulatory arrangements for farm animal welfare. A theoretical frame of reference for the analysis is found in the theory of regulatory capture. Broadly conceived, capture theory concerns the private distortion of public purposes. It describes the process in which a regulatory agency acts in the interests of the industry it is charged with regulating in a way that is inconsistent with the public interest the regulation is designed to serve (Mitnick 1980, pp. 95–96). It is a concept closely associated with notions of conflicting interests and regulatory bias, and therefore offers useful insights into some of the key issues under examination here.

The chapter begins with a brief overview of the farm animal welfare regulatory framework (Sect. 10.2). Three indicators of regulatory failure are then proposed, to provide further context for why the framework is attracting increasing criticism (Sect. 10.3). The indicators proposed include the inadequacy of current farm animal welfare standards as measured against popular normative expectations, international standards, and legislative mandates; the prevalence of serious animal welfare incidents within the agricultural sector; and the presence of regulatory process deficiencies, which are outlined in further detail in Sect. 10.6. Section 10.4 then introduces the theory of regulatory capture, highlighting relevant conditions that influence a regulator's predisposition to capture. These include poor institutional design, where an agency is delegated with pursuing a particular responsibility that conflicts with the agency's overriding mission; and "cultural capture", a process that involves a regulator adopting the norms and values of the industry it is charged with regulating (Kwak 2014, p. 79). Section 10.5 assesses the extent to which these conditions are present with respect to the Departments of Agricultures' role in administering animal welfare legislation. Key to this analysis is a consideration of the relationship between farm animal welfare and on-farm productivity – are these factors necessarily in a state of conflict, or are they mutually beneficial? Relevant scientific and economic perspectives are reviewed to argue that despite industry claims to the contrary, welfare and productivity are predominately in a relationship of conflict. While there are certainly examples of complementarity, under market conditions productivity gains are often realised at the expense of animal welfare, and subsequent animal suffering is externalised.

This finding has significant implications for the way Departments of Agriculture conceptualise animal welfare and approach their responsibilities for promoting it. The Departments' overriding mission leads them to prioritise measurable, short-term economic goals associated with promoting productive and profitable primary industries, over the more elusive, less determinate public interest in farm animal welfare. This has led the Departments to a community of interests with the regulated party where they have adopted the same instrumental approach to animal welfare as that espoused by the livestock industries. Under this approach, welfare is valued in terms of its relationship with productivity. If improved welfare contributes to productivity it is embraced and promoted by the Departments. If, however, a proposed welfare measure has a negative correlation with productivity, it is dismissed or severely compromised. Alternative viewpoints that reflect different approaches to valuing animal welfare are difficult for the Departments to embrace as the structure of their reward system does not provide sufficient incentives for them to do so. On this basis, it is argued that Departments of Agriculture have absorbed the norms and perspectives of the livestock industries amounting to a case of cultural capture. This has resulted in systematic underperformance with respect to their animal welfare regulatory responsibilities. The public interests in ensuring farm animals are protected from cruelty, and in improving welfare standards over time, has been routinely subordinated to the interests of increasing industry productivity and profitability.

When considered in light of the deficiencies in regulatory processes discussed in Sect. 10.6, a strong case for the existence of regulatory capture is established.[4] These deficiencies include the overrepresentation of industry interests in standards development processes, industry influence over the development of animal welfare science, and inadequate compliance monitoring and enforcement.

The chapter concludes by exploring options for regulatory reform to circumvent the mechanisms of capture and to improve the accountability and democratic legitimacy of the framework. These include separating the competing responsibilities by transferring animal welfare to a non-conflicted department, or to a new government entity dedicated to animal welfare; establishing a "regulatory contrarian" in the form of an Inspector-General or Ombudsman for animal welfare to monitor and audit the performance of Departments of Agriculture; and creating a tripartite system of regulation by giving animal welfare non-government organisations a more prominent role in regulatory processes.

Overall, what the analysis shows is that current governance and regulatory arrangements for farm animal welfare in Australia are procedurally unfair and democratically illegitimate. They are dominated by institutions that identify with one particular set of values towards animals and their place in the world relative to human beings. The traditional instrumental approach to farm animals is now facing sustained challenge from a growing post-material ethic of care and inclusivity, which recognises intrinsic value in farm animals independent of their material worth

[4] Figure 10.3 below outlines the causal relationship between the various stages of the capture process.

to human beings. Democratic principles demand that such alternative viewpoints are reflected in relevant governance and regulatory processes. The time has arrived to better institutionalise the public interest in animal welfare within our systems of government.

10.2 Overview of Australia's Farm Animal Welfare Regulatory Framework

The Australian Constitution sets out the respective legislative powers of the Commonwealth, and State and Territory governments. The Commonwealth has legislative authority in specific areas,[5] while the States and Territories are empowered to legislate with respect to all other matters not expressly covered.[6] As the Constitution is silent on the subject of animal welfare, State and Territory governments are vested with primary regulatory responsibility. The Commonwealth is however authorised to legislate in relation to matters that are "incidental" to its expressed powers.[7] Under the "trade and commerce"[8] and "external affairs"[9] powers, the Commonwealth has taken on regulatory responsibility for the welfare of farm animals and wildlife that are subject to international trade. As Australia is the largest exporter of live animals in the world (Farmer 2011, p. x), this area of responsibility holds great significance for animal welfare purposes. All other areas of farm animal welfare rest squarely with the States and Territories. The principal statutes for animal welfare and the responsible government departments are outlined in Table 10.1.

The substance of Australia's farm animal welfare standards is contained within subordinate legislation adopted under the principal statutes in the form of industry-based codes of practice.[10] To achieve national consistency, the codes of practice were developed through a cooperative process involving all jurisdictional governments. State and Territory ministerial endorsement for the codes of practice was provided for through the Council of Australian Government's Standing Council on Primary Industries (previously referred to as the Primary Industries Ministerial Council) (SCoPI). In 2005, Commonwealth, State and Territory governments embarked on a process of converting the mostly voluntary codes of practice into the *Australian Animal Welfare Standards and Guidelines* (Standards and Guidelines) to

---

[5] These are outlined in s 51 of the *Australian Constitution*.

[6] This is courtesy of the residual powers doctrine, which provides that any power not expressly vested in the Commonwealth Government remains within the jurisdiction of the States: See *Australian Constitution* s 107.

[7] *Australian Constitution* s 51(xxxix).

[8] *Australian Constitution* s 51(i).

[9] *Australian Constitution* s 51(xxix).

[10] Referred to as the 'Model Codes of Practice for the Welfare of Animals'; accessible at: http://www.publish.csiro.au/index.cfm

Table 10.1 Principal animal welfare legislation and administering departments (current as at 22 February 2014)

Jurisdiction	Principal legislation	Administering department
Qld	*Animal Care and Protection Act 2001*	Department of Agriculture, Fisheries and Forestry
NSW	*Prevention of Cruelty to Animals Act 1979*	Department of Primary Industries
ACT	*Animal Welfare Act 1992*	Territory and Municipal Services Directorate
Vic	*Prevention of Cruelty to Animals Act 1986*	Department of Environment and Primary Industries
	Livestock Management Act 2010	
Tas	*Animal Welfare Act 1993*	Department of Primary Industries, Water and Environment
SA	*Animal Welfare Act 1985*	Department of Environment and Natural Resources
WA	*Animal Welfare Act 2002*	Department of Agriculture and Food
NT	*Animal Welfare Act 2000*	Department of Primary Industries and Fisheries
Cth	*Australian Meat and Live-stock Industry Act 1997*	Department of Agriculture
	Export Control Act 1982	

further enhance regulatory consistency. The intended approach for the conversion is to incorporate the Standards as mandatory requirements under state law, while the Guidelines would simply be recommendations (de Witte 2009, p. 149). To-date, only Standards and Guidelines for the land transport of livestock have been finalised.[11] Standards and Guidelines for cattle and sheep farming have been drafted but are yet to be ministerially endorsed (Animal Health Australia 2014a). There are 13 further codes of practice yet to be converted (de Witte 2009, p. 149).

So despite the fragmented jurisdictional responsibility, Australia is in the process of developing a relatively consistent framework for farm animal welfare standards. Much of this can be attributed to the role of the Commonwealth Government in coordinating the national approach. A crucial feature of this coordination involved the development of the *Australian Animal Welfare Strategy* (AAWS) in 2005.[12] The AAWS outlines Australia's national policy on animal welfare and provides a strategy for achieving a number of objectives for improving the welfare animals within the country. However, the Commonwealth Government has recently withdrawn from its leadership position and provision of financial support for the AAWS. Shortly after the change of government in 2013, the new Commonwealth Minister of Agriculture announced the abolition of the advisory committees that were responsible for overseeing the implementation of the AAWS, including the AAWS

[11] Accessible here: http://www.animalwelfarestandards.net.au/land-transport/

[12] The current version is the 2010–2014 *Australian Animal Welfare Strategy*, available at: http://www.australiananimalwelfare.com.au/

Advisory Committee and six animal sector working groups (Vidot 2013). The dismantling of animal welfare capacity at a Commonwealth level also resulted in the abolition of the Department of Agriculture's Animal Welfare Branch, with its 23 staff being relocated to alternative roles mostly unrelated to animal welfare. The extent to which this will impact progress on the development of the Standards and Guidelines is currently unknown.

Enforcement arrangements for animal welfare law varies from State to State. In most jurisdictions the State based Royal Society for the Prevention of Cruelty to Animals (RSPCA) organisation plays a prominent role in providing personnel who are appointed under the legislation as inspectors to work cooperatively with livestock health and veterinary officers from the responsible departments. However, as discussed in Sect. 10.6.3, when it comes to commercial livestock, the RSPCA's role in enforcement is limited in several States.

Taken at face value, Australia's farm animal welfare framework appears to provide quite a robust national framework of protections. However, this impression is contradicted by the growing criticisms and calls for reform from political and academic sources. To provide further background to these criticisms, Sect. 10.3 considers a number of factors that may indicate shortcomings in the framework's performance.

10.3 Indicators of Regulatory Failings

Claims that Departments of Agriculture suffer from conflicting interests are not made in isolation. They attempt to explain, and give context to, a range of poor regulatory outcomes. These outcomes may be considered as indicators of regulatory failings.[13] It is important to clearly identify these indicators first as they will inform the subsequent analysis of regulatory capture. As will be explored further below, such an analysis must provide a "defensible model of the public interest" that is claimed to be subverted by the relevant capturing interests (Carpenter and Moss 2014, p. 14). The indicators will help to demonstrate how this public interest has been subverted.

In identifying the indicators it is important to ensure that they are themselves defensible, and can be measured by some objective benchmark. If they are not defensible they will be vulnerable to dismissal on the grounds that they are simply the views of a disgruntled stakeholder that has failed to get its way in the otherwise legitimate democratic processes involved in developing the framework. With this in mind, the following outcomes are proposed as some of the key indicators of failures within Australia's farm animal welfare regulatory framework.

[13] The term "regulatory failing" is used in this context to refer to a failure to achieve policy objectives, as opposed to the meaning given to a similar term – "government failure" – which in economic literature, refers to government intervention in the market that causes inefficiency and misallocation of scarce resources.

10.3.1 Inadequate Animal Welfare Standards

The first indicator of regulatory failure, and perhaps the most significant, concerns the inadequacy of the farm animal welfare standards produced by the current framework. The standards currently provide for the common forms of intensive animal farming including the use of battery cages for layer hens, individual stalls for pigs, and extremely high stocking densities for broiler chickens. They also prescribe many of the routine surgical procedures that are performed without pain relief, including castration, tail-docking, beak trimming, mulesing, dehorning, and even the spaying of cattle.[14] The negative impacts of these practices on animal welfare have been well documented by scientific research (see for instance, Rollin 1995; Benson and Rollin 2004; Pew Commission 2008).

Of course, determining "adequacy" in this context is a value laden exercise. Different stakeholders will inevitably have different views on what may be considered adequate in a given context. There are however some benchmarks and points of comparison that can provide guidance on this question. First, the standards for farm animal welfare may be assessed against popular normative conceptions. Sociological research on human-animal relations demonstrates that over the past 40–50 years, public attitudes to animals have been shifting from a traditional utilitarian[15] approach, in which animals are viewed primarily in terms of their productivity and instrumental worth, to a more compassionate and empathetic ethic of care (Franklin and White 2001, pp. 221–222; Mazur et al. 2006, p. 21; Parbery and Wilkinson 2012; Goodfellow et al. 2014, p. 40). This shift has been associated with the growth in post-material values throughout the Western world more broadly (Mazur et al. 2006, p. 21). These changing attitudes are leading to intensifying concern for animal welfare (Coleman 2007, p. 6; Parbery and Wilkinson 2012), causing a growing disparity between current livestock production methods and consumer expectations (Te Velde et al. 2002, pp. 207–211; Vanhonacker et al. 2008; Frewer et al. 2005, p. 347). One national study revealed that 52 % of Australians believe that modern farming methods relating to the production of eggs, milk, and meat are cruel (Franklin 2007, p. 19). Research from the European Union indicates that consumers are concerned about the "naturalness" of production and the overall quality of life afforded to animals (Harper and Henson 2001, p. 21). They expect animals to be provided with living conditions that allow them the ability to express their natural behaviour (Frewer et al. 2005, p. 345) and are concerned about practices that cause pain and stress (Vanhonacker et al. 2008, p. 131). The willingness of consumers to pay more for products that derive from higher welfare practices may be a measure for the discrepancy between current minimum standards prescribed by law and normative expectations. Consumer research by Taylor and Signal (2009, pp. 351–352) found that 86 % of Australians were willing to pay more to ensure farm animals

[14] For a detailed critique of current farm animal welfare standards, see Sharman (2013).

[15] The term 'utilitarian' here is used for its common meaning in terms of valuing action by its utility, not the meaning ascribed to it in the field of animal ethics, particularly by Peter Singer.

were provided with the "Five Freedoms".[16] This sociological and consumer-based research suggests that Australia's current standards for farm animal welfare are deviating from popular normative conceptions of acceptable farming practices.

A second benchmark for assessing adequacy may be the improvement of animal welfare standards in developed nations around the world. The European Union, consisting of 27 member states, passed legislation in 1999 to phase out the use of conventional battery cages by 2012.[17] In 2001, it passed legislation to prohibit the use of sow stalls by 2013.[18] In 2010, New Zealand amended its *Pig Code of Welfare* to phase out the use of sow stalls by 2015,[19] and in 2012, it amended the *Layer Hen Code of Welfare* to phase out the use of battery cages by 2022.[20] Additionally, several State jurisdictions within the U.S. have enacted similar bans and restrictions,[21] and most recently, the Canadian Government implemented a ban on the construction of new sow stalls from 1 July 2014, and a phase out of all stalls by 2024.[22] These international developments demonstrate that Australian standards are beginning to fall behind the rest of the developed world. It should be noted here that the Australian Capital Territory (ACT) has passed legislation to prohibit sow stall and battery cage production practices.[23] However, the move was more symbolic than practical as there is only one commercial egg producer and no pig farming operations within the Territory. The State of Tasmania has regulated to restrict the use of sow stalls to 10 days within a reproductive cycle.[24] It also foreshadowed a phase out of battery cages in 2012 (Tasmanian Government 2012), but no legislation has since been introduced to follow through with the announcement. There has been no indication of equivalent reforms at the national level.

A third benchmark may be found in the legislative mandates under which the standards are developed. Industry codes of practice and Standards and Guidelines are given legal effect via incorporation within State and Territory animal welfare laws. The expressed objects of these laws are generally to prevent animal cruelty

[16] The "Five Freedoms" are a universally recognised benchmark for animal welfare. They include: freedom from hunger and thirst, freedom from discomfort, freedom from pain, injury or disease, freedom to express normal behaviour, and freedom from fear and distress.

[17] *Council Directive 1999/74/EC of July 1999 laying down minimum standards for the protection of laying hens* [1999] OJ L 203, p0053–0057, Art 5.

[18] *Council Directive 2001/88/EC of 23 October 2001 amending Directive 91/630/EEC laying down minimum standards for the protection of pigs* [2001] OJ L 316 p 001–0004, Art 3.

[19] *Animal Welfare (Pigs) Code of Welfare 2010* (NZ), available at: http://www.mpi.govt.nz/Default.aspx?TabId=126&id=1846

[20] *Animal Welfare (Layer Hens) Code of Welfare 2012* (NZ), available at: http://www.biosecurity.govt.nz/files/regs/animal-welfare/2012-layer-hens-code-web.pdf

[21] See Humane Society of the United States timeline of legislative developments for farm animal welfare here: http://www.humanesociety.org/issues/confinement_farm/timelines/timeline_farm_animal_protection.html

[22] *Code of Practice for the Care and Handling of Pigs* (Canada), available at: http://www.nfacc.ca/pdfs/codes/pig_code_of_practice.pdf

[23] See *Animal Welfare Act 1992* (ACT), ss 9A and 9B.

[24] See *Animal Welfare (Pigs) Regulations 2013* (Tas), r 25A.

and to promote animal welfare.[25] A hallmark feature of the legislation is a prohibition on animal cruelty, which is generally framed in terms of causing an animal unnecessary harm.[26] At the heart of this prohibition is the principle of proportionality, which provides that any harm caused to an animal must be for a legitimate purpose, and any pain caused must not be disproportionate to that purpose (Radford 2001, pp. 247–249; for a critique of this principle, see Sankoff 2013, pp. 13–30). As enunciated by Hawkins J in *Ford v Wiley* (1889) 23 QBD 205 at [219] – "[t]he beneficial or useful end sought to be attained [by the harm causing practice] must be reasonably proportionate to the extent of the suffering caused, and in no case can substantial suffering be inflicted, unless necessity for its infliction can reasonably be said to exist." If considered against this objective standard, the legality of many of Australia's current farming practices could be brought into question, particularly when the availability of less harm causing alternatives are taken into consideration. This is precisely why State governments have sought to exempt the practices prescribed in the codes of practice from the application of animal welfare law. This demonstrates the discrepancy between the standards of care promoted by state animal welfare law, and those outlined in the codes of practice.

So while assessing "adequacy" may to some extent be a subjective exercise, there are some objective benchmarks and comparators that can provide further perspective. As the above discussion indicates, Australia's farm animal welfare standards fall short of the standards of care promoted within popular normative conceptions, international standards on animal welfare, and the legislative mandates of State and Territory animal welfare law.

10.3.2 Prevalence of Significant Animal Welfare Incidents

A second indicator that may be evidence of regulatory failure is the prevalence of animal welfare incidents within the agricultural sector in recent years. As there is no data collected nationally on such cases it is difficult to make an accurate assessment. However, anecdotal evidence from the reports of animal protection groups, the media, and government, suggest that they are quite frequent, and in the case of live animal export, they are routine.

Several high profile cases of animal mistreatment within intensive production and processing facilities have been revealed in the past 3 years. These have included separate incidents within pig (Knaus 2012), broiler chicken (Langdon 2013), and

[25] See generally, *Animal Welfare Act 2002* (WA), s 3; *Prevention of Cruelty to Animals Act 1979* (NSW), s 3; *Prevention of Cruelty to Animals Act 1986* (Vic), s 1; *Animal Care and Protection Act 2001* (Qld), s 3.

[26] See *Animal Welfare Act 1985* (SA), s 13(3)(a); *Prevention of Cruelty to Animals Act 1986* (Vic), s 9(1)(c); *Animal Welfare Act 1993* (Tas), s 8(1); *Animal Welfare Act 2002* (WA), s 19(2)(e); *Prevention of Cruelty to Animals 1979* (NSW, s 4(2); *Animal Care and Protection Act 2001* (Qld), s 18(2)(a); *Animal Welfare Act 1992* (ACT), s 8(1).

duck production facilities (Herbert 2012), and cases of cruelty in the processing of dairy calves (Fitzsimmons 2013), pigs (Mickelburough 2011), sheep (Cannane 2012), and turkeys (Levy 2013). All of these cases were reported widely in the mainstream news media.

In addition to intensive production and processing operations, anecdotal evidence suggests that welfare incidents within Australia's extensive production environment are commonplace. In giving evidence at a parliamentary inquiry into the starvation of hundreds of cattle at a university-run cattle station in the Northern Territory, a senior government veterinary officer stated that high death rates on cattle properties in the North (relative to those in Southern regions) were the industry norm (Legislative Assembly of the Northern Territory 2011a, pp. 5–6). Accordingly, he could not "support a teaching facility being prosecuted when other properties with similar situations would not be considered for prosecution." This view suggests that incidents of mass cattle starvation are not unusual in the country's North. This was supported more recently, when a Federal politician reported in January 2014, that between 5000 and 10,000 cattle were dying every day across the North of Queensland due to a prolonged drought (Neales 2014). While this figure may well have been exaggerated for political purposes, the scale of such deaths was no doubt significant.

By far the most highly reported cases of animal mistreatment have related to Australia's live export trade. At the time of writing, Commonwealth Department of Agriculture (2014a) records indicate that since 2006, there have been 49 reportable mortality events, amounting to one every 2 months. Recent incidents include 4179 sheep that died of heat stress en route to Qatar in September, 2013 (Commonwealth Department of Agriculture 2014a), and a further 1653 sheep that died in February 2014 on a Middle East-bound voyage that suffered mechanical issues (RSPCA 2014). In addition to the incidents at sea, there have been many reported cases of cruelty in the importing countries. In only 2012 and 2013, there have been 30 reported breaches of *Exporter Supply Chain Assurances System* and the *Australian Standards for the Export of Livestock* (Commonwealth Department of Agriculture 2014b). Notable cases include 20,000 Australian sheep that were brutally slaughtered in September 2012 under the order of Pakistan authorities. The shipment was originally intended for Bahrain but had been rejected due to animal health concerns (Ferguson and Masters 2012). A more recent example involved footage of Australian cattle being tortured in the streets of Gaza. Cattle were filmed by civilians as they were roped and wrestled to the ground, stabbed in the face and throat, shot at with assault rifles, and had their throats sawn open with blunt knives (Australian Associated Press 2013). Animal protection groups have compiled detailed chronologies of many other examples.[27] The frequency of these cases demonstrates that serious animal welfare incidents within the trade are routine.

[27] See, RSPCA Australia: http://www.rspca.org.au/campaigns/live-export; Animals Australia: http://www.banliveexport.com/facts/history-of-disasters.php; and Animals Australia: http://www.banliveexport.com/investigations/

While there is no nationally collated data on animal welfare incidents within the Australian agricultural sector, regular reporting by animal protection groups, the media, and government, suggest, at least anecdotally, that such incidents occur regularly.

10.3.3 Regulatory Process Deficiencies

Deficiencies in governmental processes for developing, implementing, and enforcing animal welfare standards may also be considered as indicators of regulatory failure. These deficiencies are discussed in Sect. 10.6 below as symptoms of regulatory capture. They include the overrepresentation of industry interests in standards development processes, industry influence over the development of animal welfare science, and inadequate compliance monitoring and enforcement. While these deficiencies are considered to be part of the *cause* of the failures mentioned above, they also constitute regulatory failures in themselves independent of the detrimental outcomes they create. Extensive empirical research demonstrates that people's evaluation of governmental authorities, and the decisions they make, are shaped by their views about the fairness of formal decision-making procedures (Tyler 1994, p. 811; Tyler 2005, p.xvi). The perceived impartiality and intentions of the decision-making authority are found to be key influencers of such views, which consequently, affect the legitimacy of the relevant governmental institution and the extent to which people are willing to accept its decisions (Tyler 1994, p. 811). For the reasons discussed below, current processes for the development of animal welfare standards give rise to procedural fairness concerns and reduce the public's confidence in the performance of the responsible departments.

There has been no shortage of proposed explanations for why these regulatory failings are occurring. As noted in the introduction, accusations of conflicting interests have featured prominently, but to date, there has not been a sustained theoretical examination of the substance of these claims. The remaining parts of this chapter endeavour to provide such an analysis through the application of the theory of regulatory capture.

10.4 An Introduction to Regulatory Capture

Regulatory capture is a theory familiar to a range of disciplines concerned with the performance of the State in serving the public interest. It can serve as a useful analytical tool for assessing government performance and explaining regulatory deficiencies. Broadly conceived, capture theory relates to the private distortion of public purposes. It is the process by which a regulatory agency acts in the interests of the individual, firm, or industry it is charged with regulating in a way that is inconsistent with the public interest the regulation is designed to serve (Mitnick 1980, pp. 95–96;

Briody and Prenzler 1998, p. 55). While related to the field of public corruption, regulatory capture is distinct in the sense that the influencing forces it is concerned with are not as explicit, nor illegal, and the subsequent changes in regulatory behaviour brought about by such forces, are often not as conscious or deliberate as in the classic case of corruption. These factors can make regulatory capture more difficult to identify and respond to (Carpenter 2014, pp. 57–70).

There are numerous mechanisms that may cause a regulator to deviate from serving the public interest, most of which relate to the nature of the relationship between the regulator and the regulated industry. Significant movement of key personnel between the regulatory agency and the industry – referred to as the "revolving door" – has been found to influence regulatory behaviour (Makkai and Braithwaite 1992). It can give rise to unhealthy relationships involving the granting of reciprocal privileges, particularly in circumstances where the regulator is seduced by the prospects of lucrative employment opportunities within the regulated industry. The "life-cycle" theory of capture proposes the notion that regulatory agency behaviour changes through different stages of growth. While agencies act with virility and regulatory enthusiasm upon establishment, they become passive in their dealings with industry as they mature, until finally, the agency becomes the protector of industry in an attempt to justify its continued existence (Bernstein 1955; Makkai and Braithwaite 1992, p. 62). Further identified mechanisms include asymmetric information where the regulator is overly reliant upon the industry to provide the information or funds (through fees and levies) it requires to perform its duties (McCarty 2014, p. 100; Kwak 2014, p. 75), and the existence of close ties with industry where regulators are required to maintain regular personal contact with industry representatives leading to mutual empathy, favouritism, and inappropriate regulatory concessions (Grabosky and Braithwaite 1986, pp. 205–207; Adams et al. 2003, p. 3; Baxter 2011, p. 187). However, the key mechanisms of capture considered in this analysis are those of poor institutional design and cultural capture.

10.4.1 Poor Institutional Design

The design of a government institution influences its predisposition to capture. Poor institutional design has been found to be a key contributor to capture, particularly when an agency is tasked with pursuing two conflicting objectives (Crowley 1982; Marsden et al. 2000, p. 26; Baldwin and Walker 2008, p. 187; Biber 2009; Barkow 2010; Baxter 2011, p. 194). It is certainly not uncommon for government departments – especially large conglomerate "super departments"– to be delegated with responsibilities that may at times give rise to tensions and conflicts. The classic example is an agency that has the dual-fold role of both promoting an industry, in terms of its growth and productivity, while at the same time regulating the industry to serve a public interest, such as human safety, consumer protection, or environmental conservation.

However, upon closer examination, acting as both "industry cheerleader" and "industry policeman" does not of itself create a conflict. Regulation may serve the interests of industry in creating a fair and stable environment in which to do business, and in guarding against external risks. Equally, industry productivity may in turn serve the public interest in terms of increased economic prosperity, rates of employment etc. Where the difficulties arise is in the nature of the relationship between the perceived conflicting responsibilities, especially as their respective importance is considered against the backdrop of the agency's overarching mission. If one objective is unduly prioritised over another it can affect agency performance with regard to the subordinate responsibility. Biber (2009, p. 13) has noted that agencies will systematically underperform on secondary goals that conflict with the achievement of the agencies' primary goals. In particular, agencies will pursue short term economic goals that are easy to measure at the expense of more elusive social goals in the public interest (Biber 2009, p. 13; Barkow 2010, p. 50).

It is important to point out here that institutional design of this nature is not constitutive of regulatory capture in itself. Agencies are often "pro industry" because the very purpose of their existence is to facilitate the needs of an industry. They may be instructed by the responsible minister, and the legislative mandates under which they operate, to promote industry growth and productivity. This is not capture but "electorally sanctioned pro-business governance" (Carpenter 2014, pp. 66–67). It is when industry-promoting agencies are delegated with responsibility for pursuing a public interest objective that conflicts with these economic goals that problems begin to arise. This is because such agencies are more inclined to adopt the norms and values of the industry they are designed to serve in their approach to the relevant conflicting public interest – a process referred to as "cultural capture".

10.4.2 Cultural Capture

Cultural capture (sometimes referred to as "cognitive" capture), is a process by which "those in charge of the relevant state entity internalise, as if by osmosis, the objectives, interests and perception of reality of the vested interest they are meant to regulate" (Kwak 2014, p. 78 citing Buiter 2008, p. 601). Factors that induce the capture of cultural norms include common backgrounds and experience between industry representatives and regulators, and an industry with a social purpose with which the regulators identify (Baxter 2011, pp. 182–183; Kwak 2014, p. 95). This can create difficulties in the policy formulation process because, as Kwak (2014, p. 98) explains, "when people identify with groups or adopt ideas…it is considerably harder for those people to identify the sources of their choices," and they are not open to "rational argument about the public interest." Alternative viewpoints can become difficult for the regulator to conceptualise (Lippmann 2005, pp. 9–10), and a form of "unconscious bias" towards the interests and demands of the regulated industry may emerge. As Mitnick (1980, pp. 95–96) explains:

> 'Capture' is said to occur if…quite independently of the formal or conscious desires of either the regulators or the regulated parties the basic structure of the reward system leads neither venal nor incompetent regulators inevitably to a community of interests with the regulated party.

If regulators are more readily rewarded for delivering on measurable economic goals that align with the interests of the regulated industry, then it should come as no surprise when the agency adopts the industry's norms and perspectives, and subsequently underperforms on other responsibilities that conflict with those goals.

10.4.3 The "Public Interest"

Before going on to consider the extent to which these conditions exist within Australia's farm animal welfare regulatory framework, it is necessary to offer a "defensible model of the public interest" (Carpenter and Moss 2014, p. 14). As regulatory capture necessarily involves a shift away from serving the public interest, understanding what is meant by "the public interest" is critical to determining the extent to which an agency has drifted from serving it, if it has at all. As Carpenter and Moss (2014, p. 14) note, in a pluralist society "there is no certainty in making a value claim about the common good." What matters, they argue, is that an account of the public interest is offered and defended so that "readers and future researchers can engage appropriately with the argument."

With this in mind, it is proposed that Australia's farm animal welfare regulatory framework is designed to serve a dual-fold public interest in: (1) protecting farm animals from cruelty; and (2) promoting farm animal welfare through incremental, but sustained, improvements in welfare standards over time. This model of the public interest largely reflects the objects of animal welfare laws throughout the country,[28] and is consistent with changing public attitudes towards farm animals. As discussed in Sect. 10.3.1 above, sociological research demonstrates that public concern for the welfare of farm animals is growing, fuelled by a shift from the traditional utilitarian approach, emphasising productivity and instrumental worth, to a more compassionate and empathetic ethic of care (Franklin and White 2001, pp. 221–222; Mazur et al. 2006, p. 21; Parbery and Wilkinson 2012). Consequently, the public is increasingly demanding improvements in farm animal welfare standards to meet their evolving expectations. To determine whether the current regulatory performance of Departments of Agriculture has deviated from serving this public interest, Sect. 10.5 will examine the nature of conflicting institutional priorities, and the subsequent implications for the way animal welfare is conceptualised and approached.

[28] See *Animal Welfare Act 1985* (SA), s 13(3)(a); *Prevention of Cruelty to Animals Act 1986* (Vic), s 9(1)(c); *Animal Welfare Act 1993* (Tas), s 8(1); *Animal Welfare Act 2002* (WA), s 19(2)(e); *Prevention of Cruelty to Animals 1979* (NSW, s 4(2); *Animal Care and Protection Act 2001* (Qld), s 18(2)(a); *Animal Welfare Act 1992* (ACT), s 8(1).

10.5 Conflicting Interests of Responsible Institutions

As outlined in Table 10.1 above, the administration of animal welfare legislation is delegated to the Departments of Agriculture in all jurisdictions, with the exception of the Australian Capital Territory (ACT) and South Australia (SA). The ACT is the smallest geographical jurisdiction in mainland Australia and does not have an agriculture department. In SA, while the Department of Environment and Natural Resources has formal administrative responsibility for the State's animal welfare legislation, officers from the Department of Primary Industries and Regions play a prominent role in the development of farm animal welfare policy.[29]

Departments of Agriculture exist to serve their respective agricultural sectors primarily through the promotion of profitable and productive primary industries. The prominence of these objectives is abundantly clear upon a review of departmental mission statements and corporate documents. For instance, the stated mission of Commonwealth Department of Agriculture (Commonwealth Department of Agriculture 2013a) is to "lead the development of policy advice and provide services to improve the productivity, competitiveness and sustainability of agriculture, fisheries, forestry and related industries." The vision of the Queensland Department of Agriculture, Fisheries and Forestry (Queensland Department of Agriculture and Fisheries and Forestry 2013a) is for "efficient, innovative, resilient and profitable agriculture, fisheries and forestry industries that thrive in the long term." The Department's goal is to "double Queensland's agricultural production by 2040." The vision of Western Australia's Department of Agriculture and Food (2013) is for "a progressive, innovative and profitable agriculture and food sector that benefits Western Australia." The Department states that its role is to "assist WA's $6 billion agriculture and food sector to be sustainable and profitable, with a clear focus on export-led growth." Other Department of Agriculture documents evidence similar objectives.

The true nature of a department's priorities is perhaps best illustrated by its allocation of funding and resourcing to programs and services falling under its portfolio responsibilities. Unfortunately, space does not permit a full budgetary analysis, but in general terms, animal welfare related services attract a fraction of one per cent of most Department of Agriculture funding arrangements.[30] Animal welfare services are not specifically identified in most departmental operational expenditure records. In the few cases where they are, it is in the context of promoting an industry's "social licence to operate" or facilitating market access (see for instance, Commonwealth Department of Agriculture 2013a; Queensland Department of Agriculture and Fisheries and Forestry 2013a).

---

[29] For instance, a representative from the Department of Primary Industries and Regions sits on the national Animal Welfare Committee on behalf of South Australia.

[30] Budgetary analysis undertaken by the author as part of PhD thesis due for submission in early 2015.

Upon review, the primary objectives of Departments of Agriculture are quite clear. And it must be emphasised that this chapter makes no criticism of such priorities. As noted above, some government departments are established precisely to promote and support industry growth and productivity and these objectives are entirely appropriate and democratically legitimate. The real question of concern is whether it is appropriate for Departments of Agriculture to be delegated with the responsibility for developing and administering animal welfare law.

To answer that question it is necessary to first consider the factual nature of the relationship between farm animal welfare and on-farm productivity. Do these factors conflict, or are they mutually beneficial? It is then necessary to consider what implications the answer to this question may have for the way Departments of Agriculture approach their animal welfare responsibilities.

10.5.1 Is There a Conflict between Farm Animal Welfare and On-farm Productivity?

The importance of this question to current political discussions concerning farm animal welfare cannot be overstated. It is often at the heart of debates about the scope of farm animal welfare regulation, and the position one takes on the question will often determine their position on whether the regulation is too weak or too intrusive. Those who answer in the affirmative tend to support welfare regulation and demand that it be strengthened, while those who answer in the negative often believe the regulation is superfluous and call for it to be scaled back. Unfortunately, as is the case with many controversial policy issues, the debate is often politicised and ideology tends to destroy any chance of reasoned factual discussion. The true nature of the relationship between animal welfare and productivity is often reduced to simplified and misleading claims and assumptions.

For instance, it is often asserted by representatives of livestock industries, and supportive politicians, that animal welfare and productivity are not in a state of conflict, but actually go "hand-in-hand" (see for example, AgForce Queensland 2013). This assertion is usually portrayed through two key claims. The first claim is that productivity is an indicator of good animal welfare (Rollin 2012, p. 360). This is often expressed as follows: "our animals are producing so therefore they must be happy". The veracity of this claim is ultimately a question for science. The second claim is that producers have sufficient economic incentives to provide for good states of welfare.[31] This claim is often expressed in the following form: "it's in our commercial interests to look after our animals well." This claim is ultimately a

[31] See for example, Jai Rowell MP commenting on the *Prevention of Cruelty to Animals Amendment Bill 2012* in the NSW Legislative Council on 12 September 2012: "Animals that are mistreated are not as productive as those that are not and they reproduce much less. In simple terms, unhealthy and unhappy animals produce poor-quality meat and dairy products."

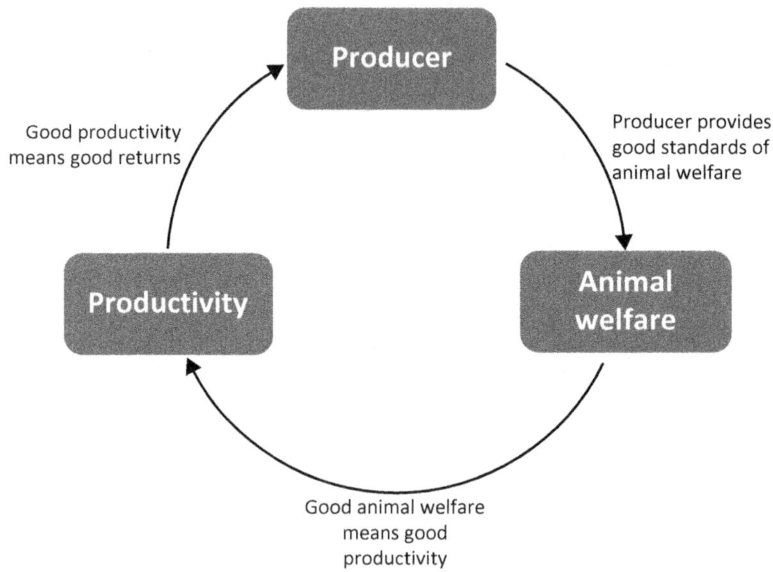

Fig. 10.1 Claimed relationship between animal welfare and productivity

question of economics. When presented together, the claims form a logic as depicted in Fig. 10.1 above.

The argument proceeds along the following lines: producer provides good standards of animal welfare, good animal welfare contributes to better productivity in terms of the animal's output (body mass, milk, eggs etc.), and better productivity results in higher returns for the producer. The argument is simple and logical, and this perhaps is why it has quite widespread appeal, permeating through public and political debates about farm animal welfare matters.[32] However, when the relevant scientific and economic literature is considered, it becomes clear that this relationship is not so straightforward and that the industry claims are actually based on a very narrow and outdated conception of animal welfare.

With respect to the first claim, animal welfare science tells us that an animal's productivity, or physical performance, is only one of several factors that need to be taken into account when making a proper assessment of an animal's state of welfare. In addition to physical performance, an animal's physiological functioning, brain state, behaviour, and even the animal's feelings should be considered to determine how a particular animal is truly coping with its environment (Broom 2008). This is important because there are many instances where an animal may be in good

[32] See for instance, South Australia, *Parliamentary Debates*, Legislative Council, 27 March 2012, 669, discussing the role of the RSPCA in farm animal welfare law enforcement; and, NSW, *Parliamentary Debates*, Legislative Council, 12 September 2012, 15402, discussing the introduction of "Livestock Welfare Panels" to the *Prevention of Cruelty to Animals Act 1979* (NSW).

physical health, but will nevertheless be experiencing a poor state of welfare. Animal welfare scientist, Professor Donald Broom (2008, p. 6) explains:

There are many circumstances where behavioural or physiological coping mechanisms are activated, indicating that welfare is poor, but the animal's health remains good. These include: situations where the coping mechanisms are successful, such as when body temperature is maintained despite extreme ambient temperatures; circumstances where failure to cope has consequences for psychological, but not physical stability, such as in the development of non-injurious pathological behaviours; and where detrimental effects upon physical stability are compensated for by management practices, such as the routine use of antibiotics.

The fact that an animal can be productive despite being in a poor state of welfare has serious implications for the way in which welfare is valued under market conditions. This brings us to the second claim. Economic research regarding the relationship between welfare and productivity provides that within an economic context, farm animals are valued simply by their productivity (see McInerney 2004). The economically rational producer will ensure that inputs for welfare are limited to the extent they contribute to productivity gains and overall higher returns. There are of course some examples of complementarity between welfare and productivity, such as in ensuring low stress handling and slaughter practices to avoid bruising and to improve meat quality (UN FAO 2001). However, in many situations the interaction is not so positive, and productivity gains are realised at the expense of welfare (in the economic literature, see Bennett 1997; Appleby et al. 2003; Appleby 2008; Lusk and Norwood 2011; Lusk 2011). To illustrate this dynamic, agricultural economist Professor John McInerney developed the following graph depicted in Fig. 10.2.

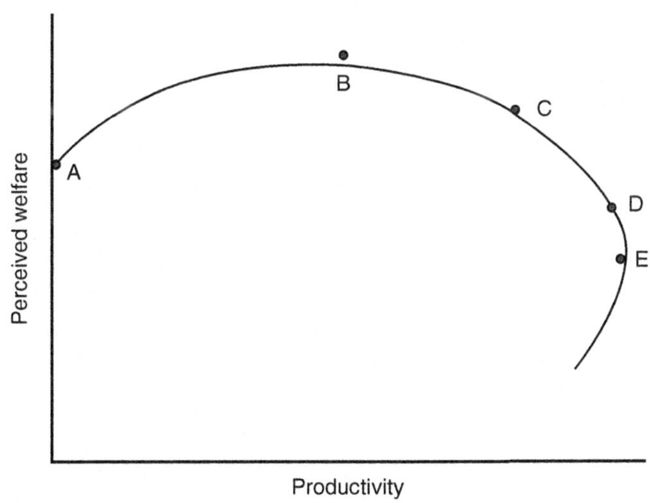

Fig. 10.2 Relationship between animal welfare and productivity (First presented in McInerney (1991). Reproduced in Bennett (1997) and McInerney (2004))

The graph shows a generalised relationship between animal welfare (as perceived by humans) and productivity within a commercial farming environment. It is not based on empirical data, but McInerney (2004, p. 19) notes that it "is now well established as providing a useful framework" and is "sufficiently robust to broadly fit any type of livestock production system." McInerney explains that while there is often a mutually beneficial relationship between welfare and productivity at lower levels of output (points A to B), as the size, efficiency, and intensity of the production system increases, welfare begins to decline (from points B through to E).[33] This is ultimately because the costs of animal welfare inputs that go beyond those related to productivity gains are not factored into the price of the end product, and the subsequent animal suffering is externalised. As McInerney (2004, p. 2) explains:

> Since animal welfare is in the nature of a nonmarket good ("externality") it carries no evident price and so farmers inevitably focus on the animals' productivity, which does provide commercial reward. Economic optimising theory demonstrates that market signals will tend to cause welfare standards to fall below the socially desirable norm.

There are, however, examples of meat and egg products that do factor in some costs of improved welfare measures. These products are distinguished precisely on the basis of their higher welfare attributes and command a price premium in return.[34] However, these input costs generally only address a limited range of welfare impacts, and in many cases relate solely to the overall system of production and do not address associated practices such as routine surgical procedures without pain relief and other welfare-compromising management practices.

Ultimately, this demonstrates that there is a conflict between farm animal welfare and on-farm productivity. Unfortunately for animals, they have the capacity to produce despite being in poor states of welfare. This leads to significant consequences under market conditions where productivity is valued and animal suffering is externalised. The traditional notion that animal welfare and productivity are one in the same no longer stands to reason when contemporary science and its sophisticated methods of assessing welfare are applied.

10.5.2 Implications for Departmental Approach to Animal Welfare Responsibilities

So what implications does this conclusion have for the way in which Departments of Agriculture approach animal welfare? First, it means the Departments have conflicting responsibilities. The promotion of profitable and productive livestock

[33] For a more detailed explanation, see, McInerney (2004, pp. 18–20).

[34] Examples include free-range and organic products, and other private accreditation schemes such as the RSPCA's Approved Farming Scheme, see: http://www.rspca.org.au/what-we-do/working-farming-industry/approved-farming-scheme

industries conflicts with the promotion and protection of farm animal welfare. Measures designed to promote welfare will often limit what a producer can do with an animal, placing restraints on the pursuit of greater productivity.[35]

When these conflicting responsibilities are considered against the backdrop of the Departments' overarching missions and reasons for existence, it is not difficult to predict which will be valued more highly and given greater priority. The structure of the reward system for Departments of Agriculture is geared towards achieving the economic goals of improving industry productivity and profitability. This has led the Departments to a community of interests with the regulated party where they have adopted the same approach to animal welfare as that promoted by the livestock industries. According to this approach, welfare is valued only in terms of its contribution to productivity. If improved welfare contributes to productivity gains, it is embraced and promoted by the Departments. If, however, welfare measures have a negative correlation with productivity, they are dismissed or severely compromised. This highly instrumental approach to animal welfare reflects the traditional utilitarian attitude to animals that has been found to be more prevalent among farmers and primary producers than within the broader Australian community (Signal and Taylor 2006; Southwell et al. 2006; Parbery and Wilkinson 2012). Alternative viewpoints that reflect different values and attitudes towards animals are not given equivalent consideration. This can also lead regulators to under appreciate risks within the environment they are regulating as their views are partial to one perception of reality (McDonnell and Schwarcz 2011, p. 1637). Departments of Agriculture have absorbed the norms and perspectives of the livestock industries in their approach to administering animal welfare legislation and in this sense they can be said to suffer from a degree of cultural capture.

Overall, this cultural capture resulted in systematic underperformance with respect to their animal welfare regulatory responsibilities. The public interest in ensuring farm animals are protected from cruelty, and in improving welfare standards over time, has been routinely subordinated to the interests of increasing industry productivity and profitability. When considered in light of the regulatory process deficiencies this has caused (outlined in Sect. 10.6 below), a strong case for regulatory capture is established. Figure 10.3 depicts the causal relationship between the stages of the capture process.

---

[35] Technological developments do have the potential to overcome this dilemma and create "win-win" solutions in which welfare can be improved without reducing productivity. However, these developments are currently context specific and do not have the wider cross species/management system application necessary to fundamentally alter the conflicting nature of the relationship between welfare and higher levels of productivity.

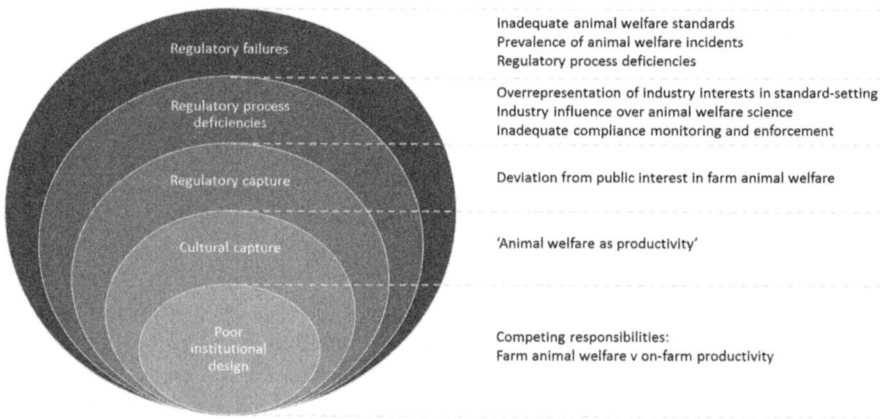

Fig. 10.3 Causal relationship between stages of capture process within Australia's farm animal welfare regulatory framework

10.6 Effects on Regulatory Processes

Australian animal law scholarship has provided much needed scrutiny of government performance in undertaking crucial regulatory functions relating to animal welfare. The specific deficiencies addressed below relate to the overrepresentation of industry interests in standards development processes, industry influence over the development of animal welfare science, and the inadequacy of compliance monitoring and enforcement functions within the agricultural sector. Baxter (2012, p. 34) has noted that regulatory capture affects governmental processes "when the degree of influence by one legitimate stakeholder in the regulatory process over another has become unbalanced." As the proceeding analysis demonstrates, the nature of each of the identified deficiencies reflects the Department of Agricultures' tendency to prioritise industry interests over the general public interest in animal welfare.

10.6.1 Overrepresentation of Industry Interests in Standards Development Process

Perceptions of procedural fairness in formal decision-making processes shape people's views about the legitimacy of the decision-making body and the decisions it makes (Tyler 1994, 2005). If people perceive the process to be unfair, whether due to bias on behalf of the decision-maker, or because they have not had an adequate opportunity to have their say, it can affect their willingness to accept the decision and to engage with future decision-making processes (Tyler 1994). Having recognised this, governments within liberal democratic societies have promoted the notion of "participatory democracy" to allow stakeholders and the broader community the opportunity to participate in government decision-making processes (Holmes 2011). Australia's process for developing farm animal welfare standards incorporates participatory mechanisms including the provision of deliberative forums for key stakeholders, and opportunities for public comment on proposed drafts (see for example, AHA 2009). Unfortunately, however, the process suffers on procedural fairness grounds due to perceived bias on the part of key decision-making institutions at various stages of the process, and the disproportionate representation of industry interests.

To achieve national consistency, Australia's standards development process is coordinated at a national level with cooperation from all jurisdictional governments. Management of the process is vested in Animal Health Australia (AHA), "a not-for-profit public company established by the Australian, State, and Territory governments and major national livestock industry organisations" (2014a). AHA's membership includes the Australian Chicken Meat Federation, Australian Dairy Farmers, Australian Egg Corporation, Australian Pork Limited, the Cattle Council of Australia, and other peak industry bodies (AHA 2014b). It exists to promote "a robust national animal health system that maximises competitive advantage and preferred market access for Australia's livestock industries" (AHA 2014a).

In 2009, AHA created a national business plan for the standards development process (Business Plan), which addresses funding, priority setting, membership of writing and reference groups, and outlines the various stages of the process. The Business Plan makes clear that control over the process is largely vested within three institutions – AHA, the national Animal Welfare Committee (consisting of representatives of the Departments of Agriculture), and the relevant livestock industries. These institutions set the priorities for what standards are to be developed, provide the funding for the process, determine whether there is need for scientific research, and commission such research if it is deemed to be required (AHA 2009).

Once the priorities have been set, AHA is responsible for establishing a standards writing group, which is made up of representatives from the three institutions mentioned above, "relevant independent science representation, invited consultants" and an "independent chair" (AHA 2009). Leading animal welfare groups, RSPCA Australia and Animals Australia, have decided not to participate in the writing groups as they have both formed the view that their involvement will have no substantive impact on the drafting process.[36] These groups do, however, participate within the stakeholder reference group, which is responsible for reviewing and providing comment on draft standards before they go out to public consultation. Once the standards have been finalised following the public consultation phase, they are submitted to a meeting of jurisdictional agriculture ministers to be formally endorsed for implementation in each State and Territory.[37]

[36] Personal correspondence with the author from RSPCA Australia staff member, 28 January 2013, and Glenys Oogjes, Executive Director, Animals Australia, 30 March 2012.

[37] Such meetings were previously undertaken under the name of the Standing Council on Primary Industries (SCoPI). However, the Commonwealth Government recently announced that this Council would be abolished as part of its budgetary reforms. The Commonwealth Minister for

The process for developing farm animal welfare standards is dominated by industry interests, if not represented through the industries themselves, then through the agency of AHA, the Departments of Agriculture, and the agriculture ministers. The control exerted by these institutions creates the appearance of a system that is heavily weighted in favour of industry interests, in which alternative viewpoints may not receive a fair hearing. Animal welfare representatives have the opportunity to provide their input to the process but the potential to actually influence substantive changes within such a forum is limited. As the Executive Director of Animals Australia, Glenys Oogjes (2011, p. 5) has observed:

> Whilst consensus is sought in the meetings of such reference groups, the reality of the dynamics of the process is that the livestock industries have an (unofficial) power of veto in decision making – if they determine that they cannot or will not accept a particular Standard, invariably the proposed Standard is varied (watered down) or becomes merely a Guideline.

The perception of bias within the process is also shared by the RSPCA. It has consistently raised concerns over AHA's role in the development process on the basis that, given its membership, "it cannot be considered an independent body in [the standards development] process" (RSPCA Australia 2008). The result is a process that lacks procedural legitimacy and ultimately produces animal welfare standards that reflect the industry status quo, and deepen the disconnect from public expectations.

10.6.2 Industry Influence over Development of Animal Welfare Science

An issue related to the standard-setting process is the development of the science upon which such standards are supposed to be based. The *Australian Animal Welfare Strategy* and the Business Plan emphasise the need for animal welfare policy to be underpinned by scientific knowledge. In 2009/10, Australia invested approximately $14.279 million in primary industries related animal welfare research, development, and extension (RD&E) (Animal Welfare RD&E Strategy 2010). The funding derived primarily from industry, government, and university sources in similar proportions of $5,195,000, $4,707,000, and $4,377,000, respectively. Much of the RD&E expenditure is coordinated through collaborative forums like industry-specific Cooperative Research Centres,[38] the Commonwealth Scientific and Industrial Research Organisation (CSIRO), and university centres such as the Animal Welfare Science Centre and the Centre for Animal Welfare and Ethics at the University of Queensland.

Agriculture has since stated that jurisdictional agriculture ministers would continue to meet but on a more informal basis (Vidot 2013).

[38] See CRC Australia: https://www.crc.gov.au/About-CRCs/Directory/Agriculture-Forestry-and-Fishing/Pages/default.aspx

The largest proportion of funding is managed by livestock industry Research and Development Corporations (RDCs). The role of industry RDCs is to "invest in R&D and innovation to improve the productivity and delivery of high quality products in order to underpin the competitiveness and profitability of Australia's agricultural, fish and forestry industries" (Commonwealth Department of Agriculture 2012). The Commonwealth Department of Agriculture is responsible for administering the legislation that governs RDCs. The RD&E expenditure of the RDCs is funded through industry levies that are matched dollar for dollar by the Commonwealth Government (Commonwealth Government 2011). Prominent livestock RDCs include Meat and Livestock Australia, LiveCorp, Dairy Australia, Australian Wool Innovation, Australian Egg Corporation, and Australian Pork Limited. In addition to RD&E activities, many of these organisations are responsible for industry marketing and representation functions.

The degree of influence and control exerted by some RDCs over animal welfare research (despite the matched contribution of public funding) has raised concerns within the scientific community. The Director of the Centre for Animal Welfare and Ethics at the University of Queensland, Professor Clive Phillips (2011) has noted the following about the RDC research process:

> [E]ssentially all funding for research into the welfare of animals that are exported live is administered by industry body Meat and Livestock Australia for the exporters' organisation LiveCorp.
>
> The industry body determines the type of work that is conducted, who does the research, how it is conducted and how it is reported.
>
> Under contracts with [Meat and Livestock Australia] or LiveCorp, researchers working for them cannot release any information about the project to anybody without permission of industry representatives. In practice that means that reports are drafted by the researchers and edited by the industry body before release. Delays in publication can be lengthy, up to one year.

Phillips also raised the concern that, due to funding pressures, "some [researchers] may be tempted to undertake work that has the objective of confirming that the status quo does not damage animal welfare, so that the industry does not have to modify its practices to meet community expectations of high welfare standards" (2011). This concern has been partly supported by subsequent empirical research conducted by van der Schot and Phillips (2012) into incidences of "publication bias" within animal welfare scientific literature. The research found that authors' assessment of animal welfare tended to support the interests of the funding agency. The data showed that the effects of new treatments in improving animal welfare were rated lower if the research was funded by industry, as compared to government, or charitable organisations. Van der Schot and Phillips (2012, p. 957) warn that this "may retard progress in animal welfare development in the animal production industries in particular", and that in light of the "changes in research funding towards more industry sponsorship, this has the potential to undermine the benefits arising from research in animal welfare."

The priorities of industry RDCs are self-evident. They exist to promote the productivity and profitability of their respective industries. Any animal welfare RD&E

commissioned by these organisations is invariably for that purpose. This induces a very instrumental approach to animal welfare RD&E in which such research is considered predominately in terms of its contribution to productivity and industry sustainability. This approach is reflected in the *National Primary Industries RD&E Framework: Animal Welfare RD&E Strategy* (2010, p. 8):

> The development of credible animal welfare policy needs to be soundly based in science, otherwise the growth and market access of Australian animal industries and the current freedoms of animal ownership may be jeopardised.

Industry RDCs are of course entitled to engage in research to pursue these objectives. The problem lies in the dedication of public funds for such purposes. As a matter of good governance, public funds that are reserved for researching an issue in the public interest should not be delegated to organisations that possess priorities which may conflict with that interest. Unfortunately, the government department responsible for administering the funding arrangements – the Commonwealth Department of Agriculture – is unlikely to perceive this to be an issue as it shares the same instrumental approach to animal welfare as that of the industry RDCs. Consequently, much of the public funding dedicated to animal welfare science is directed towards research that provides little in the way of substantive improvements to welfare standards.

10.6.3 Inadequate Compliance Monitoring and Enforcement

In addition to its detrimental influence over standard-setting processes and the development of animal welfare science, industry norms have also influenced the manner in which Departments of Agriculture approach their compliance monitoring and enforcement functions. First, the Departments have developed a compliance system that delegates the majority of inspection functions to the industry itself through formal and informal recognition of industry quality assurance (QA) schemes (Goodfellow 2013). The basic approach is that if a production facility is part of a recognised industry QA scheme, it will be subject to fewer routine inspections, or no routine inspections at all (for animal welfare compliance purposes).[39] In other words, the QA scheme is used to *substitute*, not supplement, the government inspection regime. This is despite the regulatory environment being one of high risk due to the inherent vulnerabilities of animals in having no legal (or physical) means of asserting their interests, the financial incentives for noncompliance, and sociological research indicating a possible culture of defiance to welfare regulation on the part of the farming community (Goodfellow 2013). This is evidence of the tendency for Departments of Agriculture to under appreciate risks of noncompliance within

[39] See, Edge and Bailey (2010), noting that facilities under an approved QA scheme will be "deemed to comply in the first instance."

the agricultural sector due to their strong identification with the industry's perception of reality (McDonnell and Schwarcz 2011; and discussion at Sect. 10.5.2).

At the same time as delegating inspection functions to the industry, Departments of Agriculture in several States have sought to limit the jurisdictional responsibilities of RSPCA inspectors with respect to livestock. Memorandums of Understanding between State RSPCA's and Departments of Agriculture in Queensland and Victoria provide that any complaints relating to "commercial livestock" (over ten sheep, cattle, pigs, or over 500 poultry) will be referred on to the Department for investigation (Queensland Department of Agriculture, Fisheries and Forestry 2013b; McEwen 2006). Similar arrangements appear to be in place in Western Australia where the Department of Agriculture and Food's (2014) website specifically refers to the RSPCA's role in *companion* animal compliance:

> The RSPCA receive and investigate complaints of animal cruelty and take the lead in the area of companion animal welfare. The Western Australian Government supports the RSPCA in its work pursuing companion animal welfare compliance and responsible pet ownership through a funding grant.

While no wholesale restriction of RSPCA jurisdiction exists in the State of New South Wales, there are additional requirements placed on inspectors with respect to dealing with "stock animals". Inspectors are required to seek advice from the Department of Primary Industries before charging a person in relation to an offence of failing to provide a stock animal with sufficient food, drink, or shelter.[40] Similarly, "seizure and disposal" of stock animals, in the absence of an associated prosecution, can only be authorised by the Director-General of the Department, and this can only occur after a convoluted process of establishing, and receiving advice from a "Stock Welfare Panel".[41]

A further concern relates to the Departments' approach to enforcement when noncompliance is detected. Most Departments of Agriculture take a "responsive" approach to enforcement (Goodfellow 2013), which recommends that noncompliance, "however serious" be responded to with dialogue in the first instance (Braithwaite 2011, p. 483). It recommends that punitive sanctions should only be entertained in the event of repeated breaches in defiance of prior advice or warnings issued by the regulator. This compliance-based approach to enforcement has been based on a characterisation of animal welfare offences as "technical rule violations" or "side effects of legitimate, socially value business operations" (Goodfellow 2013, pp. 204–206). As a consequence, it has created an enforcement system that implicitly accepts the *management* of animal mistreatment as opposed to its prohibition.

No clearer evidence of this approach to enforcement can be found than in the Commonwealth Department of Agriculture's regulation of the live animal export trade. As previously stated, in 2012 and 2013 alone there have been 30 reported breaches of the *Exporter Supply Chain Assurance System* and the *Australian Standards for the Export of Livestock* (Commonwealth Department of Agriculture

[40] *Prevention of Cruelty to Animals Act 1979*, s 8(4).

[41] *Prevention of Cruelty to Animals Act 1979*, Part 2B.

2014b). Several of these reports have been substantiated by the Department's own investigations, yet no fines, licence revocations, or even licence suspensions, have been imposed. The Department's only response has been to impose "extra conditions" on the approval of the exporter's next consignment (see for example, *Compliance Investigation Report 14: Sheep exported to Jordan*, Commonwealth Department of Agriculture 2014c). These additional conditions amount to nothing more than a tax, which exporters simply treat as the cost of doing business. They have no deterrent effect at all, and as a consequence, significant welfare incidents and breaches of the regulations have become routine to the point where they are now simply accepted as part of the trade. When questioned about the inevitability of further welfare incidents on national radio, the Commonwealth Agriculture Minister simply stated that "the trade will be watched over and regulated to the best of our ability but we can never make promises that…there will never be another drama in this sector. We're going to manage these issues as they occur" (ABC 2014).

This attitude of tolerance to animal welfare incidents was also clearly demonstrated by senior officers of the Northern Territory Department of Primary Industry and Fisheries when responding to a major case of neglect on a university operated cattle station in Northern Australia. The 2009 incident resulted in the death of between 200 and 800 cattle and was described as one of Australia's worst cases of animal cruelty (Murdoch 2011). Despite an Ombudsman inquiry finding that the deaths were caused by neglect and that the Territory's *Animal Welfare Act 2000* had been breached (Ombudsman 2010, p. 5), no prosecution was ever initiated as investigating officers did not feel it was warranted. In giving evidence to a subsequent parliamentary inquiry, the then Chief Veterinary Officer, Dr Brian Radunz, stated that the focus of the investigation was on the welfare of the remaining cattle and that prosecution was considered to be a "much lower priority", and that "whether to prosecute or not was a minor point" (Legislative Assembly of the Northern Territory 2011a, pp. 11–15). This was not, however, the view of another attending veterinary officer, Dr John Eccles, who thought that prosecution was warranted. Dr Eccles was relatively new to the Department but was nevertheless a senior veterinary officer with 12 years prior experience with the Australian Quarantine and Inspection Service. Despite this, Dr Eccles' views on prosecution were not supported by the Chief Veterinary Officer and he was subsequently removed from the investigation. When questioned about this removal, Dr Radunz stated that "it was done to take the heat out of the conflict, to have another person doing the investigation, another person who was experienced and skilled" (Legislative Assembly of the Northern Territory 2011a, p. 15). Dr Eccles was replaced by another veterinary officer who, Dr Radunz informed, "had been in the [area] for about five or six years", and who presumably did not express the same views about the necessity for prosecution.

In justifying his approach to the investigation, Dr Radunz likened the cattle deaths at the station to the "normal death rate" that occurs on cattle stations in Australia's Northern regions (Legislative Assembly of the Northern Territory 2011a, p. 5). He subsequently concluded: "I cannot support a teaching facility being prosecuted when other properties with similar situations would not be considered for prosecution." This view suggests that incidents of mass cattle starvation were not unusual within the Northern Territory and were simply accepted by the Department as the industry norm. The lack of enforcement culture within the Department was also made apparent when Dr Radunz stated that he could not recall any prosecutions under livestock related legislation within his 40 years in the Department (Legislative Assembly of the Northern Territory 2011a, p. 8).

Further insight into the Chief Veterinary Officer's approach to animal welfare issues was revealed in questioning about his views on the difficulties of managing cattle stations and its relevance to what would amount to a prosecutable offence under the *Animal Welfare Act* (Legislative Assembly of the Northern Territory 2011a, p. 12):

> **Dr Radunz**: What happened on the Station was that there were fires the year before, there's a failed sale of a large number of animals, so therefore there's a much larger number of animals there than would be prudent. There's probably a situation of the property being overstocked. All that can be done is to manage the balance between the feeding the animals and then, if necessary, humanely destroying the animals which either are too poor to walk out to feed or there aren't sort of enough funds to actually feed those animals. Feed those animals for a whole year.
>
> **Mr Elferink [Parliamentary representative]**: So if I understand you correctly, an active case of cruelty, somebody taking a steel pipe, for example, to a cow would represent an act of cruelty, but an act of neglect would not?
>
> **Dr Radunz**: I think neglect would be not to destroy animals which are in too poor condition that they can't stand up or they can't walk.

Dr Radunz's response implies that allowing cattle to starve to the point of disability is not sufficient to amount to a case of neglect. Rather, it is only when the responsible person then fails to destroy the animals that neglect can be established. Curiously, this is precisely what occurred at Mataranka Station but Dr Radunz nevertheless argued against the case for prosecution.

The Department's response to this incident was a classic example of a regulator adopting the norms of the regulated industry in its approach to administering a public interest statute. As the subsequent wide-spread public condemnation, (as evidenced within significant media coverage (see for example, Adams 2011; Murdoch 2011), Ombudsman (2010) and Parliamentary inquiries (Legislative Assembly of the Northern Territory, Legislative Assembly of the Northern Territory 2011a, b) evidenced, this approach was completely out of step with community expectations and the public interest in animal welfare that underpins the Northern Territory's *Animal Welfare Act*. The Northern Territory example is emblematic of enforcement problems across jurisdictions. The significant divesting of inspection responsibilities to industry, the restrictions on RSPCA jurisdiction, and the characterisation of animal welfare breaches as "technical rule violations" that need to be "managed", creates an enforcement system that often excuses cases of animal mistreatment within the agricultural sector. It represents a deviation from the public's interests and expectations in protecting farm animals from cruelty.

This examination has depicted a system that is suffering from regulatory capture. Departments of Agriculture have adopted, whether consciously or not, the norms of livestock industries in their approach to administering animal welfare legislation. In doing so, their performance has deviated from the public interest such laws are intended to serve. The protection and promotion of farm animal welfare has been routinely subordinated to the interests of productive and profitable primary industries. This has led to the development of regulatory deficiencies, which in turn, have contributed to the cases of regulatory failure outlined in Sect. 10.3. To ensure the public interest in farm animal welfare is upheld, regulatory reform needs to occur.

10.7 Options for Reform

The customary response within the regulatory capture literature to addressing undue influences is to recommend the establishment of greater accountability and transparency mechanisms. As Novak (2013, p. 48) explains, the process involves "the simultaneously mundane and heroic task of attempting to blunt the force of perennial public corruptions and private coercions by simply piling on 'all the checks and balances that human ingenuity can devise.'" There are of course myriad forms of checking and balancing devices available to create more accountable regulatory environments. Such options must be considered in light of the primary mechanisms of capture identified. To this end, poor institutional design, the catalyst for the regulatory failings discussed in this chapter, will be the primary focus of attention.

Section 10.4.1 introduced the potential problems that can arise when a department is delegated with the responsibility of pursuing two conflicting objectives, especially when one objective has clear priority over the other. As Biber (2009) has shown, agencies will systematically underperform on secondary goals that conflict with the achievement of the agencies' primary goals. In particular, agencies are found to pursue short term economic goals that are easy to measure at the expense of more elusive social goals in the public interest. The structure of the reward system for Departments of Agriculture is geared towards achieving the economic goals of improving industry productivity and profitability. Subsequently, the public interest in farm animal welfare is neglected.

Similar institutional design problems have been identified many times before in other regulatory contexts around the world. Regulatory frameworks relating to atomic energy (Barkow 2010), workplace health and safety (Gordon and Hafer 2014), financial services (Baxter 2011; Omarova 2012) minerals management (Hogue 2010), environmental protection (Biber 2009), and others have suffered from design faults leading to underperformance in key areas. To rebalance the regulatory approach, reforms have usually taken the form of one of three strategies outlined below. Of course, a detailed assessment of these options could form the basis of a separate publication. The following discussion is intended only to provide a basic introduction to the available options.

10.7.1 Separation of Competing Responsibilities: Independent Animal Welfare Authorities

Perhaps the most obvious reform for addressing issues of poor institutional design is to separate out, or decouple, the competing responsibilities (Biber 2009). This can be achieved through transferring the subordinate responsibility to either an existing agency that does not possess a conflicting organisational priority, or to a new government entity established specifically to address that responsibility. For instance, in the European Union (E.U.), responsibility for animal welfare was transferred from the Directorate-General for Agriculture, to the Directorate-General for Health and Consumers where 'protecting the health and welfare of farm animals' is prominently recognised as one of the Directorate's five key aims (EC 2012). To add a further degree of independence, scientific advice on welfare matters is provided for by an independent Panel on Animal Health and Welfare within the European Food Safety Commission (EFSA 2014).

A prominent recent example of a government creating an entirely new agency to address a public interest concern was the establishment of the U.S. Consumer Financial Protection Bureau in the wake of the 2007–2008 Financial Crisis (Kwak 2014). Following numerous congressional and other inquiries it was found that while the relevant existing regulatory institutions[42] had some consumer protection responsibilities, their incentives to pursue those responsibilities were not sufficient to ensure adequate financial consumer protections (Kwak 2014). Accordingly, the Consumer Bureau was established to fill that void and ensure consumer protection would be given the priority it required (CFPB 2014).

The establishment of independent environmental protection authorities (EPAs) provides another example of government reforms designed to separate out particular functions from traditional bureaucratic structures. In 1998, Briody and Prenzler argued for the establishment of an EPA in Queensland to address the presence of regulatory capture on the part of the Environmental Compliance Division within the then Department of Mines and Energy. Marsden et al. provided a similar critique of the Tasmanian environmental regime in 2000, identifying instances of regulatory capture with respect to the administration of new environmental regulations. Poor administrative design was a feature of the capture identified, in response to which the authors proposed the establishment of an EPA. An EPA was later established in Tasmania in 2008 (EPA Tas 2013).

Equivalent authorities could be established for animal welfare. The possible structure, scope, powers, and responsibilities of such authorities are unfortunately beyond the scope of this chapter.[43] As with the EPAs, animal welfare authorities could be based at the State level in accordance with Constitutional requirements.

[42] Including the Office of the Comptroller of the Currency, The Federal Reserve, the Office of Thrift Supervision, the Federal Deposit Insurance Corporation, and the Securities and Exchange Commission.

[43] The author has analysed these issues as part of a PhD thesis due for submission in 2015.

However, an equivalent authority could also be proposed at the federal level to deal with those areas of animal welfare responsibility governed by the Commonwealth Government. As noted in the introduction, the Australian Labor Party, when in Government in 2013, foreshadowed the proposal of an "Independent Office of Animal Welfare" at the Commonwealth level (Parke 2013). The Office was proposed precisely to address the issues of conflicting responsibilities raised in this chapter (see Parke 2013). However, when the formal proposal was announced some months later it took the form of an "Inspector-General for Animal Welfare and Live Animal Exports" (discussed further below).

Separating competing institutional objectives improves clarity in public administration and the ability to measure government performance (Baldwin and Walker 2008). If responsibility for animal welfare was transferred to independent government entities, Departments of Agriculture would be free to get on with their job of promoting productive and profitable primary industries, unencumbered by growing public demands for greater attention to be given to a conflicting objective. It should be emphasised here that such reforms would not be designed to exclude the interests of livestock industries from the animal welfare regulatory process. This would create an unworkable regulatory environment and one that would suffer from the same issues of democratic legitimacy it was designed to address. Any new entity should be required to involve industry in the process as much as possible subject only to the limits necessary to ensure the entity's independence. Achieving this balance would improve the democratic legitimacy of the current process and ensure the public interest in animal welfare was no longer neglected.

10.7.2 Regulatory Contrarian: Inspector-General or Ombudsman for Animal Welfare

Another option for reducing the effects of regulatory capture is to establish an independent government entity that has the specific task of monitoring and auditing the performance of a department. The main purpose of such an entity, as Omarova (2012, p. 641) explains, is to "help regulators overcome the various cognitive and incentive biases and 'to counteract agency inaction or ossification'." McDonnell and Schwarcz (2011) have labelled such entities 'regulatory contrarians' reflecting their role in providing a contrary perspective to prevailing bureaucratic thinking. They define contrarians by three distinguishing features: (1) "they possess persuasive authority by virtue of their position, access to media and officials, or speaking engagements and reports;" (2) "they are affiliated with, and enjoy the privileged access to, a regulatory entity but are nonetheless independent, as reflected in their budget, staffing, and/or priorities;" and (3) "they are tasked with studying the regulatory process, policy positions, and the regulated [environment] and in some way reporting on deficiencies and potential improvements." Regulatory contrarians often take the form of Inspectors-General or Ombudsman offices.

In July 2013, the Labor Government announced plans to establish an "Inspector-General for Animal Welfare and Live Animal Exports" (Fitzgibbon 2013). The proposed role of the Inspector-General was to "review and audit Australia's live animal export trade processes and systems to strengthen [the] animal welfare assurance system." The Inspector-General was to be established as an "independent, statutory office holder, reporting directly to the Minister [for Agriculture]." Unfortunately, the Bill to establish the office was never introduced to Parliament as the Australian Labor Party lost the Federal Election 2 months later. The Coalition Government has maintained a policy of opposition to the Inspector-General proposal on the basis that it would be adding another layer of "unnecessary red tape" to the live export trade (Cobb 2013).

Regulatory contrarians provide an important check on deficiencies in departmental performance. An Inspector-General or Ombudsman for animal welfare would provide a much needed contrary perspective to the prevailing Department of Agriculture views on animal welfare.

10.7.3 Tripartism

A third option for increasing transparency and accountability is that of regulatory "tripartism", an influential theory that was first proposed in 1991 by Ayres and Braithwaite as a potential restraint on regulatory capture. The theory is based on creating a more prominent role for relevant public interest NGOs in regulatory processes. It recommends three key strategies for fostering NGO involvement, including (1) granting NGOs and their members all the information that is available to the regulator; (2) giving the NGOs a seat at the negotiating table with the industry and the regulator when deals are done; and (3) granting the NGOs the same standing to sue or prosecute under the relevant regulatory statute as the regulator. Essentially, "tripartism means both opening to NGOs the smoke-filled rooms where the real business of regulation is transacted and allowing the NGO to operate as a private attorney-general" (Ayres and Braithwaite 1991, p. 441). Empowering NGOs in this way is said to create an "effective third-party guardian" of the public interest, which acts as a "built-in source of countervailing perspective on substantive policy issues and imposes structural checks on regulatory capture" (Omarova 2012, p. 642).

As set out in Sect. 10.6.1 under current arrangements, prominent animal protection groups such as the RSPCA and Animals Australia are provided with the opportunity to consult with the government and industry during the process of developing animal welfare standards. While participation on standards reference groups has some worth (even despite the disproportionate industry influence), it is unlikely that this formal process is what Ayres and Braithwaite envisaged by the "smoke-filled rooms where the real business of regulation is transacted." One of the apparent challenges of facilitating such access would be identifying precisely when and where regulatory transactions actually take place. Departmental interactions with industry are many and varied, and discussions about current or proposed regulations, and their implementation, take place on a frequent and at times informal basis. Conscious forethought and a genuine commitment to facilitating access on behalf of the responsible department would be a necessary condition for tripartism to operate effectively in practice. The prospects of Departments of Agriculture manifesting such a commitment in the context of animal welfare regulation are not great.

With respect to granting NGOs the right to prosecute, one may point to the RSPCA's role in enforcing animal welfare legislation as evidence that this feature of tripartism has been granted. However, in most jurisdictions, it is not the RSPCA as a corporate entity that is granted powers, but the inspectors who are appointed under the legislation by the responsible minister. The enforcement activities of inspectors are subject to the control and oversight of the Departments of Agriculture through conditions placed of their appointment,[44] training requirements,[45] Memorandums of Understanding regarding operational matters,[46] and routine reporting obligations.[47] As outlined in Sect. 10.6.3, the degree of control is heightened in the context of commercial livestock. The RSPCA is sometimes criticised on the basis that it does not pursue more vigorous enforcement actions within the agricultural sector including the initiation of cases of strategic significance, or "test cases." Such criticism often fails to appreciate the degree to which controlling factors operate to prevent RSPCA inspectors from taking this approach.[48] Even in the absence of more vigorous enforcement, or test cases, it is not uncommon for State RSPCAs to be threatened with having their role in law enforcement removed. Recent parliamentary debates in South Australia and Western Australia demonstrate the very real potential for such threats to materialise.[49] So while the RSPCA is granted some privileges in terms of its role in enforcing animal welfare legislation, this could not be said to amount to a freedom to operate as a "private attorney-general." In reality, the RSPCA's enforcement functions are controlled and constrained by the administering Departments of Agriculture.

The theory of regulatory tripartism presents a number of checking and balancing mechanisms that, if implemented effectively, would certainly improve the accountability and transparency of the farm animal welfare regulatory environment. The challenge, however, lies in fostering a willingness on the part of Departments of Agriculture to provide for such regulatory privileges. Unfortunately, structural and cultural factors may operate to prevent this from occurring. On that basis, any reforms directed towards promoting tripartism should probably be approached as a complementary feature to the establishment of an Animal Welfare Authority, Inspector-General, or Ombudsman.

---

[44] See for instance, *Animal Welfare Act 1985* (SA), s 28.

[45] See for instance, *Animal Care and Protection Act 2001* (Qld), s 114.

[46] See discussion in Sect. 10.6.3.

[47] See for instance, *Prevention of Cruelty to Animals Act 1979* (NSW), s 34B.

[48] This is in combination of course with traditional barriers to public interest litigation. See, McEwen 2011.

[49] See, South Australia, *Parliamentary Debates*, Legislative Council, 27 March 2012, 669; and Western Australia, *Parliamentary Debates*, Legislative Council, 24 October 2013, 5443c, each discussing proposals to abolish the RSPCA's role in farm animal welfare law enforcement.

10.8 Conclusion

This chapter has sought to provide some analytic depth to the increasing accusations of conflicting interests on the part of the Departments of Agriculture in regulating the welfare of farm animals in Australia. Such accusations are in large part substantiated. Farm animal welfare and on-farm productivity are predominately in a relationship of conflict. Consequently, Departments of Agriculture can be said to possess competing responsibilities in promoting productive and profitable primary industries while at the same time protecting and promoting farm animal welfare. This is not of itself a problem; rather, the difficulty lies in the respective priority placed on each responsibility. The Departments' overriding mission and very reason for existence, dictate against any possibility of a balanced approach. The structure of the Departments' reward system leads them inevitably to prioritise measurable economic goals associated with productive and profitable primary industries, over the more elusive, less determinate public interest in farm animal welfare. This has led the Departments to a community of interests with the regulated parties where they have adopted the same instrumental approach to animal welfare as that advanced by the livestock industries. Consequently, the Departments have deviated from serving the public interest in farm animal welfare as demonstrated by various regulatory failures and process deficiencies. These features bear all the hallmarks of a framework that is suffering from regulatory capture. And, true to form, capture has left a trail of procedurally unfair and democratically illegitimate governance and regulatory processes in its wake. Reforms are now required to correct these imbalances, to restore legitimacy, and to institutionalise the growing public interest in animal welfare within our systems of government.

References

ABC, Australian Broadcasting Corporation. 2014. *Live animal exports to Egypt resume*. 20 March. http://www.abc.net.au/news/2014-03-20/live-animal-exports-to-egypt-resume/5335184. Accessed 20 Mar 2014.

Adams, Prue. 2011. *Sorry saga*. ABC Landline. 27 July. http://www.abc.net.au/landline/content/2010/s3276556.htm. Accessed 15 May 2014.

Adams, Gary, Hayes Sharon, and Weierter Stuart. 2003. Regulatory risks: Minimising misconduct risks in agencies with regulatory functions. In *Building capacity*, 1–6. Brisbane: Crime and Misconduct Commission.

AgForce Queensland. 2013. *Agriculture – 30 issues, 30 days. Issue 13: Animal Welfare*. Media Release. 4 August. http://www.agforceqld.org.au/file.php?id=2719&open=yes. Accessed 2 Feb 2014.

AHA, Animal Health Australia. 2009. *Development of Australian standards and guidelines for the welfare of livestock – Business plan*. Available at: http://www.animalwelfarestandards.net.au/files/2011/01/Animal-Welfare-Standards-and-Guidelines-Development-Business-Plan.pdf.

AHA, Animal Health Australia. 2014a. *Australian animal welfare standards and guidelines*. http://www.animalwelfarestandards.net.au/. Accessed 2 Feb 2014.

AHA, Animal Health Australia. 2014b. *Company profile*. http://www.animalhealthaustralia.com.au/about-us/company-profile/. Accessed 2 Feb 2014.

AHA. Animal Health Australia. 2014c. *Members*. http://www.animalhealthaustralia.com.au/about-us/members/. Accessed 2 Feb 2014.

Appleby, Michael. 2008. Food prices and animal welfare. In *The animal ethics reader*, 2nd ed, ed. Susan Armstrong and Richard Botzler, 187–190. London: Routledge.

Appleby, Michael, Neil Cutler, John Gazzard, Peter Goddard, John Milne, Colin Morgan, and Andrew Redfern. 2003. What price cheap food? *Journal of Agriculture and Environmental Ethics* 16: 395–408.

Australian Associated Press. 2013. Gaza cattle torture prompts calls to ban live animal exports. *Sydney Morning Herald*, 12 December 2013. http://www.smh.com.au/environment/animals/gaza-cattle-torture-prompts-calls-to-ban-live-animal-exports-20131212-2z87t.html. Accessed 2 Feb 2014.

Australian Greens. 2013. *Ingham abuse at Turkey Abattoir: Overdue office of animal welfare needed*. Media Release. 21 March. http://greensmps.org.au/content/media-releases/ingham-abuse-turkey-abattoir-overdue-office-animal-welfare-needed. Accessed 2 Feb 2014.

Australian Labor Party. 2011. *46th national conference: national platform*. Barton: Australian Labor Party.

Ayres, Ian, and John Braithwaite. 1991. Tripartism: Regulatory capture and empowerment. *Law and Social Inquiry* 16(3): 435.

Baldwin, Richard, and Christopher Walker. 2008. The separation of policy and operations in organisations: A New South Wales case study. *Australian Journal of Public Administration* 54(2): 184–195.

Barkow, Rachel. 2010. Insulating agencies: Avoiding capture through institutional design. *Texas Law Review* 89: 15–80.

Baxter, Lawrence. 2011. "Capture" in financial regulation: Can we channel it toward the common good? *Cornell Journal of Law and Public Policy* 21: 175–200.

Baxter, Lawrence. 2012. Understanding regulatory capture: An academic perspective from the United States. In *The making of good financial regulation*, ed. Stefano Pagliari, 31–39. Surrey: Grosvenor House Publishing Ltd.

Bennett, Richard. 1997. Animal welfare, economics and policy. In *Animal welfare*, ed. Appleby Michael and Hughes Barry, 279–290. Wallingford: CAB International.

Benson, John, and Bernard Rollin. 2004. *The wellbeing of farm animals: Challenges and solutions*. Iowa: Blackwell Publishing.

Bernstein, Marver. 1955. *Regulating business by independent commission*. Princeton: Princeton University Press.

Biber, Eric. 2009. Too many things to do: How to deal with the dysfunctions of multiple-goal agencies. *Harvard Environmental Law Review* 33: 1–64.

Braithwaite, John. 2011. The essence of responsive regulation. *University of British Columbia Law Review* 44: 475–520.

Briody, Michael, and Tim Prenzler. 1998. The enforcement of environmental protection laws in Queensland: A case of regulatory capture? *Environmental and Planning Law Journal* 15: 54–71.

Broom, Donald. 2008. *Animal welfare: Future knowledge, attitudes and solutions*. Paper presented at AAWS international animal welfare conference. Gold Coast, Australia. 31 August–3 September.

Bruce, Alex. 2011. *Animal welfare Law in Australia: An integrated approach*. Sydney: LexisNexis.

Buiter, William. 2008. Central Banks and financial crises. In *Maintaining stability in a changing financial system*. Proceedings of the Federal Reserve Bank of Kansas City Economic Policy Symposium, 495–633. Jackson Hole, Wyoming, 21–23 August 2008.

Cannane, Steve. 2012. Abattoir shut down over cruelty concerns. *ABC News*. 10 February.http://www.abc.net.au/news/2012-02-09/abattoir-shut-down-over-cruelty-concerns/3821302. Accessed 14 Feb 2014.

Carpenter, Daniel. 2014. Detecting and measuring capture. In *Preventing regulatory capture: Special interest influence and how to limit it*, ed. Daniel Carpenter and David Moss, 57–69. New York: Cambridge University Press.

Carpenter, Daniel, and David Moss. 2014. Introduction. In *Preventing regulatory capture: Special interest influence and how to limit it*, ed. Daniel Carpenter and David Moss, 1–23. New York: Cambridge University Press.

Caulfield, Malcolm. 2008. *Handbook of Australian animal cruelty law*. Melbourne: Animals Australia.

CFPB, Consumer Financial Protection Bureau. 2014. *About us*. http://www.consumerfinance.gov/the-bureau/. Accessed 2 Feb 2014.

Cobb, John, and Shadow Minister for Agriculture. 2013. *Minister admits system is "strongest" but adds more red tape*. Media Release. 31 July. http://www.johncobb.com.au/MediaHub/PortfolioMediaReleases/tabid/93/articleType/ArticleView/articleId/2110/Minister-admits-system-is-strongest-but-adds-more-red-tape.aspx. Accessed 2 Feb 2014.

Coleman, Grahame. 2007. *Public perception of animal pain and animal welfare*. Paper presented at AAWS international animal welfare conference. Gold Coast, Australia. 31 August–3 September 2007.

Commonwealth Department of Agriculture. 2012. *Research and development corporations*. http://www.daff.gov.au/agriculture-food/innovation/research_and_development_corporations_and_companies. Accessed 2 Feb 2014.

Commonwealth Department of Agriculture. 2013a. *Annual Report 2012/13*. Canberra: Australian Government.

Commonwealth Department of Agriculture. 2013b. *Portfolio budget statements 2013–14*. Canberra: Australian Government.

Commonwealth Department of Agriculture. 2014a. *Investigations into mortalities*. http://www.daff.gov.au/biosecurity/export/live-animals/livestock/regulatory-framework/compliance-investigations/investigations-mortalities. Accessed 30 Mar 2014.

Commonwealth Department of Agriculture. 2014b. *Mortality investigation report 46: sheep exported to Qatar and the United Arab Emirates – September 2013*. http://www.daff.gov.au/biosecurity/export/live-animals/livestock/regulatory-framework/compliance-investigations/investigations-mortalities/report-46. Accessed 30 Mar 2014.

Commonwealth Department of Agriculture. 2014c. *Regulatory compliance investigations*. http://www.daff.gov.au/biosecurity/export/live-animals/livestock/regulatory-framework/compliance-investigations/investigations-regulatory-compliance. Accessed 30 Mar 2014.

Commonwealth Department of Agriculture. 2014d. Compliance investigation report 14: sheep exported to Jordan – February 2014. http://www.daff.gov.au/biosecurity/export/live-animals/livestock/regulatory-framework/compliance-investigations/investigations-regulatory-compliance/jordan-june-2013. Accessed 30 Mar 2014.

Commonwealth Government. 2011. *Rural research and development corporations*. About. http://www.ruralrdc.com.au/Page/About/About.aspx. Accessed 2 Feb 2014.

Crowley, R. 1982. The design of government policy agencies: Do we learn from experience? *The Canadian Journal of Regional Science* 1: 103–123.

Dale, Arnja, and Steven White. 2013. Codifying animal welfare standards: Foundations for better animal protection or merely a façade? In *Animal law in Australasia*, 2nd ed, ed. Peter Sankoff, Steven White, and Celeste Black, 151–182. Sydney: Federation Press.

de Witte, Kevin. 2009. Development of the Australian animal welfare standards and guidelines for the land transport of livestock: Process and philosophical considerations. *Journal of Veterinary Behaviour* 4: 148–156.

EC, European Commission. 2012. *Animal health and animal welfare*. http://ec.europa.eu/food/animal/index_en.htm. Accessed 14 Feb 2014.

Edge, Michelle, and P. Bailey. 2010. *The livestock management act: A new approach to livestock regulation in Victoria*. Victoria: Department of Primary Industries.

EFSA, European Food Safety Authority. 2014. *Animal welfare*. http://www.efsa.europa.eu/en/topics/topic/animalwelfare.htm. Accessed 14 Feb 2014.

Ellis, Elizabeth. 2010. Making sausages and the law: The failure of animal welfare laws to protect both animals and the fundamental tenets of Australia's legal system. *Australian Animal Protection Law Journal* 4: 6–26.

EPA Tas, Environmental Protection Authority (Tas). 2013. *Legislative context for the EPA*. http://epa.tas.gov.au/epa/legislative-context-for-the-epa. Accessed 30 Mar 2014.

Farmer, Bill. 2011. *Independent review of Australia's livestock export trade*. Available at: http://www.liveexports.gov.au/__data/assets/pdf_file/0016/2030380/independent-review-of-australias-livestock-export-trade.pdf.

Ferguson, Sarah, and Deb Masters. 2012. Another bloody business. *Four Corners*. 7 November. http://www.abc.net.au/4corners/stories/2012/11/02/3623727.htm. Accessed 14 Feb 2014.

Fitzgibbon, Joel, Minister for Agriculture. 2013. *New independent inspector-general of animal welfare and live animal exports*. Media Release. 31 July 2013. Ministry of Agriculture, Fisheries and Forestry.

Fitzsimmons, Hamish. 2013. Victorian abattoir accused of cruel treatment of unwanted dairy calves. *Lateline ABC*. 1 February. http://www.abc.net.au/lateline/content/2013/s3681709.htm. Accessed 14 Feb 2014.

Franklin, Adrian. 2007. Human-nonhuman animal relationships in Australia: An overview of results from the first national survey and follow-up case studies 2000–2004. *Society and Animals* 15: 7–27.

Franklin, Adrian, and Robert White. 2001. Animals and modernity: Changing human-animal relations, 1949–98. *Journal of Sociology* 37(3): 219–238.

Frewer, Lynn, A. Kole, S. Van de Kroon, and C. de Lauwere. 2005. Consumer attitudes towards the development of animal-friendly husbandry systems. *Journal of Agricultural and Environmental Ethics* 18: 345–367.

Goodfellow, Jed. 2013. Animal welfare law enforcement: To punish or persuade? In *Animal law in Australasia*, 2nd ed, ed. Peter Sankoff, Steven White, and Celeste Black, 183–208. Sydney: Federation Press.

Goodfellow, Jed, Melina Tensen, and Lynne Bradshaw. 2014. The future of animal welfare policy and its implications for Australian livestock industries. *Farm Policy Journal* 11(1): 39–51.

Gordon, Sanford, and Catherine Hafer. 2014. Conditional forbearance as an alternative to capture: Evidence from coal mine safety regulation. In *Preventing regulatory capture: Special interest influence and how to limit it*, ed. Daniel Carpenter and David Moss, 208–239. New York: Cambridge University Press.

Grabosky, Peter, and John Braithwaite. 1986. *Of manners gentle: Enforcement strategies of Australian business regulatory agencies*. Melbourne: Oxford University Press.

Harper, Gemma, and Spencer Henson. 2001. *Consumer concerns about animal welfare and the impact on food choice*. Reading: Centre for Food Economic Research, Department of Agricultural and Food Economics, University of Reading.

Herbert, Bronwyn. 2012. Disturbing footage prompts calls for duck farming changes. *7.30 Report ABC*. 19 June 2012. http://www.abc.net.au/7.30/content/2012/s3528922.htm. Accessed 14 Feb 2014.

Hogue, Henry. 2010. *Reorganisation of the minerals management service in the aftermath of the deepwater horizon oil spill*. Washington, DC: Congressional Research Service.

Holmes, Brenton. 2011. *Research paper: Citizens' engagement in policymaking and the design of public services*. Canberra: Parliament of Australia, Department of Parliamentary Services.

Knaus, Christopher. 2012. Police, RSPCA Probe 'Cruelty' at Piggery. *The Canberra Times*. 6 August 2012. http://www.canberratimes.com.au/act-news/police-rspca-probe-cruelty-at--piggery-20120805-23o7s.html. Accessed 14 Feb 2014.

Kwak, James. 2014. Cultural capture and the financial crisis. In *Preventing regulatory capture: Special interest influence and how to limit it*, ed. Daniel Carpenter and David Moss, 71–99. New York: Cambridge University Press.

Langdon, Alison. 2013. Roam Free. *60 Minutes*. 28 March 2013. http://sixtyminutes.ninemsn.com.au/article.aspx?id=8633776. Accessed 14 Feb 2014.

Legislative Assembly of the Northern Territory. 2011a. Council of Territory Co-operation, Sub-committee – Animal Welfare Governance, public hearing – meeting no. AWG09. 3 August 2011.

Legislative Assembly of the Northern Territory. 2011b. Council of Territory Co-operation, Sub-committee – Animal Welfare Governance, public hearing – meeting no. AWG09. 30 June 2011.
Levy, Megan. 2013. Torture for fun: Police given shocking Abattoir fotage. *Sydney Morning Herald*. 21 March 2013. http://www.smh.com.au/national/torture-for-fun-police-given--shocking-abattoir-footage-20130321-2ggm2.html. Accessed 14 Feb 2014.
Lippmann, Stephen. 2005. Public airwaves, private interests: Competing visions and ideological capture in the regulation of U.S broadcasting, 1920–1934. *Research in Political Sociology* 14: 111–150.
Lusk, Jayson. 2011. The market for animal welfare. *Agriculture and Human Values* 28: 561–575.
Lusk, Jason, and Bailey Norwood. 2011. Animal welfare economics. *Applied Economic Perspectives and Policy* 33(4): 463–483.
Makkai, Toni, and John Braithwaite. 1992. In and out of the revolving door: Making sense of regulatory capture. *Journal of Public Policy* 12(1): 61–78.
Marsden, Simon, Kathy Gibson, and Carl Hollingsworth. 2000. Tasmania's environmental improvement programs and the "brown issues": Environmental accountability or regulatory capture? *Environmental and Planning Law Journal* 17(1): 24–32.
Mazur, Nicole, Cecily Maller, Heather Aslin, and Robert Kancans. 2006. *Australian animal welfare strategy stakeholder analysis phases 1–4*. Canberra: Bureau of Rural Sciences.
McCarty, Nolan. 2014. Complexity, capacity, and capture. In *Preventing regulatory capture: Special interest influence and how to limit it*, ed. Daniel Carpenter and David Moss, 99–124. New York: Cambridge University Press.
McDonnell, Brett, and Daniel Schwarcz. 2011. Regulatory contrarians. *North Carolina Law Review* 89: 1629–1682.
McEwen, Graeme. 2006. *Legal challenges in animal welfare*. Paper presented at a meeting of the Victorian Bar. Melbourne, Australia. 15 November 2006.
McEwen, Graeme. 2011a. Strategic litigation and law reform. *Journal of Animal Law* 7(91): 104.
McEwen, Graeme. 2011b. *Animal law: Principles and frontiers*. Barristers Animal Welfare Panel.
McInerney, John. 1991. *Economic aspects of the animal welfare issue*. Paper presented at the Society of Veterinary Epidemiology and Preventative Medicine Annual Meeting, London.
McInerney, John. 2004. *Animal welfare, economics and policy: A report on a study undertaken for the farm and animal health economics division of DEFRA*. Available at:www.archive.defra.gov.uk/evidence/economics/foodfarm/reports/documents/animalwelfare.pdf.
Mickelburough, Peter. 2011. Abattoir Shut Amid Animal Cruelty Claims. *Herald Sun*. 25 November. http://www.heraldsun.com.au/news/victoria/abbattoir-shut-amid-animal-cruelty--claims/story-fn7x8me2-1226206386636. Accessed 14 Feb 2014.
Mitnick, Barry. 1980. *The political economy of regulation*. New York: Columbia University Press.
Murdoch, Lindsay. 2011. Witness tells of mass animal grave. *Sydney Morning Herald*. 25 April, 2011: http://www.smh.com.au/environment/animals/witness-tells-of-mass-animal-grave-20110424-1dt35.html. Accessed 20 Feb 2014.
National Primary Industries Research, Development and Extension Framework. 2010. Animal Welfare RD&E Strategy. Canberra: National Primary Industries Research.
Neales, Sue. 2014. Bush speaks out on drought's secret toll. *The Australian*. 1 February 2014.http://www.theaustralian.com.au/business/bush-speaks-out-on-droughts-secret-toll/story-e6frg8zx-1226815292496. Accessed 14 Feb 2014.
Novak, William. 2013. A revisionist history of regulatory capture. In *Preventing regulatory capture: Special interest influence and how to limit it*, ed. Daniel Carpenter and David Moss, 25–49. New York: Cambridge University Press.
Omarova, Saule. 2012. Bankers, bureaucrats, and guardians: Toward tripartism in financial services regulation. *The Journal of Corporations Law* 37(3): 621–674.
Ombudsman, Northern Territory. 2010. Report of investigation into the treatment of cattle and horses at Charles Darwin University Mataranka Station. Available at: http://www.ombudsman.nt.gov.au/publications-reports/public-reports/.

Oogjes, Glenys. 2011. Australian land transport standards and guidelines: Is the new review process providing protection for transported farm animals? *Australian Animal Protection Law Journal* 6: 5.

Parbery, Peter, and Roger Wilkinson. 2012. *Victorians' attitudes to farming*. Victoria: Department of Primary Industries.

Parke, Melissa, MP. 2013. Independent office of animal welfare. Commonwealth, *Parliamentary Debates*, House of Representatives. 11 February 2013. Canberra: Hansard.

Pew Commission. 2008. *The welfare of animals in concentrated animal feeding operations*. Washington, DC: Pew Commission on Industrial Farm Animal Production.

Phillips, Clive. 2011. Animal exports: How the industry controls research to shut down debate. *The Conversation*. 9 June, 2011. Available at: https://theconversation.com/animal-exports-how-the-industry-controls-research-to-shut-down-debate-1750.

Queensland Department of Agriculture, Fisheries and Forestry. 2013a. *Strategic plan 2013–17*. Brisbane: Queensland Government.

Queensland Department of Agriculture, Fisheries and Forestry. 2013b. *Annual report 2012–13*. Brisbane: Queensland Government.

Queensland Department of Agriculture, Fisheries and Forestry. 2013c. Activity agreement between State of Queensland acting through the Department of Agriculture, Fisheries and Forestry and RSPCA Queensland. 1 July 2013. Brisbane: Queensland Government.

Radford, Mike. 2001. *Animal welfare law in Britain: Regulation and responsibility*. Oxford: Oxford University Press.

Rollin, Bernard. 1995. *Farm animal welfare*. Iowa: Iowa State University Press.

Rollin, Bernard. 2012. An opinion on the role of ethics in modern animal production systems. In *Sustainable agriculture*, ed. Christine Jakobsson, 358–495. Uppsala: Baltic University Press.

RSPCA Australia. 2008. *Submission to the draft land transport standards*. 26 May 2008.

RSPCA Australia. 2014. *Live exports: Risky then, risky now, risky in the future*. http://www.rspca.org.au/media-centre/news/2014/live-exports-risky-then-risky-now-risky-future. Accessed 14 Feb 2014.

Sankoff, Peter. 2013. The protection paradigm: Making the world a better place for animals? In *Animal law in Australasia*, 2nd ed, ed. Peter Sankoff, Steven White, and Celeste Black, 1–30. Sydney: Federation Press.

Sharman, Katrina. 2010. Regulation of the treatment of farm animals. In *Animal law in Australia and New Zealand*, 212–221. Sydney: Deborah Cao/Lawbook Co.

Sharman, Katrina. 2013. Farms animals and welfare law. In *Animal law in Australasia*, 2nd ed, ed. Peter Sankoff, Steven White, and Celeste Black, 61–84. Sydney: Federation Press.

Signal, Tania, and Nicola Taylor. 2006. Attitudes to animals: Demographics within a community sample. *Society and Animals* 14(2): 147–157.

Southwell, Angela, Amarylise Bessey, and Barbara Barker. 2006. *Attitudes towards animal welfare*. Canberra: TNS Social Research.

Tasmanian Government. 2012. *Improving animal welfare in the pig and poultry industries*. Media Release. 17 May 2012. Available at: http://www.farmpoint.tas.gov.au/farmpoint.nsf/news/88697F65EE238E86CA257A050001F0A8. Accessed 14 Feb 2014.

Taylor, Nik, and Tania Signal. 2009. Willingness to pay: Australia consumers and "on the Farm" welfare. *Journal of Applied Animals Welfare Science* 12(4): 345–359.

Tyler, Tom. 1994. Governing amid diversity: The effect of fair decisionmaking procedures on the legitimacy of government. *Law and Society Review* 28(4): 809–831.

Tyler, Tom. 2005. *Procedural justice*. Vermont: Ashgate.

UN FAO, Food and Agriculture Organisation, and Humane Society International. 2001. *Guidelines for humane handling, transport and slaughter of livestock*. Available at: http://www.fao.org/docrep/003/x6909e/x6909e00.htm#Contents.

van der Schot, Agnes, and Clive Phillips. 2012. Publication bias in animal welfare scientific literature. *Journal of Agricultural and Environmental Ethics* 26(5): 945–958.

Vanhonacker, Filiep, Wim Verbeke, Els Van Poucke, and Frank Tuyttens. 2008. Do citizens and farmers interpret the concept of farm animal welfare differently? *Livestock Science* 116: 126–136.

Velde, Te., Noelle Aarts Hein, and Chees Van Woerkum. 2002. Dealing with ambivalence: Farmers' and consumers' perceptions of animal welfare in livestock breeding. *Journal of Agricultural and Environmental Ethics* 15: 203–219.

Vidot, Anna. 2013. Federal Government Scraps Animal Welfare Advisory Group. *ABC Rural*. 8 November 2013. http://www.abc.net.au/news/2013-11-08/animal-welfare-committee-scrapped/5079284. Accessed 14 Feb 2014.

Western Australia Department of Agriculture and Food. 2013. *Annual report 2013*. Perth: Government of Western Australia.

Western Australian Department of Agriculture and Food. 2014. *Animal welfare roles and responsibilities*. https://www.agric.wa.gov.au/animalwelfare/animal-welfare-roles-and-responsibilities?page=0%2C1. Accessed 14 Feb 2014.

Chapter 11
Blackfish and Public Outcry: A Unique Political and Legal Opportunity for Fundamental Change to the Legal Protection of Marine Mammals in the United States

Joan E. Schaffner

Abstract Since 2010, stories involving cetaceans, including Tilikum, Lolita, and eighteen unnamed beluga whales in Russia, have raised public awareness in the United States of their fate. Americans are learning that cetaceans are incredibly complex, intelligent, and social creatures, with severely inadequate legal protections under U.S. law and that maintaining them in captivity does great harm to them. This situation provides an opportunity to change the legal paradigm for these animals. This chapter first presents an overview of current U.S. laws designed to protect cetaceans in the wild and in captivity and explain why these laws are inadequate. The chapter then discusses why the recent conflation of events concerning these famous cetaceans may provide the best vehicle for changing the legal paradigm for animals. Finally, the chapter analyzes a variety of legal approaches available to provide greater protections for cetaceans in the U.S. and argues that because the traditional utilitarian approach cannot properly protect cetaceans in the wild or captivity the time may be ripe for a deontological approach.

11.1 Tilikum's Tragic Tale

It is November 1983. I am a two-year-old orca about 11.5 ft long swimming in the cold, rough waters of Berufjördur, Iceland alongside my mom, siblings, and some 30 pod mates. Life is good. I still secure an occasional "drink" from mom but now most of my sustenance is fresh fish. We may swim up to 200 km a day, foraging for food, playing, socializing and communicating with our unique calls developed by our pod. My mom is one of the maternal leaders of our pod about 35 years of age

J.E. Schaffner (✉)
George Washington University Law School, Washington, DC, USA
e-mail: jschaf@law.gwu.edu

D. Cao, S. White (eds.), *Animal Law and Welfare - International Perspectives*, Ius Gentium: Comparative Perspectives on Law and Justice 53,
DOI 10.1007/978-3-319-26818-7_11

and destined to live to perhaps 80 years of age. I expect to remain within my pod for my lifetime, if lucky for at least 60 years. Wait! I hear something unfamiliar … loud noises from above are bursting in the sea … something is approaching … I am tangled, caught in this net and cannot escape! I cry out for mom. She is trying to help me but is unable to reach me. I am being pulled up and away from my family. I am flailing, trying to bite through the netting but the more I flail the more tied-up I become. I see one of my pod mates killed as I am brought to the top of the sea.

I am lucky to have survived but I have no idea what to expect. I am carried away to a concrete tank where I wait for what seems like an eternity. I swim slowly in circles and hear no familiar sounds. Eventually I am transported to Sealand in Canada and find myself in a pool created by netting tied to floating docks in the marina - 100 ft by 50 ft and 35 ft deep. The water stinks from boat sewage and I find it difficult to maneuver. Two other female orcas are here but I do not know them and they do not like me. They are fighting over who will be the matriarch but they both are "superior" to me. They ram into me and bite me to keep me in my place. Meanwhile the humans here begin to train us to perform in ways that are completely unnatural. These humans begin calling me "Tilikum" which apparently means "friend" in Chinnok. If I am their friend why did they rip me from my mother and family, force me into these small horrible enclosures with other orcas that bully me, and train me to perform in ways that are completely foreign? At night they crowd us in a 26-ft diameter metal-sided pool. It is awful….

Several years have passed and it is February 1991. I am full grown, weigh about 2000 lb, and am about 22 ft long. This pen allows me no freedom to move about as I would in the ocean nor to engage in natural behaviors. We continue to "perform" several times a day. We just finished another show for these humans. I do not understand what they find entertaining. All of a sudden one of the humans has fallen into the pool and we swim toward her and pull her down under the water. She struggles to get away but we surround her and keep her trapped under the water. After a while she stops moving and eventually she is removed from the pool. Soon thereafter I am taken to another place, Sea World in Florida. The space is bigger but life remains the same. I miss my family and life in the wild. Will this ever end?

It is now 20 years later and we have finished another show. My trainer is lying by the pool and I seize the opportunity to grab her arm and pull her into the pool. The events of 1991 come flooding back to my mind as I maul this human like the other one. Soon she too becomes motionless and is removed from the pool. Maybe this time they will send me back to my family. No such luck. They have removed me from the shows and I now just languish in a small pool … swimming in circles. I am almost 30 years old, my teeth are worn and broken, and I have lesions on my skin. I am viewed as an elder now. In fact I expect I will not live too many more years. I have no companionship although I have sired some thirteen calves. I hope that they have fared better than I. I wonder when my lonely and tragic life will end … and I thought I was the lucky one as compared to my pod mate who was killed some 30 years ago on that horrific November day in 1983 (Zimmerman 2010).[1]

[1] Based loosely on Tilikum's story as reported by Tim Zimmerman (2010).

11.2 Tilikum's Story Presents Historic Legal Opportunity

This tragic story of Tilikum may set the stage for an historic opportunity to alter the legal paradigm for at least certain non-human animals. Orcas have been held in the United States for public display since 1965 and during that time some 130 have died in captivity. Today forty-two captive orcas remain alive world-wide, twenty-six of whom are owned by Sea World. Since 2010, stories involving cetaceans, including Tilikum, Lolita, and eighteen unnamed beluga whales in Russia, have raised public awareness in the United States of their fate. Americans are learning that cetaceans are incredibly complex, intelligent, and social creatures with severely inadequate legal protections under U.S. law and that maintaining them in captivity does great harm to them. This situation provides an opportunity to change the legal paradigm for these animals.

This chapter first presents an overview of current U.S. laws designed to protect cetaceans in the wild and in captivity and explain why these laws are inadequate. The chapter then discusses why the recent conflation of events concerning these famous cetaceans may provide the best vehicle for changing the legal paradigm for animals. Finally, the chapter analyzes a variety of legal approaches available to provide greater protections for cetaceans in the U.S. and argues that because the traditional utilitarian approach cannot properly protect cetaceans in the wild or captivity the time may be ripe for a deontological approach.

11.3 U.S. Legal Protections for Cetaceans[2]

11.3.1 Protection in the Wild

11.3.1.1 The Marine Mammal Protection Act

Cetaceans arguably enjoy more legal protections in the United States than many other species (except humans of course!); yet these protections are severely inadequate. The Marine Mammal Protection Act (MMPA) was enacted in 1972 with

[2] Protection for cetaceans at the international level is primarily governed by the International Whaling Commission (IWC). The IWC was created under the International Convention for the Regulation of Whaling, signed in 1946, and "is the global intergovernmental body charged with conservation of whales and the management of whaling" with 89 government members. (IWC, Home, http://iwc.int/home. Accessed 25 June 2015). Since 1986 there has been a moratorium on whaling for commercial purposes although aboriginal subsistence whaling is allowed. Norway and Iceland do not abide by the moratorium and continue their commercial take of whales (IWC, Commercial Whaling, http://iwc.int/commercial. Accessed 25 June 2015). While permits may be issued for scientific research, only Japan and Iceland extensively utilize such permits (IWC, Scientific Permit Whaling, http://iwc.int/permits. Accessed 25 June 2015). IWC conservation initiatives are aimed as severely depleted populations and also target the primary causes of whale mortality in the wild, including "collisions with ships, entanglements in fishing gear, . . .

Congressional acknowledgment that "man's impact upon marine mammals has ranged from … malign neglect to virtual genocide. These animals, including whales, porpoises, seals, sea otters, polar bears, manatees, and others, have only rarely benefitted [sic] from our interest; they have been shot, blown up, clubbed to death, run down by boats, poisoned, and exposed to a multitude of indignities, all in the interests of profit or recreation, with little or no consideration of the potential impact of these activities on the animal populations involved." (HR Rep. 92-707, 1972, p. 4144) Lolita's capture in 1970 in Penn Cove, near Puget Sound, off the coast of Washington state, was one of the more horrific examples of such destruction and exploitation and provided the impetus for enactment of the MMPA. At four years old, Lolita's captors ripped her from her mother, the matriarch of the L-Pod, for public display in Miami, Florida's Seaquarium, after pounding the whale pods with bombs, herding nearly 100 orcas into a makeshift sea pen, using nets to separate the young from the adults, and drowning four calves and one adult female as they desperately tried to reach each other (Lolita Petition 2013, p. 5).

Noting that "marine mammals have proven themselves to be resources of great international significance, aesthetic and recreational as well as economic" (16 U.S.C. § 1361), Congress directed the Secretary of Commerce via the National Oceanic and Atmospheric Administration's (NOAA), National Marine Fisheries Service (NMFS) for cetaceans and seals, and the Secretary of the Interior, for all other marine mammals, to establish limits on the taking of all marine mammals and to issue permits for their taking when appropriate to serve the primary purpose of the Act—to maintain the optimum sustainable population of marine mammal species and population stocks in order to maintain the health and stability of the marine ecosystem. Interestingly the Committee commented that the split enforcement between the two agencies was neither helpful nor useful but the Committee was hopeful that the Department of Natural Resources would soon be created to facilitate the implementation and enforcement of the Act. Notably, in 2015, the split authority remains with NOAA, in the Commerce Department, enforcing the MMPA for cetaceans.

The mission of the Department of Commerce is to promote "job creation, economic growth, sustainable development and improved standards of living for all Americans, … competitiveness in the global marketplace, … and effective management and monitoring of our nation's resources and assets to support both environmental and economic health" (U.S. Dept. Commerce 2014). NOAA's Fisheries mission is "[s]tewardship of living marine resources for the benefit of the nation through their science-based conservation and management and promotion of the health of their environment" (NOAA Fisheries Strategic Plan 2015). The mission statements of these agencies demonstrate that their primary goals are to promote human economic interests and to protect the environment as a resource for human

disease, pollution or habitat loss" (IWC, Conservation and Management, http://iwc.int/conservation. Accessed 25 June 2015). The United States has been an active member since the establishment of the IWC in 1948.

use. The mission statements do not mention the welfare of marine mammals as individual sentient beings.

The MMPA prohibits the take—"harass, hunt, capture, or kill, or attempt to harass, hunt, capture, or kill" (16 U.S.C. § 1362)—of a marine mammal without a permit. Permits may issue for "scientific research, public display, or enhancing the survival or recovery of a marine mammal species of stock" (16 U.S.C. § 1371(a)(1); 1374(c)) or for incidental, e.g. "unintentional, but not unexpected," takes of "small numbers" of animals having no more than a "negligible impact" on the species during commercial fishing operations (16 U.S.C. § 1371(a)(2)) and non-fishery commercial activities (16 U.S.C. § 1371 (a)(5)). No permit may issue for public display if the marine mammal was "(1) pregnant at the time of taking; (2) nursing at the time of taking, or less than eight months old, whichever occurs later; (3) taken from a species or population stock which the Secretary has, by regulation, designated as depleted; or (4) taken in a manner deemed inhumane by the Secretary"[3] (16 U.S.C. §1372 (b)). Additionally, the applicant for a public display permit must show they offer a program of education or conservation based on professionally recognized standards, have a license to exhibit under the Animal Welfare Act (AWA), and maintain facilities open to the public on a regularly scheduled basis (16 U.S.C. § 1374(c)).

Although the basic guidelines are designed to severely limit takes from the wild and that when allowed they are closely monitored, humane, and performed only when necessary to protect marine mammals, a case study into a recent permit request to import 18 beluga whales from Russia for public display demonstrates the inadequacy of these provisions. In 2006, two beluga whales were captured from the Sakhalin-Amur River region of the Russia Sea of Okhotsk by a local team working for the Utrish Dolphinarium Ltd. and transported over 4000 miles to the Utrish Marine Mammal Research Station on the Russian coast of the Black Sea to await import to the United States by the Georgia Aquarium, Inc. (GAI) for public display. In 2010 they were joined by another 11 belugas and, in 2011, by yet another five belugas, all similarly captured and transported.

On 15 June 2012, GAI filed their permit application with NMFS to import these beluga whales. GAI claimed the import of the 18 belugas was to "enhance the North American beluga breeding cooperative by increasing the population base of captive belugas to a self-sustaining level and to promote conservation and education." (Georgia Aquarium 2012, p. 1) GAI planned to transport the 18 belugas on simultaneous flights of approximately 30 h in duration through Liege, Belgium to Atlanta, Georgia, USA. Three of the belugas would remain at GAI and the remaining belugas would be transferred to partner facilities throughout the United States, including Sea World in Miami and in San Diego. Notably, since 1972, 56 beluga whales have died at the facilities in which these belugas would live—37 had been wild caught

[3] These limits do not apply to permits other than those for public display. Moreover, the Secretary may allow a take for public display when factors (1) or (2) are present if for the protection or welfare of the animal.

and 19 had died shortly after birth in captivity—25 of the deaths have occurred since 2000 (Animal Welfare Institute 2012, p. 6). On 22 October 2015 Maris, a 20-year old beluga whale who had lived at GAI for ten years, died suddenly only four months after giving birth to a calf who died two weeks later (Faith 2015).

On 12 October 2012, NMFS held a public meeting to inform interested parties and receive public comment regarding the proposed import. Throughout the public comment period NMFS received almost 9000 public comments (Payne 2013, p. 13). The concerns raised fell into several categories including: the humaneness of the capture, transport and captivity of the belugas; the status of the species in the wild; the education and/or research components of the display; the commerce in wild belugas; and the failure of the applicant to consider alternatives, including the import of already captive belugas, specifically the rescue and relocation of several belugas from Marineland of Canada, and/or the use of artificial insemination techniques.

On 5 August 2013 NMFS denied the permit (Weiting 2013). GAI has filed a lawsuit challenging NMFS' denial (*Georgia Aquarium v. Pritzer*, 2013). The Georgia Aquarium filed suit challenging the denial of the permit as arbitrary and capricious and in September 2015 the case was dismissed (*Georgia Aquarium v. Pritzker*, 2015). While the denial was cause for celebration for marine mammal advocates, the basis of the denial had little to do with concern for the welfare of the individual animals. NMFS determined that "[t]horough consideration of the ongoing beluga capture operation and information available regarding the population status in the Sea of Okhotsk suggests that the level of total removal, including past and present live capture operations, have likely contributed to an adverse impact on this population" (Weiting 2013, p. 1), thus the import was inconsistent with the purposes of the MMPA. Of note, NMFS also found five of the belugas were captured at approximately 1.5 years of age and thus likely were nursing at the time of capture in violation of the permit criteria[4] (Weiting 2013, p. 2). Interestingly, the statutory nursing limitation on the take for public display has no "resource management" explanation but was a response to the public's interest in protecting vulnerable and helpless young animals and has been interpreted broadly to include both obligatory and convenience nursing by at least one judge (*AWI v. Kreps* 1977, 1012). While maintaining the sustainability of the species in the wild is important, NMFS' primary rationale for denial of the permit is the environmental/ecosystem impact of these animals as resources for human use. The numerous concerns raised by the public regarding the welfare of the individual animals were summarily dismissed by NMFS.

Most notably, NMFS dismissed public concerns raised about the humaneness of the capture (e.g. take) and transport. Recall that NMFS must deny a permit for public display[5] if the animal was taken in an "inhumane" manner. When is a take "inhumane?" Dr. Donald Broom defines "humane" as "treatment in such a way that their welfare is good to a certain high degree. The welfare is either above the threshold, in which case the treatment is humane, or it is not" (Broom, Chap. 3, this vol-

[4] This would preclude the import of these five animals only.

[5] Notably, the take limitations apply only for public display permits, leaving all other allowed takes of marine mammals without protection for the animals' welfare during the take.

ume). Thus, the term "humane" sets an objective standard that is either met or it is not. In contrast, the MMPA defines "[h]umane in the context of taking a marine mammal [as] that method of taking which involves *the least possible degree of pain and suffering practicable* to the mammal involved" (16 U.S.C. § 1362(4)). This definition establishes no objective standard. Notably, the 1972 House Report stated that the term "humane" "is not a simple concept and involves factors such as minimizing trauma to groups of highly intelligent, social animals such as whales and porpoises where the taking of any member may be distressing to the group. In many cases, *where an animal may not be taken humanely the bill will prevent that animal from being taken at all*" (H.R. Rep. 92-707, 1977). This understanding of what was meant by "humane" was never codified. Thus, regarding the capture of the belugas, NMFS stated that the method need only involve "the least possible degree of pain and suffering practicable," that the method used was similar to wild captures previously permitted by NMFS, and thus the take was "humane" regardless of whether some argue that the capture techniques are *per se* inhumane (Payne 2013, p. 12).

Regarding the transport, NMFS noted that the legal standard demands that the transport "not present any *unnecessary* risks to the health and welfare of the marine mammals" (Payne 2013, p. 7). In consultation with the United States Department of Agriculture (USDA) Animal and Plant Health Inspection Service (APHIS) and FWS CITES policy specialist for the transport of live animals, NMFS recommended the transport option that would be shorter, involve the least number of stops, and have fewer animals on board for each of three separate transports such that the attention received by each animal would be maximized and would eliminate transport container changes at Liege airport. As this option was the "best" option available, NMFS deemed the transport "humane" (Payne 2013, p. 7). In essence, it is "humane" because it is the lesser of the evils proposed by the various transport options. Notably there was no discussion of objectively humane standards for take or transport as that is not legally required.

In responding to public comments to the permit request to import the Russian beluga whales, NMFS dismissed the concerns raised about the inhumane nature of holding the belugas in captivity for public display as beyond the purview of NMFS (Payne 2013, p. 17). The statute expressly provides for public display permits with the standards of care in captivity governed by the AWA under the purview of APHIS. But, how does the public display of marine mammals promote the purpose of the MMPA? Marine mammal advocates and scientists have argued that holding cetaceans in captivity for public display purposes severely harms the individual animals, does not promote the protection of the species, and thus is antithetical to the goals of the MMPA.

Studies of the effects of captivity on orcas such as Tilikum and Lolita have noted several harms to these animals (Rose 2011). First, captivity shortens the lives of orcas. Captive orcas' annual mortality rate is approximately two and a half times that of their wild counterparts. In the wild, female orcas have a mean life expectancy of 46 years and a maximum life span of 80–90 years. The males may live 60–70 years with a mean life expectancy of 31 years. However, the maximum life span of captive orcas matches only the average mean life expectancy of wild orcas, with the majority of captive orcas dying before their early 20s. Captive-born orcas fare no better than wild-born orcas and despite so-called improvements in orca husbandry, the survival rates for orcas has not improved (Rose 2011, p. 2). The most common causes of death of captive orcas are pneumonia, septicemia, and other infections. It is believed that a contributing factor to these infectious deaths is immunosuppression caused by chronic stress, psychological depression, and boredom suffered by captive orcas (Rose 2011, p. 5).

Second, captive orcas suffer from severely poor dental health with damaged, broken and worn teeth, often with the pulp exposed. Such abrasion comes not from prey with abrasive morphology but rather from gnawing on concrete walls or steel gates out of aggression or boredom. In contrast, wild orcas' teeth show little or no wear and virtually no broken teeth. Poor dental health is a cause of several medical conditions, including heart disease and pneumonia (Rose 2011, p. 6).

Third, orcas in the wild are rarely aggressive towards each other; however, in captivity they become extremely aggressive. Placing incompatible orcas from different pods together in very close surroundings with no ability to escape creates an environment where the orcas bully each other resulting in lacerations and other wounds to the animals. In addition, captive female orcas often behave abnormally towards their calves, rejecting them, acting aggressively towards them, or merely failing to properly care for them, unlike their maternal wild counterparts. Further, although there has been no recorded case of a wild orca killing a human being; captive orcas have killed four humans. This aberrant and aggressive behavior suggests abnormal psychological and social development of captive orcas[6] (Rose 2011, pp. 7–8). Other well-documented behaviors also suggest abnormal development including: highly repetitive non-purposeful behaviors such as swimming in circles (Defran and Pryor 1980), vomiting (Sweeney 1990), physical trauma and self-mutilation, (Sweeney 1988) unresponsiveness, excessive submissiveness, and hypersexuality (towards humans and other cetaceans) (Stewart and Marino 2009). The research is clear, captivity seriously harms cetaceans. As one cetacean scientist concluded:

> Considering orca natural history, it is unsurprising that orcas do not thrive in captivity. They are kin-bonded creatures with a long dependency period on the mother, and life-long family ties to her, their siblings, and more distant relations. When in captivity they are kept in artificial social groups with no resemblance to those in nature. They are cooperative predators, whose home ranges are hundreds if not thousands of square kilometers in size and who can and often do swim almost 200 kilometers in a day. When in captivity, they are made to exist inside a comparatively small concrete enclosure, less than one ten-thousandth of normal habitat size. Captivity cannot adequately provide for such large, social, wide-ranging predators. (Rose 2011, p. 9)

Sea World and others claim that the conservation and educational benefits of captive breeding and public display justify orca captivity (Georgia Aquarium 2015). From a conservation standpoint, orcas as a species generally are not endangered or

---

[6] Also, interestingly, all captive male orcas have fully collapsed dorsal fins, while only 1–5 % of male orcas in some populations have fully collapsed dorsal fins (Rose 2011, pp. 5–6).

threatened.[7] The threat they face in the wild is not from a lack of reproductive success but rather from habitat degradation. Moreover, "no captive-born animals are destined for release to the wild" (Rose 2011, p. 15 n.105). These factors fail to satisfy any common definition of conservation breeding. As for educational benefit, the public displays of orcas and other cetaceans distort the public's understanding of marine life and is secondary, at best, to the entertainment value of the unnatural performance. The complex social and cognitive abilities of these animals cannot be captured by watching the captive animals perform (Whale and Dolphin Conservation Society 2015).[8] In fact, if the public truly understands the complex social and cognitive capacities of cetaceans, they likely may find their captivity disturbing. There is no legitimate evidence that public displays of whales and dolphins are educational (Marino et al. 2010). In this era of international travel and communication via the internet, it can be cheaper and far more interesting and educational to travel to their natural habitat to see them or to watch videos of these magnificent mammals in the wild (Kirby 2012).

11.3.1.2 Endangered Species Act

Marine mammal species determined to be threatened or endangered are protected under the Endangered Species Act (ESA). The ESA was described by the Supreme Court as "the most comprehensive legislation for the preservation of endangered species ever enacted by any nation" (*TVA v. Hill*, 1978, p. 180).[9] A species is listed as "endangered" when it is presently in danger of extinction throughout all or a significant portion of its range, and "threatened" if likely to become endangered in the foreseeable future. The term "species" includes a "distinct population segment" (DPS) of a vertebrate species which is determined by considering the "(1) discreteness of the population segment in relation to the remainder of the species/taxon and, if discrete; (2) the significance of the population segment to the species/taxon" (8 Fed. Reg. 47277, 47278). The agency considers several factors when evaluating whether to list the species as endangered or threatened including: loss of habitat, overutilization, "disease or predation, inadequacy of existing regulations, and all other natural and man-made factors affecting the species' continued existence" (Liebesman and Petersen 2010, p. 13). The ESA prohibits the "take" of any

[7] However there are certain groups of orcas who are endangered. In fact, the capture of Lolita which resulted in the death of at least four orcas and the taking of Lolita and seven other orcas, along with other factors, caused such devastation to the Southern Resident population that the Southern Resident Killer Whale Distinct Population Segment is classified as endangered. 70 Fed Reg 69903.

[8] Jacques Cousteaus has said, "There is about as much educational benefit to be gained in studying dolphins in captivity as there would be studying mankind by only observing prisoners held in solitary confinement." (Palmer 2010)

[9] Notably, while Congress has substantially changed the statute since this landmark decision, it is still viewed as "a powerful tool for species protection and conservation" (Liebesman and Petersen 2010, p. 9).

endangered species, defining "take" broadly as "to harass, harm, pursue, hunt, shoot, wound, kill, trap, capture, collect, or attempt to engage in any such conduct" (16 U.S.C. § 1538(a)(1)(B)). NMFS is tasked with implementing and enforcing the ESA with regard to cetaceans.

As with the MMPA, the ESA is designed to protect human interests in maintaining species as a resource for human use and protects the species not the individual animal. This is clearly evidenced by the situation of Lolita, held captive in Miami for 45 years, although she is a member of an endangered DPS of killer whales. In November 2005, NMFS issued a final rule listing the Southern Resident killer whale (SRKW) DPS, (the J, K, and L pods) as endangered (70 Fed Reg 69903). Hunters seeking orcas for captive public display had targeted SRKW largely because their preferred range was within the shallow inland waterways off the coast of Oregon, Washington, and Vancouver Island. The loss of Lolita to captivity and the 46 other orcas who died for public display, seriously depleted the SRKW from a population high of 96 members and resulted in the endangered listing. "The latest count [of the SRKW population] revealed a total of 85 wild members, down from 90 in 2006" (Lolita Petition 2013, pp. 4–5). Nevertheless, Lolita, the lone survivor, was excluded from protection under the ESA because NMFS declared in the final rule that all SRKW "placed in captivity prior to the listing" including their "captive born prodigy" were not included in the listing (70 Fed. Reg. 69903, 69911).

Advocates petitioned NMFS to list Lolita as endangered as a member of the SRKW, arguing that her differential treatment from the other members of her family is illegal under the ESA (Lolita Petition 2013). They argue first that the ESA mandates that listing decisions be based solely on "the best available scientific and commercial data" (16 U.S.C. § 1533 (b)(1)(A)) concerning the biological status of the species. Thus economic considerations, such as profits made off the public display of Lolita, are precluded from consideration. Second, the ESA does not contemplate separate legal status for captive-held animals. For example, the ESA defines an endangered or threatened species based on the number of species found within the species' "range." The Agencies interpreting "range" have consistently interpreted "range" to mean the "natural range of species *in the wild*" (78 Fed. Reg. 33790, 33792). Further, sections 9 and 10 of the ESA that provide certain exemptions for animals held in captivity demonstrate that Congress intended captive-held animals to have the same legal status as their wild counterparts, otherwise the exemptions would be unnecessary. Third, the exemptions permitted under section 9 of the ESA for captive animals as of the date of the species listing do not apply when the holding in captivity is in the course of a commercial activity as Lolita clearly is held. Thus, Lolita should be protected under the ESA just as her family members in the wild are protected. In response to this petition, NMFS published a final notice in February 2015 including Lolita in the endangered species listing for the SRKW (80 Fed. Reg. 7380).

11.3.2 Protection in Captivity: The Animal Welfare Act

Lolita's life in captivity is tragic. Lolita is a 20-ft long, 7000 lb orca. For 45 years she has been confined at Miami Seaquarium to a barren shallow tank, with a horizontal dimension of 35 ft, unshaded from the extreme mid-day Florida sun and with no companion of her own species since the orca, Hugo, died in 1980 after repeatedly ramming his head into the tank walls. In the wild, Lolita would be swimming with her mother, Ocean Sun, and L-pod family over many miles daily, communicating, foraging, socializing, and diving to the depths of the ocean able to escape the extreme sun (Lolita P's Opposition 2012, pp. 6–7).

Miami Seaquarium is a licensed exhibitor under the AWA. Congress enacted the AWA[10] in 1970 to "ensure that animals intended for use in research facilities or *for exhibition purposes* or for use as pets are provided *humane* care and treatment" (7 U.S.C. § 2131(1)). Congress directed the Secretary of Agriculture to promulgate standards for the "*humane* handling, care, treatment, and transportation of animals by … exhibitors" (7 U.S.C. § 2143(a)(1)) including "*minimum* requirements for handling, housing, feeding, watering, sanitation, ventilation, shelter from extremes of weather and temperatures, [and] adequate veterinary care" (7 U.S.C.§ 2143(a)(2)(A)). Often when confronted with opposition by advocates arguing that it is inhumane to have cetaceans in confined tanks for public display, the owners point to the fact that their operations are governed by the AWA which establishes "humane" guidelines for their care and thus cetacean captivity is not only legal but humane. Lolita's situation demonstrates the falsity of this premise.

On 22 August 2012 several plaintiffs filed suit against the USDA, the agency responsible for implementing and enforcing the AWA, challenging their renewal of Seaquarium's exhibitor license. (Lolita Complaint 2012) In February 2012, in anticipation of Seaquarium's license renewal, the plaintiffs had submitted voluminous materials to the USDA demonstrating that Seaquarium was in violation of several AWA standards along with "sworn declarations from world-leading cetacean experts and an expert in zoo design" detailing the physical and psychological harms caused Lolita by these violations. (Lolita Complaint 2012, p. 17) Nevertheless, in April 2012 the USDA renewed their license.

The plaintiffs, challenging the USDA's findings, argued that all exhibitors must demonstrate compliance with relevant AWA standards before being issued a license (7 U.S.C. § 2133) and failure to comply by the regulations and standards are grounds for denial of the license (9 C.F.R § 21(e)). The standards for marine mammal exhibitors require a primary enclosure that "must be constructed and maintained so that the animals contained therein are provided sufficient space, both horizontally and

[10] The name of the statute is not scientifically proper. As Donald Broom notes, "Welfare is a characteristic of an individual animal whilst animal protection is a human activity. Welfare includes both the ease of coping, or difficulty of coping, and any failure to cope. It varies over a range from very good to very poor and can be evaluated scientifically" (Broom, Chap. 3 this volume).

vertically, to be able to make normal postural and social adjustments with adequate freedom of movement in or out of the water" (9 C.F.R. § 3.104(a)). For cetaceans, the primary enclosure must provide a pool of water that has a "minimum horizontal dimension (MHD) (9 C.F.R. § 1.1)[11] … two times the average adult length" (9 C.F.R. § 3.104(b)) of the species. The "average length of an adult orca is 24 ft," thus "Lolita's tank must have a MHD of 48 ft" (Lolita Complaint 2012, p. 12). Moreover, the regulations require that marine mammals be provided appropriate natural or artificial shelter "from the weather or from direct sunlight" (9 C.F.R. § 3.103(b)) and that animals who are "primarily social in the wild … be housed with at least one compatible animal of the same or biologically related species" (9. C.F.R. § 3.109).

Lolita's confinement arguably violates each of these conditions. First, Lolita's tank has a MDH of only 35 ft (25 % less than the minimum standard). Second, the tank is situated as to leave her exposed to the sun during the mid-day when the Miami sun is the hottest. Third, although she lives with white-sided dolphins, they are not sufficiently related to her to provide adequate social interaction for her (Lolita Complaint 2012, p. 16).

The USDA did not dispute these allegations. Instead, they argued that plaintiffs' complaint must be dismissed because violations of AWA regulations are not grounds for denying the renewal of a license. The USDA distinguishes *issuing* a license from *renewing* a license, arguing that once issued the license remains valid and in effect upon renewal; the license is not reissued (Lolita D's Motion 2012, p. 9). For renewal, the only requirements are that a licensee timely *certify* compliance (not *demonstrate* compliance) with the AWA, submit an annual report, and pay the required fees (9 C.F.R. §§ 2.2(b), 2.6, 2.7). In fact, the USDA noted that even a licensee with unpaid fines under the AWA must be granted renewal (Lolita D's Reply 2012, p. 6). The USDA explained that the administrative renewal process is not designed as an enforcement tool, rather, a separate enforcement proceeding is required. Notably, the USDA's enforcement authority under the AWA is completely discretionary and unreviewable by a court. No private entity may challenge the agency's lack of enforcement of the AWA (Lolita D's Reply 2012, p. 7).

The plaintiffs argued in response that to require renewal of licenses even when the licensee is in violation of the regulations and worse, has outstanding fines based on past violations, is "unreasonable and contrary to the overall purpose of the AWA" (Lolita P's Opposition 2012). The purpose of the AWA is to insure that animals under the care of licensees are humanely treated. To renew licenses when there are clear violations of the regulations that detail only the *bare minimum* standards for care, fails horribly to protect the animals. Moreover, what is the purpose of requiring a "certification" of compliance from the licensee when the certification can be demonstrably false and not provide a basis for denying the renewal of the license?

The USDA's only response was that they have "repeatedly rejected requests to transform the administrative renewal process into an enforcement proceeding"

[11] MHD is defined as "the diameter of a circular pool of water, or in the case of a square, rectangle, oblong or other shape pool, the diameter of the largest circle that can be inserted within the confines of such pool of water" (9 C.F.R. § 1.1).

(Lolita D's Reply 2012, p. 6). The agency explained that enforcement of the AWA is based on random, unannounced inspections using a risk-based assessment to determine minimum inspection frequency. After inspection, licensees are given time to come into compliance noting that a "cooperative system has been more effective than enforcement actions for each citation. It is unrealistic and counterproductive to make license renewal contingent on not having any citations" (Lolita D's Reply 2012, p. 7). In contrast, the renewal process assists in record keeping and tracking of AWA licensees, not enforcement. The USDA explained that "Congress expressly left the agency with the discretion to choose when and how often to investigate and inspect licensees ... for compliance" (Lolita D's Reply 2012, p. 9). Apparently such discretion includes deliberately failing to enforce even the most blatant violations that have harmed an intelligent and social marine mammal for 45 years, and allowing such violation to continue indefinitely. The plight of Lolita and all cetaceans in captivity for public display suggests that woefully inadequate laws often are worse than having no law at all as they provide cover for those who exploit animals and lead the public to believe that the laws protect the interests of these animals when in fact they receive very little protection. In July 2014, the trial court agreed with the USDA allowing them to renew licenses even if they know the applicant is violating AWA regulations and in June 2015 a federal appeals court upheld the decision (*Animal Legal Defense Fund v. USDA* 2015).

Notably, as discussed above, Lolita is now listed as endangered under the ESA. However, it is unclear how, if at all, this will affect her treatment. In fact, Miami Seaquarium has no current plan to move Lolita. The ESA does not prohibit the possession of an endangered animal who was lawfully captured, even if such possession is solely for commercial purposes. Ironically, NMFS has indicated that continued possession of Lolita in captivity would likely not be prohibited while her release into the wild could be prohibited as a "take" under the ESA (NOAA Fisheries, West Region, 2015, Questions & Answers). Marine mammal advocates have challenged these views. First, holding Lolita in a facility in violation of the AWA arguably results in "harm" to Lolita which in turn constitutes a "take" under the ESA. If found, Seaquarium would require a permit under the ESA to retain possession of Lolita but such permit excludes holding the animal in captivity in the course of a commercial activity. Thus, arguably Seaquarium would be denied the permit as they hold her in the course of a commercial activity. Second, releasing her into the wild would not likely cause harm to the DPS from the introduction of disease or harmful genetic mixing of the population because only Lolita will be released and she currently does not suffer from any disease and is genetically related to the DPS. Moreover, her reintegration into the L25 sub-pod of which she is a member could provide substantial benefit to the DPS by adding a post-reproductive female who can provide alloparenting and other caregiving roles to orphaned teenage males and facilitate study into the threats facing the DPS (Mashuda 2014).

11.4 The "Perfect Storm" for Cetaceans and U.S. Law

Cetaceans, especially at this time in history, provide a unique opportunity for legal change to better protect animals' interests. First, there has been an international dialogue seeking rights for Cetaceans since at least May 2010 when the "Declaration of Rights for Cetaceans: Whales and Dolphins" (Cetacean Rights 2010) was adopted by the Helsinki Group in Helsinki, Finland. Ironically, just three months earlier, Tilikum killed his trainer during a show at Sea World in Orlando. This incident generated a media frenzy that raised public awareness of the inhumanity and danger of keeping orcas captive in small tanks for mere entertainment and profit. Three years later, the movie *Blackfish* told Tilikum's story to American audiences while the government of India, Ministry of Environment of Forests, banned all dolphinaria in their country finding "cetaceans … are highly intelligent and sensitive, and various scientists who have researched dolphin behavior have suggested that the unusually high intelligence[,] as compared to other animals[,] means that dolphin[s] should be seen as 'non-human persons' and as such should have their own specific rights and [it] is morally unacceptable to keep them captive for entertainment purposes" (Government of India 2013). Soon thereafter, NMFS denied Georgia Aquarium's permit request to import the 18 wild-caught beluga whales from Russia to the United States for public display. This was the first time in the history of the MMPA that a public display permit for marine mammals was denied. Meanwhile Bikers 4 Orcas, a group of bikers and others who care for the well-being of orcas, has a Facebook page with over 5500 likes and planned the ultimate Freedom Ride world-wide on 14 June 2014 to bring attention to and protest the keeping of orcas in captivity for public display, including a ride in Miami specifically for Lolita. In March 2014, California Assemblyman Bloom introduced the Orca Welfare and Safety Act that would outlaw keeping an orca in captivity, using an orca for performance or entertainment purposes, artificially inseminating captive orcas, and importing orcas or orca semen from other states (California AB-2140 2014; Kirby 2014a). The bill was tabled pending an interim study. In August 2014, Sea World's stock price dropped 35 % to a record low (Kirby 2014b). In October 2015, the California Coastal Commission approved Sea World's request to expand its orca habitat but banned Sea World from breeding any of their 11 captive whales (O'Neill 2015). Consistent with the theory of deliberative democracy developed by the German philosopher Jurgen Habermas and others, that posits that public discourse is essential to encouraging change in the law (Sankoff 2012), this extensive public attention focused on cetaceans and their plight in captivity sets the stage for a public discourse that may inform American legislative and judicial decision-makers and provide an impetus for legal change.

Second, scientists have studied whales and dolphins extensively and have determined that they have "large, complex brains especially in areas involved in communication and cognition … [with a] level of self-awareness similar to that of human beings" (DW 2013; Marino 2008, 2011). Cetaceans have highly developed emotional processes and feel empathy, guilt, embarrassment, and pain. Orcas form "complex societies with dynamic social roles in intricate networks, many with distinctive cultural attributes in vocal, social, feeding, and play behavior... Orcas produce dozens of community, clan, and pod-specific call types, and there is evidence that calls evolve over time and in parallel when shared between separate but associating pods... The complex and stable vocal and behavioral cultures of orcas appear to have no parallel outside humans" (Tilikum Complaint 2011, paras. 12–16). Marine scientists have argued that "[o]ur current knowledge of cetacean brains and cognitive abilities demands we develop a new ethic of respect and coexistence with [whales and dolphins]" (White 2013, p. 224; Rendell and Whitehead 2001). The cetaceans' complex cultures, a characteristic very similar to the human characteristic often used to support treating non-human animals differently from humans, provides substantial support for including these mammals, with humans, in "an extended moral community" (White 2013, p. 223). Thus, in a society founded on equality principles where a fundamental premise of fairness and justice is to treat like players alike, cetaceans are uniquely capable of being among the first non-human species to be included in our moral community and be granted basic fundamental legal rights.

Third, as the American public hears the stories of Tilikum, Lolita, and others and learns of their complex social and cognitive skills, Americans begin to understand and experience what their lives are like. Consistent with the feminist method of consciousness-raising, storytelling constructs a social reality in which others may own their experience and set the stage for change (MacKinnon 1989). Although these animals are not our "companions," we engage with them under circumstances where we can develop ties that enhance our understanding of their individual nature and needs and better appreciate their plight and desire for release from captivity. In fact, a 2012 survey of Americans on orca captivity revealed that only about 25 % of Americans favor public display of orcas (Humane Society of the U.S. 2012). Additionally, in comparison to the entrenched human interest to use animals for food, clothing, or research, use of animals for mere entertainment is more easily overcome, especially once we understand the harm done to these animals who are so similar to ourselves. Finally, unlike domestic animals who depend upon humans for their well-being and in turn require expenditure of resources to properly care for them, cetaceans are free-roaming animals (when not captured, of course) and thus granting them basic rights, for example, rights to life and bodily liberty, does not impose a resource burden on humans for their care.

Fourth, the public display of cetaceans raises serious public and employee safety concerns for humans. Consistent with a pluralist view of the political process that aligning multiple and diverse interest groups—e.g. those for cetaceans and humans alike—facilitates legal change (Anderson 2011) recent events align animal advocates with public and employee safety advocates, facilitating legal transformation for cetaceans. Tilikum killing his trainer in front of hundreds of spectators at Sea World Park in Orlando provided one vivid example of the severe distress and resulting aggression caused by his captivity, along with the threat he poses to human safety. In fact, this tragedy resulted in a finding of two violations of the Occupational Safety and Health Act (OSHA) (*Secretary of Labor v. Sea World* 2012). First, the

Occupational Safety and Health Review Commission ruled that Sea World violated OSHA section 1910.23(d)(1)(iii) by not maintaining a railing along the stairway at the orca tank which could result in an employee falling into the tank and being killed by the whales. For this "high" gravity violation Sea World was fined $5000.00 (*Secretary of Labor v. Sea World* 2012, p. *35). Second, the Commission affirmed Sea World's willful violation of the OSHA general duty clause and upheld an abatement established by the Secretary that "would prohibit animal trainers from working with killer whales, including 'waterwork' and 'drywork,' unless the trainers are protected through the use of physical barriers or through the use of decking systems [e.g. a floor raising mechanism], oxygen supply systems or other engineering or administrative controls that provide the same or greater level of protection for the trainers" (*Secretary of Labor v. Sea World* 2012, *11, *32). The fine for the willful violation of the general duty clause, classified as a "very high" gravity violation, was a mere $7000.00 penalty (*Secretary of Labor v. Sea World* 2012, p. *35).

Sea World appealed this decision to the D.C. Circuit Court of Appeals arguing that the Secretary exceeded their authority under OSHA and failed to establish two elements of a General Duty Clause violation. Specifically SeaWorld argued that it did not subject its employees to a "recognized hazard" because any risk inherent in a business activity cannot constitute a "recognized hazard." Further, the abatement imposed by the Administrative Law Judge is not feasible as it undermines the intrinsic premise of their business model because physical contact between the trainers and the killer whales is critical to Sea World's core business (*Sea World v. Perez* 2014). The DC Circuit denied the appeal. The court found that substantial evidence supported the finding that both "waterwork" and "drywork" with killer whales were recognized hazards even if such hazard is an inherent aspect of the activity. Moreover, the abatement is feasible and does not change an essential aspect of their business as there will still be human interactions and performances with killer whales. They merely must take place under increased safety measures. In fact, Sea Word on its own ceased "waterwork" with all of its killer whales and imposed distance between the trainers and whales during "drywork" (*Sea World v. Perez* 2014). Thus the fines imposed were upheld. However, the combined penalties of $12,000 are woefully inadequate to affect Sea World's operations given their multi-million-dollar quarterly profits[12] (Garcia 2013) and may provide incentive for employee safety advocates to join animal advocates in their efforts to prohibit the use of cetaceans for public display.[13]

[12] Sea World reported 2013 third-quarter profits at $120 million, a 30 % increase over 2012 third quarter. In 2014, Sea World reported attendance dropped 4.3 % (Kirby 2014b).

[13] Sea World has filed yet another lawsuit, now against the OSHA investigator claiming she, inter alia, violated ethical rules and federal law by sharing confidential information with the producers of *Blackfish* (Hightower 2014).

11.5 Approaches to Secure Greater Legal Protections for Cetaceans

The stage is set to improve the plight of cetaceans under U.S. law. Legal change for cetaceans may include substantially improving their standard of care while in captivity, prohibiting their captivity for public display, and perhaps even altering their legal status from "things" to "persons."

The law is a primary means of defining and enforcing societal ethics. As Verrinder et al. explains, "two main frameworks of ethics are utilitarianism founded on our interdependence and therefore our need to balance interests to create the greatest well-being of all concerned, and deontology founded in defining our duty to the fundamental principles which protect each individual's interests" (Verrinder et al., Chap. 4 this volume).

U.S. law governing animals is based on a utilitarian paradigm. The law views free-roaming animals as resources for our use and allows the "take" of such animals for a variety of uses so long as their species are not nearing extinction. It allows a variety of uses of both domestic and captive wild animals when their use to humans is determined to be justified. The law sets minimal standards for the animals' care during such use although arguably, in many instances, such standards do not adequately meet the animals' needs. Moreover, the standards that are established utilize terms such as "unnecessary" which requires a balancing of interests. In fact, as we have seen, even the word "humane," arguably a term that would set an objective standard, is defined in the MMPA as that which involves "*the least possible degree of pain and suffering practicable* to the mammal involved" (16 U.S.C. § 1362(4)). As we have seen, for cetaceans in captivity, these standards are severely inadequate and may not be enforced as enforcement is left to the discretion of the agency with no oversight by the courts. Non-human animals have no legal right to life or liberty under U.S. law.

In contrast, a deontological-based paradigm for animals would provide greater protections for animals by granting at least certain animals basic legal rights to life and liberty, independent of humans' interest in using them for their own purposes. Admittedly such a transformation will, if at all, evolve slowly over time. Incremental laws that follow a deontological approach, in contrast to those that follow a utilitarian approach, are those that establish a bright-line rule prohibiting an activity essential to an exploitative institution and providing no alternative form of exploitation (Francione 2008, pp. 114–15).

Legal change for cetaceans in the U.S. may be accomplished through regulatory, legislative, and/or judicial means. Regulatory change is largely at the discretion of the government agency responsible for enforcement of the statute. Individuals, including NGOs, may affect regulation by challenging regulations as arbitrary and not in compliance with the law or by informally working with agency personnel to persuade them to improve upon the standards that govern animal welfare. Legislative change requires persuading congress that the law should change and is a highly political process. Large industries, including those engaged in animal entertainment, fishing, etc., spend substantial money lobbying congress to protect their economic interests in the exploitation of animals. As a result, lobbyists for animals must overcome immense obstacles to secure greater statutory protections for animals and often are wary to seek amendment to a statute for fear that industrial lobbyists may use the amendment process to introduce changes that will do more harm than good for the animals. Finally, advocates may use the courts to improve the legal protection and status of animals through the common law. Although the common law is created by judges and thus arguably judges have absolute authority to alter the common law, judges have differing views regarding the scope of such authority.[14]

Thus, there are several approaches to securing greater legal protection for cetaceans ranging from regulatory amendment to improve the standards set to enhance their welfare while in captivity to a grant of legal rights to bodily liberty and integrity. A first step to improving the lives of cetaceans in captivity would be to amend the AWA regulations and enhance the standards that govern cetaceans in order to better meet their physical, social and cognitive needs. This utilitarian strategy, however, is ineffective as a practical matter because cetaceans' physical, social and cognitive needs cannot properly be met if they are maintained in a pool or aquarium. Moreover, generally there are insufficient resources and incentives to properly enforce the laws that exist as demonstrated by the USDA's continued renewal of Seaquarium's AWA license to display Lolita despite Seaquarium's continued regulatory violations. Thus, under this approach cetaceans would remain in severely inadequate conditions suffering from physical and emotional harm.

The second approach is for the United States to follow the lead of India and outlaw the public display of cetaceans by removing the permit provision for public display from the MMPA. In fact, as stated above, California became the first U.S. jurisdiction with public display of orcas to propose a bill to ban orcas in captivity for public display within its state (California AB-2140 2014). Such an amendment to the MMPA would fit the deontological paradigm as it establishes a bright-line rule—prohibiting public display permits for marine mammals—that prohibits an activity essential to an exploitative institution, and provides no alternative form of exploitation. Nevertheless, under this approach cetaceans still would not be protected from being held in captivity under MMPA permits for scientific research or to enhance the survival or recovery of a species or stock, nor are they protected from incidental takes in the wild.

Among deontological approaches to alter fundamentally how the law views cetaceans, and move towards establishing certain fundamental legal rights for them, are those seeking a classification of legal "personhood" for cetaceans through the

---

[14] The highest courts of Maryland and Texas demonstrated divergent views of their authority to alter the common law. The Maryland Court of Appeals altered long-established tort liability principles and held that pit bull dogs are inherently dangerous and thus a person with control over such dog would be held strictly liable, stating "the *Court certainly has the power to change the common law*" (*Tracey v. Solesky*, 2012, p. 640). In contrast, the Texas Supreme Court refused to grant non-economic damages for the wrongful killing of a companion dog although relational affection for such animal is "unquestionable" claiming "the *legislature* is best equipped" to alter damages recoverable in such cases (*Strickland v. Medlen*, 2013, pp. 185, 196).

courts. Such strategies are revolutionary and controversial yet may succeed in the foreseeable future. One attempt, while creative, was unsuccessful. In October 2011 suit was filed against Sea World seeking a declaration by the court that five wild-captured orcas—Tilikum, Katina, Corky, Kasatka and Ulises—were held by Sea World in violation of the Thirteenth Amendment of the U.S. Constitution prohibiting slavery (Tilikum Complaint 2011). The complaint alleged that "Plaintiffs were forcibly taken from their families and natural habitats, are held captive at Sea World San Diego and Orlando, denied everything that is natural to them, subjected to artificial insemination or sperm collection to breed performers for Defendants' shows, and forced to perform, all for Defendants' profit. As such, Plaintiffs are held in slavery and involuntary servitude" (Tilikum Complaint, 2011, para. 1). The case clearly was one of first impression as the U.S. Constitution has never been applied to non-humans.

The Thirteenth Amendment of the U.S. Constitution states "neither slavery nor involuntary servitude ... shall exist within the United States." While the language of the Amendment does not reference "human" or "person," the Amendment originally was designed to end the ownership of African American persons by white slave owners after the Civil War. However, the plaintiffs argued that although the *slavery of animals* was not contemplated by the framers of the amendment, *slavery* was contemplated, and the application of what constitutes slavery may change over time to adapt to changing circumstances and societal views.

The Supreme Court over the years has consistently affirmed the notion that the constitutional text is subject to changing application as it is "intended to endure for ages to come, and consequently, to be adapted to the various crises of human affairs" and noting that its nature requires "that only the great outlines be marked, its important objects designated, and the minor ingredients which compose those objects, be deduced from the nature of the objects themselves" (Tilikum P's Opposition 2012, p. 5). Justice Brennan has stated that "the genius of the Constitution rests not in any static meaning it might have had in a world that is dead and gone, but in the adaptability of its great principles to cope with current problems and current needs. What the constitutional fundamentals meant to the wisdom of other times cannot be the measure of the vision to our time" (Tilikum P's Opposition 2012, p. 7).

One example of the expanding application of the constitutional text is the Court's privacy cases interpreting the term "liberty" under the Fourteenth Amendment substantive due process clause. Over the years the Court has interpreted citizens' right to "liberty" from government interference beyond that which the framers had contemplated, striking as unconstitutional laws that criminalized the possession of contraceptives, abortion, and private, consensual sodomy. In 2003, Justice Kennedy writing for the majority of the Court explained that the framers "knew times can blind us to certain truths and later generations can see that laws once thought necessary and proper in fact serve only to oppress. As the Constitution endures, persons in every generation can invoke its principles in their own search for greater freedom" (Tilikum P's Opposition 2012, p. 8).

A second example is the Court's expanding application of the Fourteenth Amendment equal protection clause. Originally the Court interpreted "equal protection" narrowly upholding laws that prohibited interracial marriages and required the separation of the races. Years later the Court rejected this interpretation based on new research and evolving societal views of what is just (Tilikum P's Opposition 2012, pp. 9–10). Similarly, here, although the framers were blind at the time to envisioning animals as enslaved, today, equipped with new research into the nature of orcas and other cetaceans, along with evolving societal views concerning the just treatment of non-humans, the court should find that Sea World's treatment of these plaintiffs violates the fundamental principle of slavery proscribed by the Thirteenth Amendment.

The Plaintiffs' argument, although creative, was not persuasive to the court. Unwilling to look beyond the Amendment's plain and ordinary meaning, historical context, and judicial interpretations, the court found that the Thirteenth Amendment "only applies to 'humans' and therefore affords no redress to Plaintiffs' grievances" (*Tilikum v. Sea World* 2012, p. 1262). In support, the court noted that both "historic and contemporary sources reveal that the terms 'slavery' and 'involuntary servitude' refer only to persons" and the Supreme Court in one of its earliest applications of the Amendment made such application clear as well (*Tilikum v. Sea World* 2012, p. 1263). Moreover, President Lincoln's Emancipation Proclamation in 1863 referred to "persons" and thus excluded non-human animals. Finally, the court held that while fundamental constitutional concepts of due process and equal protection have evolved over the years, this has not been the case with the Thirteenth Amendment. The court stated that the "Thirteenth Amendment targets a single issue: the abolition of slavery within the Unites States. … As slavery and involuntary servitude are uniquely human activities, as those terms have been historically and contemporaneously applied, there is simply no basis to construe the Thirteenth Amendment as applying to non-humans" (*Tilikum v. Sea World* 2012, p. 1264).

Notions of liberty and equality underlie a second creative approach to establish legal "personhood" for animals. Steven M. Wise has spent many years searching for the characteristic that entitles humans to legal personhood and the basic fundamental rights to bodily liberty and bodily integrity. He has determined that "practical autonomy" is that characteristic. Thus to the extent a species has "practical autonomy" that species should be granted legal personhood. He describes "practical autonomy" as having three elements—"cognitively complex enough to want something, … able to act intentionally to achieve one's desires," and consciousness (Wise 2013, p. 1283).

Cetaceans are imbued with practical autonomy and are self-conscious, placing them at the level of human beings. According to Wise this makes cetaceans among the best candidates for legal personhood. Wise explains that, at this time, sentience should not be deemed sufficient for legal personhood because many animals are sentient and thus such a claim would likely be rejected by a court due to the dramatic practical and political consequences of such a ruling. Wise believes that by limiting the initial cases to species who are self-aware he is more likely to persuade courts to grant them legal personhood.

Wise and the Nonhuman Rights Project have chosen to use the common law rather than the Constitution to seek the right of bodily integrity and liberty for these self-aware nonhumans (Wise 2010). Wise explains that when interpreting a document, whether it be a rule, statute, or the Constitution, judges are more constrained as their task is to interpret the language of the text in light of the intent of the drafters. However, the common law is judge-made law, and thus judges have the power to interpret and change the law as they see fit. "Great common law judges understand that the common law is a flexible living organism that changes as morality changes and scientific facts and experiences accrue" (Wise 2013, p. 1288). Thus, a great common law judge with the authority to make law is an important ingredient when arguing for a paradigmatic shift in how the law views non-humans.

Wise believes that the common law writ of habeas corpus, the claim that allowed Judge Mansfield in the eighteenth century to transform James Somerset from a legal thing to a legal person, will be the vehicle to transform cetaceans from legal things to legal persons in the twenty-first century. The common law writ of habeas corpus provides for the release of anyone unlawfully detained by another (Wise 2010, pp. 8–9). Utilizing the "Normative Model" of equality, Wise argues that self-aware autonomous animals share the relevant human characteristic that grants humans the right to legal personhood. Wise explains that normative equality "requires that the criteria used to decide which entities are sufficiently different to be treated differently fit certain moral criteria. It rejects differentiations that burden a plaintiff in a manner that reflects deeply personal social stereotypes that are biologically immutable and changeable only at unacceptable personal costs, and it prohibits classifications that consider morally irrelevant traits" (Wise 2013, p. 1287). This theory requires the judge to recognize the similarities between humans and self-aware autonomous animals who share the trait deemed sufficient to be granted legal personhood while ignoring the many morally irrelevant differences. The result under such a theory would be the grant of habeas corpus and an associated finding of legal personhood for a captive self-aware autonomous mammal. In December 2013, the Nonhuman Rights Project filed its first three cases in New York State on behalf of four chimpanzees[15]—Tommy, Kiko, Hercules, and Leo—held captive in that state. All three petitions were denied on the grounds that the chimpanzees are not legal persons (Nonhuman Rights Project, Press Release 2013) and were appealed. In April 2015 Manhattan Supreme Court Justice Barbara Jaffe issued an Order to Show Cause in Hercules' and Leo's case. Hercules and Leo are confined in a biomedical research lab at Stony Brook University. The University must demonstrate a legally sufficient reason for their detainment of the chimpanzees or the judge will issue a writ of habeas corpus and effectively find that Hercules and Leo are legal persons because only

[15] The choice of the first lawsuits involved many criteria, including the right jurisdiction based on their laws, the philosophy and background of their justices, and the available alternatives for the animal if released (Nonhuman Rights Project, About Us, http://www.nonhumanrightsproject.org/about-us-2/. Accessed 3 December 2015). Currently there are no safe havens for captive orcas who cannot be released into the wild; thus they filed on behalf of chimpanzees rather than an orca.

legal persons may have a writ of habeas corpus issued on their behalf. Judge Jaffe, in a thoughtful decision, explained that such a dramatic change in the law, granting habeas corpus to a non-human animal, was best made by the Court of Appeals, and denied the writ (*The Nonhuman Rights Project v. Stanley*, 2015).

11.6 Conclusion

The stage is set for a transformative moment under U.S. law to significantly improve the lives of cetaceans and perhaps to establish their legal personhood. The stories of Tilikum, Lolita, and the eighteen beluga whales have captured the attention, empathy and concern of the American public as individual self-aware beings and not mere resources for human consumption. Their stories demonstrate that the so-called conservation and educational benefits derived from their captivity do not justify the serious physical and emotional harms they suffer from such captivity, that current U.S. law fails miserably to protect their most basic needs in captivity and in the wild, and that they share with humans complex cognitive and social abilities that demand they be granted basic rights to bodily liberty and bodily integrity.

References

Anderson, Jerry, L. 2011. Protection for the powerless: Political economy lessons for the animal welfare movement. *Journal of Animal Law and Policy* 4: 1–63.

Cetacean Rights. 2010. *Declaration of rights for cetaceans: Whales and dolphins*. http://www.cetaceanrights.org/. Accessed 25 June 2015.

Defran, R.H., and K. Pryor. 1980. The behavior and training of cetaceans in captivity. In *Cetacean behavior: Mechanisms and functions*, ed. L. Herman, 319–364. Malabar: R.E. Krieger.

Francione, Gary. 2008. *Animals as persons*. New York: Columbia University Press.

Garcia, Jason. 2013. Sea world posts record sales and profits for third quarter. *Orlando Sentinel*, 13 November 2013. http://articles.orlandosentinel.com/2013-11-13/business/os-seaworld-earnings-third-quarter-20131113_1_seaworld-orlando-seaworld-executives-seaworld-president. Accessed 25 June 2015.

Georgia Aquarium. 2015. Beluga whale research. https://www.georgiaaquarium.org/conserve/research/beluga-whales. Accessed 25 June 2015.

Hightower, Kyle. 2014. Sea world complaint alleges misconduct by federal employee who investigated trainer death. *Star Tribune*. 28 February 2014. http://www.startribune.com/politics/national/247843941.html. Accessed 25 June 2015.

Humane Society of the United States, 2012. *Press release: Poll reveals Americans oppose keeping orcas in captivity for public display*. http://www.humanesociety.org/news/press_releases/2012/07/poll_reveals_americans_oppose_orca_captivity_071612.html. Accessed 25 June 2015.

Karimi, Faith. 2015. Maris the beluga whale dies suddenly at Georgia Aquarium. CNN. 23 October 2015. http://www.cnn.com/2015/10/23/us/georgia-marisbeluga-whale-dies/. Accessed 3 Dec 2015.

Kirby, David. 2012. The real sea world: Seeing wild orcas is better … and cheaper. *Takepart*. 26 November 2012. http://www.takepart.com/article/2012/11/26/real-sea-world. Accessed 25 June 2015.

Kirby, David. 2014a. Sea world's worst nightmare: Calif. lawmaker to propose ban on orcas in captivity. *Takepart*. 6 March 2014. http://www.takepart.com/article/2014/03/06/seaworlds-worst-nightmare-calif-lawmaker-propose-ban-orcas-captivity. Accessed 25 June 2015.

Kirby, David. 2014b. Will sea world's financial woes force it to free the whales? *Takepart*. 14 August 2014. http://www.takepart.com/article/2014/08/13/will-seaworlds-financial-woes-force-it-give-captive-whale-shows-us. Accessed 25 June 2015.

Liebesman, Lawrence, and Rafe Petersen. 2010. *Endangered species deskbook*. Washington, DC: Environmental Law Institute.

MacKinnon, Catherine A. 1989. *Toward a feminist theory of the State*. Boston: Harvard University Press.

Marino, Lori. 2008. Cetacean brains. In *The encyclopedia of neuroscience*, vol. 2, ed. Squire Elsevier, 807–810. LaJolla: Academic.

Marino, Lori. 2011. Brain structure and intelligence in cetaceans. In *Whales and dolphins: Cognition, culture, conservation and human perceptions*, ed. P. Brakes and M.P. Simmonds, 115–128. Oxon/New York: Earthscan.

Marino, Lori, et al. 2010. Do zoos and aquariums promote attitude change in visitors? A critical evaluation of the American zoo and aquarium study. *Society and Animals* 18: 126–138.

Mashuda, Steve. 2014. *Comment on the proposed amendment to the endangered species act listing of the Southern resident killer whale distinct population segment*. http://www.orcanetwork.org/Main/PDF/EarthjusticeCommentsLolitaInclusion032714.pdf. Accessed 25 June 2015.

NOAA Fisheries, West Coast Region. 2015. *Southern resident killer whale – Lolita – included in endangered listing*. http://www.westcoast.fisheries.noaa.gov/protected_species/marine_mammals/killer_whale/lolita_petition.html. Accessed 25 June 2015.

NOAA Fisheries Strategic Plan. 2015. Our mission. http://www.nmfs.noaa.gov/om2/mission.html. Accessed 25 June 2015.

Nonhuman Rights Project. 2013. Press Release, Nonhuman Rights Project advances to New York Appellate Courts in Three Chimpanzee Rights Cases. 10 December 2013. http://www.nonhumanrightsproject.org/wp-content/uploads/2013/12/NhRP-Press-Release-12-10-13.pdf. Accessed 3 Dec 2015.

O'Neill, Ann. 2015. Sea World can expand tank but not breed whales, board rules. CNN. 11 October 2015. http://www.cnn.com/2015/10/08/us/seaworld-orcatank-expansion-plan-hearing/. Accessed 3 Dec 2015.

Palmer, C. 2010. Marvel at wild animals—From a distance. *CNN*. http://www.cnn.com/2010/OPINION/02/25/palmer.killer.whales.captivity/. Accessed 25 June 2015.

Rendell, L.E., and H. Whitehead. 2001. Culture in whales and dolphins. *Behavioural and Brain Sciences* 24: 309–324.

Rose, Naomi. 2011. *Killer controversy: Why orcas should no longer be kept in captivity*. Humane Society International and the Humane Society of the United States. http://www.hsi.org/assets/pdfs/orca_white_paper.pdf. Accessed 25 June 2015.

Sankoff, Peter. 2012. The animal rights debate and the expansion of public discourse: Is it possible for the law protecting animals to simultaneously fail and succeed? *Animal Law* 10: 281–320.

Stewart, K.L. and Lori Marino. 2009. *Dolphin-human interaction programs: Policies, problems, and practical alternatives*. Animals and Society Institute Policy Paper.

Sweeney, J.C. 1988. Specific pathologic behavior in aquatic mammals: Self-inflicted trauma. *Soundings Newsletter of the International Marine Animal Trainers Association* 13(1): 7.

Sweeney, J.C. 1990. Marine mammal behavioral diagnostics. In *CRC handbook of marine mammal medicine: Health, disease, and rehabilitation*, ed. L.A. Dierauf, 53–72. Boca Raton: CRC Press.

U.S. Department of Commerce. 2014. *About the department of commerce, Mission statement*, http://www.commerce.gov/about-department-commerce. Accessed 25 June 2015.

Whale and . 2015. *Captivity*. http://us.whales.org/wdc-in-action/captivity. Accessed 25 June 2015.

White, Thomas I. 2013. Review of whales and dolphins: Cognition, culture, conservation and human perceptions. *Journal of Animal Ethics* 3: 222–224.

Wise, Steven M. 2010. Legal personhood and the nonhuman rights project. *Animal Law* 17: 1–11.

Wise, Steven M. 2013. Nonhuman rights to personhood. *Pace Environmental Law Review* 30: 1278–1290.

Zimmerman, Tim. 2010. The killer in the pool. *Outside*. 30 July 2010. http://www.outsideonline.com/outdoor-adventure/nature/The-Killer-in-the-Pool.html?page=all. Accessed 25 June 2015.

Executive Materials

Animal Welfare Institute. 2012. Comment on GAI petition. https://awionline.org/sites/default/files/uploads/documents/ML-AWI-BelugaComments10-29-1-2.pdf. Accessed 25 June 2015.

Endangered and Threatened Wildlife and Plants: Endangered Status for Southern Resident Killer Whales, 70 Fed Reg 69903 (18 November 2005).

Georgia Aquarium, Inc. 2012. Application for permit to import certain marine mammals for public display under the Marine Mammal Protection Act. http://www.nmfs.noaa.gov/pr/permits/sci_res_pdfs/17324_final_application.pdf. Accessed 25 June 2015.

Government of India, Ministry of Environment & Forests, Central Zoo Authority. 2013. Circular, policy on establishment of dolphinarium, F. No. 20-1/2010-CZA(M). 17 May 2013.

Listing Endangered or Threatened Species: 12-Month Finding on a Petition to Delist the Southern Resident Killer Whale, 78 Fed. Reg. 47277 (5 August 2013).

Listing Endangered or Threatened Species: Proposed Amendment to the Endangered Species Act Listing of the Southern Resident Killer Whale Distinct Population Segment, 79 Fed. Reg. 4313 (27 January 2014).

Payne, Michael P. 2013. Report on the application for a public display permit (File No. 17324): Recommendation for Denial, Memorandum to Donna S. Weiting, Director, Office of Protected Resources, NMFS. http://www.nmfs.noaa.gov/pr/permits/sci_res_pdfs/17324_denial_letter_final.pdf. Accessed 3 December 2015.

PETA, et al. 2013. Before the Secretary of Commerce. Petition to include the Orcinus Orca known as Lolita in the ESA Listing of the Southern resident killer whales. 23 January 2013. http://www.nmfs.noaa.gov/pr/pdfs/petitions/killerwhale_lolita.pdf. Accessed 25 June 2015. [Lolita Petition].

Endangered and Threatened Wildlife and Plants; 12-Month Findings on Petitions to Delist U.S. Captive Populations of the Simitar-horned Oryx, Dama Gazelle, and Addax, 78 Fed. Reg. 33790 (5 June 2013).

Weiting, Donna S. 2013. Letter to Bill Hurley, Georgia Aquarium. http://www.nmfs.noaa.gov/pr/permits/sci_res_pdfs/17324_denial_letter_final.pdf. Accessed 3 December 2015.

Cases Cited

Animal Legal Defense Fund, et al. v. USDA, Case3:12-cv-04407-SC, Complaint for Declaratory and Injunctive Relief (N.D. Cal. Aug. 22, 2012) [Lolita Complaint].

Animal Legal Defense Fund, et al. v. USDA, Case3:12-cv-04407-SC, Defendant's Motion to Dismiss or in the Alternative for Summary Judgment (N.D. Cal. Aug. 22, 2012) [Lolita D's Motion].

Animal Legal Defense Fund, et al. v. USDA, Case3:12-cv-04407-SC, Plaintiff's Opposition to Defendant's Motion to Dismiss or in the Alternative for Summary Judgment (N.D. Cal. Aug. 22, 2012) [Lolita P's Opposition].
Animal Legal Defense Fund, et al. v. USDA, Case3:12-cv-04407-SC, Defendant's Reply in Support of Motion to Dismiss or in the Alternative for Summary Judgment (N.D. Cal. Aug. 22, 2012) [Lolita D's Reply].
Animal Legal Defense Fund, et al. v. USDA, 2015 WL 3653162 (11th Cir. June 15, 2015).
Animal Welfare Institute v. Kreps, 561 F.2d 1002 (D.C. Cir. 1977).
Georgia Aquarium, Inc. v. Pritzer, et al, Order, Civil Action No 1:13-cv-03241-AT (N.D. Ga., Sept. 28, 2015). http://www.nmfs.noaa.gov/pr/permits/ga_court_decision_092815.pdf. Accessed 3 Dec 2015.
SeaWorld of Florida, LLC v. Perez, 748 F.3d 1202 (D.C. Cir. 2014).
Secretary of Labor v. Sea World of Fl, Decision and Order, OSHRC Docket No. 10-1705 (June 11, 2012).
Strickland v. Medlen, 397 S.W3d 184 (2013).
Tennessee Valley Authority v. Hill, 437 U.S. 153 (1978).
The Nonhuman Rights Project v. Stanley, Decision and Order, Index No. 152736/15 (Jaffe, J., Supr. Ct. N.Y., N.Y. Cnty, 29 July 2015). http://www.nonhumanrightsproject.org/wp-content/uploads/2015/07/Judge-Jaffes-Decision-7-30-15.pdf. Accessed 3 Dec 2015.
Tilikum v. Sea World, 842 F. Supp. 2d 1259 (S.D. Cal. Feb. 8, 2012).
Tilikum, et al. v. Sea World Parks & Entertainment., Inc., Complaint for Declaratory Relief, No. 11-cv 11CV2476JM WMC, 2011 WL 5077854 (S.D. Ca. Oct. 25, 2011) [Tilikum Complaint].
Tilikum et al. v. Sea World, Plaintiffs' Opposition to Defendant's Motion to Dismiss, No. 11-cv-2476 JM WMC, 2012 WL 1506067 (S.D. Cal. Jan. 13, 2012) [Tilikum P's Opposition].
Tracey v. Solesky, 427 Md. 627 (as amended Aug. 21, 2012).

Legislation

Animal Welfare Act, 7 U.S.C. § 2131-2156 (2013)
California AB- 2140 (Bloom), California Captive Orca Welfare and Safety Act (amended Mar. 28, 2014).
Endangered Species Act, 16 U.S.C. §§ 1531-1544 (1982)
House Report (Merchant Marine and Fisheries Committee) No. 92-707, Marine Mammal Protection Act of 1972, 1972 U.S.C.C.A.N. 4144, 92nd Cong. 2nd Sess. 1972.
Marine Mammal Protection Act, 16 U.S.C. § 1361, et. seq. (2007)
Marino, Lori. 2010. Marine Mammals in Captivity: What Constitutes Meaningful Public Education? Before the Subcomm. On Insular Affairs, Oceans, and Wildlife of the H. Comm. on Natural Resources, 111th Cong. http://naturalresources.house.gov/uploadedfiles/marinotestimony04.27.10.pdf. Accessed 3 December 2015.
U.S. Const. amend. XIII, sec. 1.

Chapter 12
Wildlife Crimes and Legal Protection of Wildlife in China

Deborah Cao

Abstract The chapter focuses on crimes against wildlife as illustrated by ivory trade in China and its wildlife law in an attempt to identify some of the problems in the current legal protection regime for wildlife. It argues that wildlife crime is a serious crime, and transnational wildlife crime is a growing menace to the animal and human world and requires international actions. Despite the legal framework and laws at the international and domestic levels in many countries including China, African elephants, rhinos and other endangered wildlife are facing extinction due to the growing demand and trade for such animal products in Asia and the ineffective protection system. A fundamental change of attitude and conception in wildlife protection is required, that is, wildlife and animals in general need to be protected irrespective of their species and wildlife must not be seen as resources for human exploitation as illustrated by China's ivory case. A more practical and immediate urgent measure is a comprehensive ban of all trade of ivory products in China and elsewhere. Wildlife are victims of human crimes and deserve our protection, and the definition of crimes against wildlife needs to be expanded to include harms done to them either legally or illegally in order to safeguard and enhance animal rights and interests in our increasingly globalized world.

12.1 Introduction

One night in August 2011, a container ship loaded with copper ores, sailing from across the ocean from Tanzania, arrived in Shishi, Fujian Province, China. When the container underwent customs inspection, 524 pieces of elephant tusks were detected hidden in the ores. The recipient of the goods, a Mr Chen, was subsequently arrested. Ten days later, another container ship arrived in Shishi from Nigeria with goods for the same Mr Chen. Another 702 raw tusks were found concealed in the cashews being transported on the ship. Altogether Chen tried to smuggle 2154 tusks in four

D. Cao (✉)
Law Futures Centre, Griffith University, Brisbane, QLD, Australia
e-mail: d.cao@griffith.edu.au

D. Cao, S. White (eds.), *Animal Law and Welfare - International Perspectives*,
Ius Gentium: Comparative Perspectives on Law and Justice 53,
DOI 10.1007/978-3-319-26818-7_12

consignments, about 7.68 tons, representing an equivalent to over 1000 elephants killed. Chen was a Chinese government licensed ivory carving master and producer, with a licensed ivory shop in Shanghai and a licensed ivory carving workshop in Fujian. According to Chinese media reports, about 800–900 ivory smuggling seizures were made annually in recent years in China.

The essay focuses on crimes against wildlife as illustrated by ivory related crimes in China and its wildlife law. It argues that wildlife crime is a serious crime, and transnational wildlife crime is a growing menace to the animal and human world that requires international actions. Despite the legal framework and laws at the international and domestic levels in many countries including China, African elephants, rhinos and other endangered wildlife are facing extinction due to the growing trade for such animal products in Asia and the ineffective protection system. Admittedly, there is no easy or simple solution, but a fundamental change of attitude and conception in wildlife protection is required, that is, wildlife and animals in general need to be protected irrespective of their species and wildlife must not be seen as resources for human exploitation as illustrated by China's ivory case. A more practical and immediate measure is a comprehensive ban of ivory trade in China and U.S. as promised by the Chinese and U.S. governments in 2015. Wild animals are victims of human crimes, and the definition of crimes against wildlife needs to be expanded to include harms done by humans either legally or illegally in order to safeguard and enhance animal rights and interests in our increasingly globalized world.

12.2 Crime Against Animals as a Serious Crime

Crimes against animals are being gradually recognized as a serious crime. In the past few years, crimes against wildlife have started to draw international attention. As reported, in 2014, the FBI makes animal cruelty a top tier felony in the U.S. The U.S. government successfully co-sponsored a resolution at the 2013 U.N. Commission on Crime Prevention and Criminal Justice encouraging U.N. Member States to classify wildlife trafficking as a "serious" crime as defined in the U.N. Convention against Transnational Organized Crime, to facilitate further international cooperation among states that have ratified or acceded to the treaty, and lead to increased penalties for traffickers.[1]

According to WWF (2013), in the last few years, illegal wildlife trade has exploded to meet increasing demand for elephant ivory, rhino horns, and tiger products, particularly in Asia, involving crime syndicates with wildlife being trafficked much like drugs, weapons or humans.[2] Wildlife trafficking is believed to be the fifth most profitable illicit trade in the world. Both the WWF and the United Nations

[1] http://www.whitehouse.gov/the-press-office/2013/07/01/fact-sheet-us-efforts-combat-wildlife-trafficking

[2] http://worldwildlife.org/pages/stop-wildlife-crime

Office on Drugs and Crime (UNODC) estimate that the total value of the illegal global wildlife trade is around ten billion U.S. dollars annually, excluding timber and marine wildlife.[3] In recent years, organized crimes have become involved. It is reported that wildlife poaching is creating "a lucrative, bloody money stream for criminal networks," and "the revenues generated by the killing of wild animals and the sale of their parts support local insurgencies and terrorist activities, and promote political instability."[4] Worryingly, there is no effective system of deterrent to wildlife crimes. Such criminal enterprises and participants are rarely arrested, prosecuted, convicted or punished. Perpetrators of wildlife crimes, unlike in other major types of criminal activities, often operate with impunity, making the trade a low-risk and high-profit business.[5]

In recognition of the dire and aggravating situation, in November 2010, five international organizations, UNODC, Secretariat of the Convention on International Trade in Endangered Species of Wild Fauna and Flora (CITES), INTERPOL, World Customs Organization (WCO) and the World Bank joined forces to form the International Consortium on Combating Wildlife Crime (ICCWC),[6] to support governments, national wildlife and forest law enforcement agencies and sub-regional networks in a collaborative attempt to protect the world's natural resources from criminal exploitation. Encouragingly, on 1 July 2013, President Obama issued an Executive Order on Combating Wildlife Trafficking in direct response to the current surge in wildlife trafficking.[7] It is noted in the Executive Order:

> … wildlife trafficking is a multi-billion dollar illicit business that is decimating Africa's iconic animal populations. Many species – most notably elephants and rhinoceroses – now face the risk of significant decline or even extinction. Like other forms of illicit trade, wildlife trafficking undermines security across nations. Well-armed, well-equipped, and well-organized networks of poachers, criminals, and corrupt officials exploit porous borders and weak institutions to profit from trading in illegally taken wildlife.[8]

It is too early to tell whether this message on the need to treat wildlife crime as a serious crime on a par with drug and arms trafficking will produce sufficient awareness and actions to combat such crimes internationally. In China, the focus of this essay, certain wildlife crimes have been classified as serious crimes and attract heavy penalties including life imprisonment and also the death penalty until recently.

---

[3] http://worldwildlife.org/pages/stop-wildlife-crime and http://www.cites.org/eng/news/sg/2013/20130926_wildlife_crime_unodc.php

[4] http://www.nytimes.com/2013/01/09/opinion/the-poaching-of-elephants-and-rhinos.html?ref=elephants;http://www.nytimes.com/2012/12/30/world/africa/to-save-wildlife-and-tourism-kenyans-take-up-arms.html; http://www.nytimes.com/2013/01/01/world/africa/ruthless-smuggling-rings-put-rhinos-in-the-cross-hairs.html; see alsohttp://dotearth.blogs.nytimes.com/2012/09/04/a-searing-look-at-ties-between-africas-wars-and-the-ivory-trade/

[5] http://www.worldwildlife.org/pages/stop-wildlife-crime

[6] http://www.cites.org/eng/prog/iccwc.php

[7] http://www.whitehouse.gov/the-press-office/2013/07/01/fact-sheet-us-efforts-combat-wildlife-trafficking

[8] http://www.whitehouse.gov/the-press-office/2013/07/01/fact-sheet-us-efforts-combat-wildlife-trafficking

Nevertheless, wildlife crimes are increasing in China, especially crimes related to illegal ivory trade, to be discussed next.

12.3 Relevant Laws for Wildlife Protection in China

The basic legal regime relating to wildlife exists in China, consisting of constitutional provision, national laws, subordinate administrative regulations, provincial and local wildlife implementing regulations, and binding interpretations and opinions by the highest court and relevant legal and administrative authorities (see further discussion by Chang, 2006, pp. 197–201, and Cao 2013, and 2015), unlike the situation for domestic animals in China.

The most important law related to animals in China is the Wildlife Protection Law (1988 as amended in 2004) (WPL). It is the basic law on wildlife in China, setting out the guiding principles and framework for wildlife management, utilization and protection. There are subordinate regulations and administrative measures such as Regulation on the Implementation of the Land Wildlife Protection (1992). There are local regulations at the provincial level designed to implement the national laws, authorizing local authorities and specifying local conditions. China is a signatory to the Convention on the International Trade in Endangered Species of Wild Fauna and Flora (CITES) and other international and bilateral treaties related to wildlife. Interestingly, the WPL (1988) states that if any international treaty concerning the protection of wildlife, concluded or acceded to by China, contains provisions differing from those of this Law, the provisions of the international treaty shall apply, unless China has made reservation about such provisions (Art. 40). There are also interpretations and opinions by the highest court in China, the Supreme People's Court, on the specific provisions of the above mentioned national laws and on the applications of the laws in court trials in particular in determining appropriate sentencing and punishment. These include the Interpretation of the Supreme People's Court on the Application of Laws for Criminal Cases concerning the Destruction of Wildlife Resources (2000); the Supreme People's Court's Interpretation on the Application of Law in Smuggling Cases (2000). Part of the Interpretation deals with the smuggling of endangered or protected wildlife. Importantly, the Criminal Law (1997) has prohibitions of criminal acts related to wildlife and smuggling of wildlife and their products (see Cao 2015).

In pursuance of the WPL (1988), a Special State Protection Animal List (1988) was compiled and published in 1989 by the central Chinese government. It originally listed over 250 species of animals under special state protection and more species have been occasionally added to the list. It has two classes of protection, Class I and Class II. Panda, tiger, Yangtze alligator, leopard, Tibetan antelope, golden monkey, among others are in Class I, and other wildlife, such as pangolin, black bear, brown bear, gecko gecko, are in Class II. In 1993, the SFA also classified the animal species not native to China but listed in the CITES Appendix I and Appendix I, including rhino, kangaroo, crab-eating macaque, ostrich, African elephant, zebra and others and their body parts or derivatives as Class I or Class II state protected animals categories. These animals are given the same class of protection as the state protected Chinese indigenous wildlife (for detailed discussions, see Cao 2015).

In the Chinese language, *yesheng dongwu* as used in Wildlife Life Protection Law refers to both "wildlife" and "wild animals", and in this chapter, the two English expressions are used interchangeably. The term "wildlife" protected under WPL (1988) refers to "the species of land and aquatic wildlife that are rare and valuable or endangered and the species of land wildlife that are beneficial or of important economic or scientific value" (Art. 2). Furthermore, the Supreme People's Court pronounced in its Interpretation for the Application of Law in Smuggling Cases (2000) that the term "rare and valuable animals" refers to in Article 151 of the Criminal Law (1997) include the wildlife in the first and second class state protection categories as listed the Wildlife List of State Special Protection and such wildlife raised and bred, and the wildlife in CITES Appendix I and Appendix II and such wildlife raised and bred. The Supreme People's Court Interpretation on the Application of Law for Criminal Cases of Wildlife Resources Destruction (2000) also stipulates that "rare and valuable, or endangered wildlife" referred to in Article 341 of the Criminal Law (1997) include wildlife in the first and second class state protection categories as listed the Wildlife List of State Special Protection and such wildlife raised and bred, and wildlife in CITES Appendix I and Appendix II and such wildlife bred and raised (see Cao 2015).[9]

In terms of penalty and punishment of wildlife crimes, the WPL (1988) sets out the overall criminal liability and legal responsibility for acts of violation of the law in relation to smuggling offences (Art. 36) for importing and exporting protected wildlife or their products. The relevant specific penalties and punishments are mainly found in the Criminal Law (1997), and the relevant regulations and court interpretations. In the Criminal Law (1997), the crime of smuggling includes the smuggling of protected wildlife:

> Article 151 …Whoever smuggles cultural relics, gold, silver or other precious metals, the export of which is forbidden by the State, or rare and valuable animals or their products, the import and export of which are forbidden by the State, shall be sentenced to fixed-term imprisonment of not less than five years and not more than ten years and concurrently sentenced to a fine; if the circumstances are particularly serious, the person shall be sentenced to fixed-term imprisonment of not less than ten years or life imprisonment and property confiscation; if the circumstances are minor, the person shall be sentenced to fixed-term imprisonment of not more than five year and concurrently sentenced to a fine.

Previously, Art 151 contained the words "death penalty" for cases with particularly serious circumstances. The "death penalty" wording was deleted in the 2011 amendment to the Criminal Law (1997). In the Criminal Law (1997), there are a number of wildlife related offences (for detailed discussions of these offences, see Cao 2015):

---

[9] http://www.forestry.gov.cn/slga/2569/content-467597.html

> Article 341 Whoever illegally hunts, catches or kills rare and valuable or endangered wildlife under special state protection, or illegally buys, transports or sells rare and valuable or endangered wildlife under special state protection or their products, shall be sentenced to fixed-term imprisonment or criminal detention of not more than five years and concurrently sentenced to a fine; if the circumstances are serious, the person shall be sentenced to fixed-term imprisonment of not less than five years and not more than ten years and concurrently sentenced to a fine; if the circumstances are particularly serious, the person shall be sentenced to fixed-term imprisonment of not less than ten years, and concurrently sentenced to a fine or confiscation of property.

Also relevant are Interpretation of the Supreme People's Court on the Application of Laws for Criminal Cases concerning the Destruction of Wildlife Resources (2000) and the Interpretation of the Supreme People's Court on the Application of Law in Smuggling Cases (2000).

In terms of implementation and enforcement, there are administrative and law enforcement agencies throughout China, and some of them have been established as required under its CITES treaty obligations. For instance, China has the Inter-agency CITES Enforcement Coordination Group at the national and provincial levels to coordinate the domestic wildlife enforcement and international enforcement cooperation. Police, in particular, forest police in every province in charge of wildlife crime related matters, and customs and other law enforcement departments are responsible for investigating and arresting suspects for possible violation.

In sum, there is a basic legal framework for wildlife management and protection in China. There are laws, subordinate regulations and measures, and enforcement agencies empowered with the tasks and responsibilities for wildlife protection. China has some of the toughest penalties for wildlife crimes.

12.4 Issues in Elephant Protection

Over the past 5 years or so, China has emerged as the single largest conduit for illegal ivory trade in the world (EIA 2014). One recognized fact is that the resurgence of wildlife trafficking and poaching in the case of elephants and rhinos is directly linked to and fuelled by China's economic boom in recent years and the growing Chinese population in Africa. It was reported that the majority of the illegal ivory, as much as 70 %, is flowing to China. Of all the ivory trade in the world, around 90 % is sold to China.[10] In 2011, more than 150 Chinese citizens were arrested across Africa for smuggling ivory.[11] A growing demand for ivory products in China has seen the rising of ivory price in China and other countries in East and Southeast Asia (see Gettleman 2012; Levin 2013).[12] As a result of the increasing lucrative trade, organized criminal syndicates are believed to be involved in

[10] http://art.people.com.cn/n/2014/0224/c206244-24445159.html

[11] http://art.people.com.cn/n/2014/0224/c206244-24445159.html

[12] http://www.nytimes.com/2013/03/17/opinion/sunday/slaughter-of-the-african-elephants.html

elephant poaching and large scale ivory shipments from Africa, abetted by corruption at key stages in the smuggling chain (EIA 2014). According to reports and investigations, East Asian nationals, particularly the Chinese, drive the modern ivory trade, and ivory traffickers are primarily Asian organized criminal networks, spanning Asia and Africa (BornFree USA 2014). They operate within a networked black market that has ties to various other forms of illicit activity. Chinese traffickers are present in virtually every single African range state, and operate at nearly every point along the ivory supply chain (BornFree USA 2014).

Admittedly, there is no easy or simple solution that can immediately end the massive and brutal killing of elephants for ivory trade. However, a number of issues are worth considering in the legal protection regime of elephants and wildlife in general. They include lenient punishment for offenders and lack of enforcement of existing laws, legalized trade of ivory products in China and elsewhere, and the legal status of wildlife as resources for exploitation.

12.4.1 Punishment for Ivory Related Crimes

As often pointed out, the lack of comprehensive international legal norms dealing with wildlife offences means that domestic law is primarily responsible for determining the nature, scope and consequences of wildlife offences (see Cao 2014 and Favre, Chap. 5 this volume). Thus, what is needed is more effective and universal international law and legal norms for transnational wildlife crimes and comprehensive wildlife laws at the national level in terms of criminal sanction. As pointed out, CITES enforcement is problematic as it depends on each individual state for implementation of its treaty obligations (Kelch 2011). Member states are required to impose criminal penalty for trade in contravention of CITES. In many countries and jurisdictions, trade in endangered species of animals may be criminal, attracting criminal sanctions (see Bilchitz, Chap. 7 this volume, for wildlife protection laws in South Africa).

In China, as mentioned before, under the Chinese law, smuggling, killing, buying, selling and transporting protected wildlife and their products are crimes that can attract lengthy jail terms, ranging from jail time under 5 years or suspended sentences to life imprisonment and also the death penalty until recently. China says that it has jailed hundreds of illegal ivory dealers in the past decade and around 30 ivory smugglers were sentenced to life imprisonment (Levin 2013).

In recent times, a number of major cases involving ivory smuggling convictions in China were reported widely in the Chinese media (see also Cao 2015). For instance, in 2010, a court in Shantou, Guangdong Province, sentenced Ke Chengcai and Ke Jiayou to death with 2 years reprieve (this was before the 2011 amendment of the Criminal Law), and Xu Jiewen to life imprisonment, for smuggling 770 elephant tusks from the Philippines. The total amount of tusks involved was 2.1 tons, valued at CNY165 million (just under 27 million U.S. dollars).[13] The Chen case mentioned in the beginning of the essay involved a government licensed and accredited ivory carving master and trader in Fujian Province. In 2013, Chen was convicted of smuggling around 7.68 tonnes of ivory out of Kenya, Tanzania and Nigeria, one of the largest ivory smuggling cases in China. Chen was sentenced to life imprisonment in the first instance but reduced to 15 years in jail on appeal. His two accomplices were sentenced to 15 and 7 years respectively. The ivory involved was worth over CNY378 million (about 55 million U.S. dollars). In 2013, the two suspects from Guangdong were convicted and sentenced to 14 and 12 years imprisonment for smuggling 1.04 tons of ivory from Africa. They were from a legal ivory manufacturer and trader in Guangdong. They appealed but the original sentences were upheld.[14]

In another major ivory case in 2013, Gong and four others in Anhui Province smuggled over three tons of ivory from Japan to China from 2010 to 2012, with an estimated value of CNY135 million (about 20 million U.S. dollars). Gong was convicted of illegally buying and selling protected animal products and was sentenced to 15 years imprisonment and confiscation of CNY3 million worth of property; an accomplice was sentenced to 13 years with a confiscation of property, and three more involved were sentenced to 10, 6 and 5 years jail respectively with fines and confiscations. Two others involved were sentenced to 3 and 5 years jail with suspended sentences and with fines and confiscations. Still another person was sentenced to five and a half years (see also Cao 2015).[15] In a recent case, in September 2015, the Qingdao City Intermediate People's Court sentenced an owner of an antique shop to 15 years imprisonment for smuggling 18 elephant trunks to China through the postal system via Japan.[16]

Notwithstanding these lengthy jail terms and severe punishments, it is true that punishment is not uniform or guaranteed (Wagman and Lieberman 2011, p.173). Despite the legal stipulation and lengthy jail terms in many cases involving the smuggling of ivory in China, a disturbing trend has emerged over the past couple of years of lenient sentencing in some Chinese courts, especially in Shanghai. For instance, many cases that involved significant amounts of ivories or tusks that would be classified as particularly serious cases under the current law, and involved values of smuggled ivory over CNY250,000. They would have normally attracted minimum 10 years imprisonment under the law, but they were given lenient treatment and minimum punishment (see Cao 2015).

Lenient sentencing is a common problem in wildlife crimes including ivory related crimes in most jurisdictions in the world. Although CITES demands criminal punishment for wildlife crimes, many countries do not always enforce their laws or penalize CITES violations (Kelch 2011). For most of the reported convictions of wildlife related crimes from 1997 to April 2013 for many jurisdictions around the

[13] http://news.qq.com/a/20100102/000644.htm

[14] http://zh.bendibao.com/news/2013115/38914.shtm

[15] See http://bwwz.forestry.gov.cn; http://www.148china.com/display.asp?id=2082

[16] http://www.legaldaily.com.cn/index/content/2015-09/08/content_6259277.htm?node=20908

world as documented in the data compiled by TRAFFIC (2013), common sentencing includes suspended sentences, fines, community service, home confinement, with a small number of short jail terms of 1 to 3 years in Europe, Americas and Asia except China and a small number of cases in African countries in the last year or so. For instance, in 2012, two New York ivory dealers pleaded guilty to selling and offering for sale about one ton of illegal ivory, including jewellery and carved tusks. Under plea agreements, the convicted criminals and their companies were ordered to forfeit the ivory and fined with the money donated to elephant conservation.[17] In another case, in 2010, a New York court sentenced a person to 33 months imprisonment and fined USD25,000 for importing 71 elephant ivory carvings into the U.S. from Nigeria and Uganda, concealed inside the cavities of wooden and metal handicrafts. In another ivory smuggling case in 2011, a piano importer was sentenced to 3 years probation for illegally importing 855 elephant ivory key tops, totalling 1710 pieces of ivory, and was a fine of USD17,500. One rare custodial sentencing from the U.S. is case involving a Canadian citizen who was sentenced by a U.S. court in 2008 to 5 years imprisonment and fined USD100,000 for illegally importing ivory from Cameroon. The sentence is the result of an international investigation by special agents of the U.S. Fish and Wildlife Service, wildlife officials from Canada and the U.S. Attorney's Office in Cleveland, Ohio. The person pleaded guilty on four federal felony charges (two felonies for Lacey Act violations and two felonies for smuggling) for activities relating to the illegal trade in raw African elephant ivory.[18]

The TRAFFIC (2013) report indicates that since 2010, more lengthy jail terms are being handed down in countries in southern Africa, most of them relating to rhino poaching. For instance, in 2012, a Thai national was sentenced by a court in South Africa to 40 years in jail for his role in organizing illegal rhinoceros poaching expeditions. The sentence is the longest ever imposed for poaching in South Africa and follows a year-long investigation.[19] Also in South Africa, in August 2012, two people were each sentenced to 29 years imprisonment for killing a rhinoceros cow and a calf. They were each sentenced to 10 years imprisonment for killing an adult female rhinoceros, and another 10 years each for killing a calf. They were further sentenced to an effective 4 years each on charges of possession of illegal firearms and 2 years for illegal possession of ammunition as well as 3 years each for trespassing. Similarly, in a case from Zimbabwe, in June 2012, three people were sentenced to 17 years imprisonment for rhino poaching.[20] In 2012, two farmers were sentenced

[17] District Attorney, New York County media release, 12 July 2012

[18] The U.S. Lacey Act (16 U.S.C. §§ 3371–3378) makes it a federal violation to import, export, transport, sell, receive, acquire, or purchase in interstate or foreign commerce any fish, wildlife or plants, taken, possessed, transported, or sold in violation of any law or regulation of any State.

[19] It was reported that South African authorities currently issue permits to what it terms 'bona fide' hunters for trophy hunting. The SA Department for Environmental Affairs says 'a hunting client may only hunt one White Rhinoceros within a specific calendar year.'

[20] www.zoutnet.co.za/details/17-08-2012/21_years_in_jail_for_two_rhino_poachers/14483, 17 August 2012; www. newsday.co.zw/article/2012-06-29-poachers-jailed-17-years, 29 June 2012

to 18 years and 10.5 years imprisonment respectively by a Zimbabwe court. Also in Zimbabwe, in 2012, two people were sentenced to 21 years in jail each for killing and dehorning two Black Rhinoceroses. In Kenya, in January 2012, three Mozambican nationals were each sentenced to 25 years imprisonment after being found guilty of poaching rhinos in Kruger National Park as well as of possessing weapons and ammunition.

Regarding ivory cases, despite a surge of elephant killings in Africa, many ivory smugglers escaped with minimal fines and were set free. Some of the lengthy jail times in African countries in recent years came in the wake of both aggravating poaching and trafficking of ivory. For instance, in 2013, Kenya increased penalties to up to 15 years in jail and/or a fine of KSH10 million (approximately USD118,340), raising the severity and treatment of wildlife crime to the level of organized crime in Kenya. Previously, the offense was punishable with up to 10 years imprisonment and/or KSH500,000 (USD5917).[21] In early 2014, a court in Kenya fined a Chinese man named Tang a record 20 million shillings (USD230,000) for smuggling ivory, under the new anti-poaching laws. Tang was caught with an elephant ivory tusk weighing 3.4 kg in a suitcase on his way from Kenya back to China. If he is unable to pay the fine, he will spend 7 years in prison. Also in 2013, a Kenya court sentenced a Chinese ivory smuggler in a landmark ruling in which a Ms Chen was jailed for 31 months for trying to smuggle 6.9 kg of ivory she had disguised as 15 bags of macadamia nuts on her flight to Hong Kong.[22] In another serious case, in March 2014, a court in Tanzania jailed a Chinese ivory smuggler for 20 years. Yu Bo pleaded guilty to unlawful possession of 81 elephant tusks and the skins of two pangolins.[23] Yu Bo was arrested trying to smuggle 81 tusks — the equivalent of about 40 elephants, during a Chinese naval visit to Tanzania. Convicted of ivory smuggling and fined $5.6 million, he was sentenced to 20 years in prison — the only person brought to justice in eight major cases going back to 2009, during which time 26.5 tons of Tanzanian ivory was seized (EIA 2014).

In September 2014, one of the toughest penalties imposed in South Africa for the illegal possession of ivory was handed down when a Chinese national named Liang was sentenced to 10 years jail and a R5 million (about USD456,700) fine for possessing one ton of poached elephant tusks. Three years of his jail sentence would be suspended provided he pay the R5 million fine within a year. The court heard he was probably part of an international ivory smuggling syndicate working as a courier and exporter. Liang was arrested in September 2012. As reported, the nerve cavities

[21] http://www.awf.org/news/awf-applauds-passage-kenyan-parliament-motion-increase-penalties-wildlife-crime

[22] http://www.scmp.com/news/world/article/1298759/kenya-jails-chinese-ivory-smuggler-two-and-half-years

[23] http://www.africadaily.net/reports/Tanzania_jails_Chinese_ivory_smuggler_for_20_years_999.html

of some of the tusks he was carrying were still moist with blood and tissue, indicating 34 elephants had been killed.[24]

As we have seen, punishments for ivory related crimes vary in different jurisdictions, and there is a tendency of leniency across the board and a lack of enforcement of law (see also Cao 2014, 2015). There is a welcome development in Africa of more severe penalties for such crimes. Such severity needs to prevail in other jurisdictions such as Europe and the U.S., as a strong deterrent. Wildlife crimes, especially transnational wildlife crimes, need to be taken seriously in all jurisdictions. Perpetrators of such crimes need to be given the full force of the law irrespective which country they are from or where they commit the crimes. Moreover, the scope and definition of crimes or harms against animals need to be expanded, to include acts of cruelty against animals that are legalized. In this regard, I define *crime against animals* as acts or omissions that are against the law and subject to criminal prosecution and sanctions; and *international or transnational crime against animals* as acts or omissions that are a form of abuse and harms to animals regardless of legality, including those committed by individuals and those committed or facilitated by the state, corporations or other national or international entities (see Cao 2014; see also White 2011). Certain uniformity and extensive cooperation among nations are required to combat wildlife crimes.

One could argue that the severe criminal penalties for ivory smugglers in China have not succeeded in deterring criminals over the years. There are indeed many problems with the Chinese criminal justice system and judicial process. As mentioned earlier, there is a tendency of leniency in sentencing in Chinese courts in the past few years. Furthermore, the criteria for assessing the severity of ivory related cases have not changed from the early 2000s while many things and conditions in China have considerably changed. More important is the lack of enforcement of the law or insufficient enforcement of the law. The punishment may be severe when offenders are brought justice, but most such offenders are not caught, prosecuted or punished.

Another related aspect is the lack of awareness of poaching and other problems associated with ivory and lack of awareness of the law and the crime by the Chinese society as a whole. Most Chinese would not regard buying, selling and consuming ivory products as a negative or criminal matter. After all, when the buying, selling and manufacturing of ivory products are legal with the support from the government and ivory carving is regarded as a high art form and cultural heritage, most ordinary people naturally would not consider such activities as criminal. It was found that many Chinese did not know that the taking of elephant tusks would involve the killing the animals until recently when publicity awareness campaigns were launched and promoted in Chinese official and social media. All of these contributed to the ineffectiveness of the existing law in China (see Cao 2015).

[24] http://www.iol.co.za/capetimes/heaviest-jail-term-for-ivory-smuggler-1.1747233#.VGSpNzSUcWY

12.4.2 Wildlife as Resources for Utilization

As pointed out, the current surge in illegal ivory trading in China is a consequence of several connected factors; the creation of a parallel legal domestic market for ivory in China by CITES decisions, the role of the Chinese government and industry in stimulating demand for ivory products, and failure to stop the flow of smuggled ivory to mainland China (EIA 2014). All of these are closely linked or derive from the basic philosophy and practice in China that animals, including protected endangered animals, are valuable resources to be utilized for human benefit. In China's case, the main objectives of Chinese wildlife protection include the exploitation and utilization of wildlife including protected and endangered species. The Chinese Constitution states that "the State ensures the rational use of natural resources and protects rare and valuable animals and plants (*zhengui de dongwu*)" (Art. 9), the only mention of animals in the Constitution. It means that natural resources including animals are utilized, and rare and valuable animals are afforded protection subject to utilization (see Cao 2014, 2015). Accordingly, the main objectives of the WPL (1988) are to promote wildlife protection, rescue, domestication and rational utilization of utilizing wildlife resources and maintaining ecological balances (Art.1).

More specifically, first, the WPL (1988) provides that wildlife resources are owned by the State, and the State protects the lawful rights and interests of units and individuals engaged in the development or utilization of wildlife resources according to law (Art. 3). Second, the WPL (1988) reconfirms the state's support for wildlife domestication and utilization in "actively domesticating and breeding wildlife species, rationally developing and utilizing wildlife resources" (Art. 4). To achieve this end, the WPL (1988) establishes a licensing system for such purposes and provides detailed stipulation on the mechanisms for animal business operations. It requires that anyone who intends to domesticate or breed wildlife under special state protection shall obtain a license (Art. 17).

Regarding the commercial and other uses of wildlife, the law explicitly prohibits the sale and purchase of wildlife under special state protection or such products (Art. 22 of the WPL (1988)), but the same provision provides for the licensing scheme whereby trade of such animals is allowed when such trade is for scientific research, domestication and breeding, exhibition or other special purposes (Art. 22). This law of supporting and legitimizing the trade of protected animal and their products is not empty words. They have been actively pursued and implemented in China for various purposes, and one of the major purposes is for commercial exploitation for profit. For instance, in pursuance of the WPL (1988), the Regulations for the Implementation of Terrestrial Wildlife Protection (1992) set out procedures and rules for domestication and breeding operations of protected wildlife. Such protected animals and their products can be legally purchased under this system. In 2003, the State Forestry Administration (SFA) issued a notice establishing a "pilot marking" scheme for utilization of wildlife products. Such "markings" are given to approved businesses and their wildlife products, which are then allowed for transportation and trade. Under this scheme, the list of pilot entities with the marking has expanded from the initial three to over 400. Gradually, in the past decade, ivory production and retail outlets, some tiger bone wines, possession and trade of wild animal pelt products, production and trade of traditional Chinese medicine containing leopard bones and tiger bones, and the trade of tiger skins, leopard skins and their products (from those bred in captivity) have all been added to the list, legalizing their production and trade for commercial and other purposes (see Cao 2015).[25]

The thinking and practice of regarding animals as resources to be utilized for human benefit is not limited to China, but prevailing around the world. In the existing international wildlife protection framework, generally speaking, wildlife including endangered species is not banned from trading. Instead, international treaties and laws including CITES regulate the trade of wildlife and legitimize exploitation of animals. CITES is the most important international treaty on wildlife and is regarded as the most successful one as well. However, as rightly pointed out, the success and progress achieved under CITES need to be viewed against the backdrop of the prevailing human-centred approach to the treatment of animals in our societies. CITES does not prohibit trading of wildlife including endangered species. As in other realms of exploitation of animals, the supposed interests of humans overshadow the interests of animals in CITES as well as in other legal regulatory regimes (Kelch 2011). CITES aims to ensure that international trade in specimens of wild animals and plants does not threaten their survival. It views wild fauna and flora as possessing value from aesthetic, scientific, cultural, recreational and economic points of view, and "accepts the legitimacy of trade in animals, including those that are endangered, and this necessarily legitimizes the exploitation of all these animals" (Kelch 2011, p. 234). For CITES and other international agreements on animal management and conservation (and many domestic laws related to wildlife), their main concerns are with human use of animals, not with protecting them for their own sake, not recognizing their intrinsic value as sentient lives. Accordingly, exploitation often tramps protection. As pointed out, the interests of humans, whether economic, cultural, political, religious, or social, and the long-term interests of humans, easily outweigh the interests of animals (Kelch 2011, p. 235). Despite the worldwide efforts to enforce laws to protect endangered animal species, "CITES may be becoming nothing more than a brightly harnessed show pony in an international economic game of ecological exploitation" (Kelch 2011, p. 231).

In relation to China, another example of exploitation of endangered animal products for human benefit and profit is CITES's sales of seized ivory to China and Japan in recent times, aiding and abetting the worsening of elephant poaching in

---

[25] For instance, see the SFA notice on inclusion of markings of products made from tiger and leopard skin, and medicinal tablets containing Tibetan antelope horns and pangolin body parts and various products containing body parts of CITES listed snakes for legal trading purposes: http://www.forestry.gov.cn/portal/main/govfile/13/govfile_1102.html; see also http://eia-international.org/wp-content/uploads/EIA_Table-of-relevant-laws-and-notifications-in-China_Feb-25-2013.pdf

Africa and trafficking in Asia (EIA 2010, 2012, 2014). The CITES approved ivory sales in 1999 and 2008 were believed to be the major drivers to the subsequent drastic increase in poaching and illegal ivory trade related to the Chinese demand. As pointed out (IFAW 2012), the ivory sales have failed to reduce ivory prices and control illegal trade, and the influx of "legal" ivory further challenges market control and enforcement efforts, providing loopholes for smuggled ivory to be freely traded under the cover of legal domestic markets. The CITES sale spurred production and trade of ivory products in China and stimulated the demand for ivory from a growing class of wealthy consumers that covets ivory products as collectables and investment vehicles.[26] The regulated ivory trade has provided unscrupulous Chinese carvers and collectors with the ideal legal camouflage to buy and sell contraband tusks (Levin 2013).

12.5 Protected Wildlife Trade Ban

In recent times, there have been proposals and debates in Western countries on whether ivory trade should be legalized in order to save African elephants.[27] China has supported and argued for the trade legalization and such trade has always been legal within China. For many years China has been pushing for legalizing international trade of ivory.[28] However, in more recent times, there have been some dramatic changes in the official attitude and policy. Encouragingly, in early 2015, a temporary moratorium of import of African ivory carvings was imposed by the government.[29] According to the ban, for 1 year from 26 February 2015, the Chinese government stops issuing import permits for carvings obtained since 1975.[30] In May 2015, the Chinese government also announced the decision to phase out the legal, domestic manufacture and sale of ivory products. More significantly, in September 2015, during the Chinese President's visit to the U.S., the two countries, the two largest ivory markets in the world, announced that they would enact a nearly complete ban on the import and export of ivory. The ban would cover "significant and timely restrictions on the import of ivory as hunting trophies" as well as unspecified

[26] See a report on the failure of the sale http://newswatch.nationalgeographic.com/ 2013/05/30/china-ivory-prosecution-a-success-exposes-fundamental-failure/Bryan Christy in A Voice for Elephants on May 30, 2013. See also http://www.vanityfair.com/culture/features/2011/08/elephants-201108

[27] See http://voices.nationalgeographic.com/2014/09/15/opinion-can-elephants-survive-a-continued-ivory-trade-ban/

[28] For instance, in July 2012, the CITES Management Authority of China in its submission on the control of trade in ivory in China to a CITES standing committee meeting in Geneva http://www.cites.org/eng/com/sc/62/inf/E62i-08.pdf

[29] http://www.nytimes.com/2015/02/27/world/asia/china-bans-import-of-ivory-carvings-for-one-year.html

[30] http://www.forestry.gov.cn/portal/main/govfile/13/govfile_2136.htm

"significant and timely steps to halt the domestic commercial trade of ivory."[31] The official statement also says that recognising the importance and urgency of combating wildlife trafficking, the two countries commit to take positive measures to address this global challenge, and will further cooperate in joint training, technical exchanges, information sharing, and public education on combating wildlife trafficking, and enhance international law enforcement cooperation in this field, and will cooperate with other nations in a comprehensive effort to combat wildlife trafficking.[32]

The latest developments and position taken by the Chinese government are positive and encouraging, different from its earlier insistence in the legal trade of ivory. This should also make some in the West and other countries rethink about their argument and advocacy for the legalization of ivory trade and the legalized killing of endangered animals for trophy hunt and other commercial purposes. In China's case, I always argue that the root cause of the worsening poaching and criminal activities is the very fact that ivory trade is legal in China. The biggest issue is that ivory is traded in China as a valuable commodity. Ivory trading itself, legal and illegal, is the problem. In China, ivory products are seen predominantly in terms of its economic value and status symbol, with its cultural or artistic value being secondary. It is the monetary value of ivory, not cultural icon, that is driving the growing production and demand. "No market value" is the key to the current ivory crisis. Ivory may continue to maintain its artistic value as part of the Chinese cultural heritage, but that culture and heritage of ivory carving does not depend on its economic value or market price. The ban of ivory trade is a crucial first step by the Chinese and U.S. governments. More crucial will be the enforcement of such ban, especially in China.

Finally, what are the possible solutions to the worsening wildlife crimes, especially crimes against elephants and rhinos and other endangered species? It seems that a combination of measures and new forms of intervention are needed, including, for instance, ban of the trade of such animal and their body parts, tough law and effective legal enforcement, and the use of administrative, civil, and criminal enforcement tools in tandem, and the involvement of multiple agencies and community if we can take a leaf from the development in environmental laws (c.f. White 2014, pp. 452–465). In particular, apart from special law enforcement agencies (such as forestry police in China and animal NGOs) should also be given certain power to intervene and investigate suspected wildlife crimes, taking a leaf from the roles of the RSPCA in the U.K. and Australia in relation to domestic animal cruelty cases. Above all, political will and financial resources are needed for the tough law to work.

To end the chapter in a positive note, and as an example of a combination of measures, it was reported, over the past two decades, the number of Asian elephants

---

[31] https://www.whitehouse.gov/the-press-office/2015/09/25/fact-sheet-president-xi-jinpings-state-visit-united-states

[32] https://www.whitehouse.gov/the-press-office/2015/09/25/fact-sheet-president-xi-jinpings-state-visit-united-states

in Yunnan Province, China, has roughly doubled, to nearly 300, thanks to government financed feeding programs, wildlife education efforts and the tough law on elephant killing, and as reported, since 1995 when four people were executed for killing an elephant in China, only a handful of Chinese domestic elephants have been poached for their tusks (Jacobs 2014).[33]

References

Born Free U.S.A. 2014. *Out of Africa: Mapping the global trade in illicit elephant ivory.*

Cao, Deborah. 2013. Legal protection of animals in China. In *The global guide to animal protection*, ed. A. Linzey, 234–235. Chicago: University of Illinois Press.

Cao, Deborah. 2014. Crimes against animality: Animal cruelty and criminal justice in a globalized world. In *The routledge handbook of international crime and justice studies*, ed. Bruce A. Arrigo and Heather Y. Bersot, 169–190. London: Routledge.

Cao, Deborah. 2015. *Animals in China: Law and society*. London: Palgrave Macmillan.

EIA. 2010. *Open season*. London: EIA.

EIA. 2012. *Blood ivory: Exposing the myth of a regulated market*. London: EIA.

EIA. 2014. *Vanishing points: Criminality, corruption and the devastation of Tanzania's elephants*. London: EIA.

Gettleman, J. 2012. Elephants dying in epic frenzy as ivory fuels wars and profits. *The New York Times*, September 4, 2012, A1.

IFAW. 2012. *Making a killing: A 2011 survey of ivory markets in China.*

Jacobs, A. 2014. In land that values ivory, wild elephants find a safe haven. *The New York Times*, April 26, 2014.

Kelch, T.G. 2011. *Globalization and animal law*. Alphen aan den Rijn: Wolters Kluwer.

Levin, D. 2013. From elephants' mouths: An illicit trail to China. *The New York Times*, March 1, 2013.

Wagman, B.A., and M. Liebman. 2011. *A worldview of animal law*. Durham: Carolina Academic Press.

White, R. 2011. *Transnational environmental crimes*. London: Routledge.

White, R. 2014. What is to be done about environmental crime? In *The Routledge handbook of international crime and justice studies*, ed. Bruce A. Arrigo and Heather Y. Bersot, 445–467. London: Routledge.

[33] Similarly, China also has very tough law enforcement for crimes relating to panda killing, and the death penalty was also applied in some instances in the recent past before capital punishment was taken out of the criminal law for such crimes. It has served as a strong deterrent and has worked for pandas. For discussions of the tough law regarding panda protection in China, see Cao (2015).

Index

D. Cao, S. White (eds.), *Animal Law and Welfare - International Perspectives*,
Ius Gentium: Comparative Perspectives on Law and Justice 53,
DOI 10.1007/978-3-319-26818-7

CPSIA information can be obtained at www.ICGtesting.com
Printed in the USA
BVOW06*0325130216

436605BV00004B/9/P

9 783319 268163